D1256181

Neuroengineering the Future

NEUROENGINEERING THE FUTURE

Virtual Minds and the Creation of Immortality

By

Bruce F. Katz, Ph.D.

INFINITY SCIENCE PRESS LLC
Hingham, Massachusetts
New Delhi, India

Infinity Science Press LLC
11 Leavitt Street
Hingham, MA 02043
Tel. 877-266-5796 (toll free)
Fax 781-740-1677
info@infinitysciencepress.com
www.infinitysciencepress.com

This book is printed on acid-free paper.

Library of Congress Cataloging-in-Publication Data

Katz, Bruce F., 1959-
Neuroengineering the future: virtual minds and the creation of immortality / by Bruce F. Katz.
p. cm.
Includes bibliographical references.
ISBN 978-1-934015-18-6 (hardcover : alk. paper)
1. Biomedical engineering. 2. Neural networks (Neurobiology) I. Title.
R856.K38 2008
610.28—dc22
2008017792
8 9 10 4 3 2 1

Our titles are available for adoption, license or bulk purchase by institutions, corporations, etc. For additional information, please contact the Customer Service Dept. at 877-266-5796 (toll free).

To Ayça, for her support, and my parents, Cynthia and Jordan, for theirs.

Table of Contents

Part 2. Neural Engineering: The State of the Art

Preface

I am not the first, and certainly will not be the last, to stress the importance of coming developments in neural engineering. This field has all the hallmarks of a broad technological revolution, but larger in scope and with deeper tentacles than those accompanying both computers and the Internet. What justifies this strong claim? The brain is not merely just another organ in the body, albeit more complex by many orders of magnitude than any other; it is the gateway through which all sensation must pass, and it is only via the brain's conscious emanations, thoughts, that anything whatsoever can be experienced. To modify the brain, then, is to modify not only how we perceive but what we are, our consciousnesses and our identities. The power to be able to do so cannot be overstated, and the consequences scarcely be imagined, especially with our current unmodified evolutionarily provided mental apparatuses.

Here are just a few of the topics that we will cover, the realization of any of which more than justifies the claim that this will be a technology like no other:

i) Brain-machine interfaces to control computers, exoskeletons, robots, and other devices with thought alone;
ii) Mind-reading machines that will project the conscious contents of one's brain onto a screen as if it was a movie;
iii) Devices to enhance intellectual ability and to increase concentration;
iv) Devices to enhance creativity and insight;
v) And finally, and most significantly, mechanisms to upload the mind to a machine, thus preserving it from bodily decay, and bodily death.

The book is divided into three sections. The first develops the neurophysiological as well as philosophical foundations on which these advances may be made. The second describes the current state of the art, and neuroengineering developments that will be with us in the near term. The final part of the book speculates on what will happen in the long-term, and what it will like to be a postevolutionary entity. Corresponding to these three divisions are three approaches to reading this work. The newcomer to the brain and the philosophy of mind will want to concentrate on the first section, and then approach the rest of the book as interest determines. The technologist will be inclined to focus on the second section, although the first part of the book should not be spurned. There are important issues centering on consciousness and identity, for example, that will ultimately determine how neurotechnology develops. The futurist will naturally be drawn to the final section, but in their case it is all the more imperative that the initial development be mastered, especially the chapters with a philosophical bent. The uploading of the soul to a machine is not just a matter of creating the proper technology; it is first and foremost figuring out what it means to have a soul.

What this book is *not* about is as significant as its contents. First, this is not about genetic or other biological-based enhancements of the human species. As important as these developments will prove in the coming years, they are outside the ken of my knowledge, and I cannot comment intelligently upon them. More to the point, it is likely that neuroengineering will end up eclipsing these developments in the long run. To take memory as a paradigmatic example, there may very well be drugs on the market in the next few decades that will preserve memory and may even enhance the creation of new memories. But any developments along these lines will be limited by inherent and unavoidable biological constraints. As discussed in Chapter 2, the necessity of having a protective skull also limits the brain's ability to grow, and with a fixed number of neurons there can only be so many memories—this is a strict and unavoidable mathematico-physical fact. On the other hand, there seems to be no limit to the size that memory could grow if the brain was augmented by external storage devices. Of course, there are serious challenges ahead in achieving this, including how to interface this device with the brain, how to access the memories on the device once it is connected, and how to integrate recalled memories with other cognitive functions, but these, in the end, are technical problems, not problems in principle. They will be overcome, haltingly at first, and then more consistently, and then the fact a fish oil extract when taken with one's morning coffee wards off memory decay by 12% will fade into insignificance.

The second thing this book is not about is the ethics of neuroengineering. Again, this is not to say that there are extremely important issues that must be confronted. To take just a few examples:

i) If we are able to engineer bliss, how will we prevent addiction to this state?
ii) The mind of a murderer is uploaded into a machine, but in such a way as to lose the memory of his crime. Is he still responsible?

iii) Suppose "Big Electrode" develops a killer app, say a special form of deep brain stimulation (see Chapter 7) that allows it to control people's thoughts. Who will regulate such a technology, and will we be able to do it before it is too late?
iv) Finally, suppose we are able to become immortal through uploading. Given limited resources, how will we decide who is lucky enough to undergo this, and who does not?

The developing field of neuroethics is currently attempting to answer these questions, but it is unlikely that it will be able to keep pace with coming developments. The problem is that we falsely believe that we have resolved many ethical dilemmas whereas in reality we have simply produced reasonable compromises that meet the demands of current circumstances, and it is only by a wholesale return to fundamental premises of ethics that we will begin to make progress.

Should this then limit our movement in this direction? I, along with many others, say no. The fact of the matter is that the body is made for one thing and one thing only. It is a conduit for genetic material. Our current governess, evolution, does not operate at the level of the individual; it cares not a whit for him or her. Its aims are not our aims, its goals not ours. If we are to be happy, truly happy, if we are to think great thoughts and not thoughts about aches, pains, itches, and mortgages, if we are to discover the laws that govern the Universe, what it is made of, and not merely be shot around like a pinball from life event to life event, we must, to some extent, subvert these very laws. We must take nature by the throat rather than vice versa.

Or course we will make mistakes along the way, some of them bad, some catastrophic. The trajectory on which we will be thrust by becoming self-modifying agents is radically different than the gentle course evolution has placed before us. But in the long run, it can only to be our advantage to have a choice about the only thing that is truly our own, the contents of our consciousnesses. In summary, it is not that the ethics of neuroengineering is superfluous; it is just that these concerns will asymptotically approach minimal importance as time progresses. When we do have control, not just to be blissed out, floating on a cloud in a blindingly beautiful virtual sky, but to be able to concentrate intensely for 100 or 1000 hours without sleep until we solve a particularly thorny problem in mathematics or metaphysics, we will want to seize that opportunity and there will be no good reason not to. This book describes how we might be able to make this happen.

There is another objection along these lines, subtler, but just as powerful, and often voiced. This is the objection from naturalness. Is the world as it is structured now not sufficient, is not the course of a life as it currently unfolds not beautiful, and does not existence already contain within it an energetic mystery sufficient for all our needs, and if so, why gild the lily? To which I can only heartily agree but with an important caveat. How many moments in a day is this beauty apparent, and how many of us have even these few moments? Yes, the body is a thing of grace and beauty, except. Except when it is cold, when it is tired, when it is hungry, when it is bloated, when it is working in a cubicle on a report that it is already late that it doesn't care about in the first place, when the kids are not fed and the dog still needs to be walked, when it is hungry and it is coming down with the flu and just missed the 6:37 because of that report and it will have to wait another hour in the station. It is only in those sparse punctate moments when it is lounging, for example, on some Caribbean beach and the first piña colada of the late afternoon starts to kick in that the body is truly convinced that it was placed on the earth not for penance but as a celebration of what surrounds it. Neural-based technology, then, is not about having more—we have enough, and we *have* had enough every since we invented fine silks, cigars, and baseball and largely eradicated childhood diseases like polio. It is about fully drinking in what is already here. Contra the naturalists, we were not constructed to do this by biology, but this does not mean that this is not what we can become.

Nature also never intended for us to live much beyond our childbearing years. To the transhumanist, death is not so much a tragedy as an absurdity. The idea of halting in midstream simply because a pump stops working or because a few cells start to act up is a fate that no one would submit to given the choice. We would not accept this in our car or house or electronic devices, nor should we accept it in the case of our most important material possession, the body. As will be argued, the only sure way of preventing entropic decay is to extract the soul from its current housing, the brain, and render it into a purely digital form. The endgame of neuroengineering is not necessarily the cessation of death, though, but the ability to choose the timing and nature of termination. As we will see, this includes *inter alia* the absorption of the smaller, birth mind into a larger and more comprehensive composite identity.

Writing a book like this is a bit like being Gromit in "The Wrong Trousers" where he must furiously layout the train tracks just ahead of his speeding carriage. In particular, many of the developments discussed in the third part of the book depend on a concrete exposition of how the self could be virtualized. The literature on identity is by no means sparse, but there is little on the precise conditions necessary for identity to be preserved across separate media, as opposed to generalities regarding the possibility or lack thereof of this transfer. As in many areas, until one needs to actually implement an idea, one can afford the luxury of vagueness; this was not an option open to me if I was to give a recipe for identity transfer, as I do in Chapter 11, for example. Thus, I often had to shore up conceptual difficulties before proposing technological developments to come. Overall though, I have to admit it was a pleasure speculating on these matters, and my only regret is that my journey was not aided by neural-enhancing devices still beyond the current technological horizon.

ACKNOWLEDGMENTS

I wish to offer heartfelt thanks to all humans, transhumans and others sentient beings who have made this work possible by their prior efforts. Thinking about thought is not easy and I would still be at metaphysical square one if not for their sacrifices. I would especially like to thank my colleagues on the fNIR project Allon Guez and Youngmoo Kim, and all of the crew that helped collect the data under the supervision of Diem Bui. My friends in the aesthetics community also provided valuable intellectual support over the years, and I thank them and wish to mention in particular Colin Martindale and Paul Hekkert. I am also grateful to Anthony Freeman for shepherding my consciousness paper over rocky terrain and three anonymous reviewers; this paper became the backbone for much of the book. I would also like to thank ISS Inc. and Hitachi Medical Systems for their generous loans of fNIR equipment, and Bruno Dubuc for the use of images from his excellent site on the brain. Finally, I would also like to thank David Pallai at Infinity Science Press for his patience and guidance.

Part 1 Foundational Material

Chapter 1 INTRODUCTION

The year is 2054, and Leonard Moscowitz is dying a slow death. Science has long since conquered the ravages of cancer with the clever use of nanobots—tiny machines that course through the bloodstream, seeking out and destroying malignant cells and sparing the healthy ones. Artificial hearts have become routine, and many people even choose to have their heart replaced as a preventative procedure, long before problems arise. Unfortunately, Moscowitz, an accountant from Long Island, a loving husband and father of two daughters, suffers from the chronic wasting disease ALS, or Lou Gehrig's disease as it is named after its most famous victim. ALS has proven more resistant to medical advances than its more prominent counterparts, cancer and heart disease.

And yet, there is a hopeful glint in the eyes of his wife of 35 years, and in the eyes of his daughters. True, Moscowitz's body is ravaged, and he breathes only with the help of a respirator. He has not been able to swallow without aspirating for six months, and is kept alive via a feeding tube. But his mind is more or less intact. Functional magnetic resonance imaging (fMRI), a technology invented in the twentieth century and perfected in the first two decades of the twenty-first century, has shown only minor neurological deficits, and those are chiefly in the cerebellar and motor cortices.

ALS primarily affects peripheral neurons and spares the central nervous system. In the past, this has meant that the afflicted helplessly watch on as their body deteriorates and eventually gives out entirely. Such would have been Moscowitz's fate also, had it not been for a new radical procedure, a procedure that the attending neurologist has come to discuss with the family. As is turns out, Moscowitz's neurologist is no ordinary brain specialist, but one with expertise in a relatively new area, that of brain-machine interfaces. He has not come here to augment or otherwise alter Moscowitz's mental life, but to transfer it in its entirety to a new and more viable medium. He has come, you see, to upload the essence of Leonard Moscowitz into a machine. He approaches the patient's bed with his electronic clipboard in hand, and begins the session.

"Mr. Moscowitz, I have a number of questions I need to ask you. I know that you have given consent previously, but we now need your final approval. Nod your head if you agree to the following. First, I Leonard Moscowitz agree to the termination of my bodily existence."

He gives a reluctant but firm nod.

"I also understand that this is a new procedure, and it is possible that the transfer will not work. Approximately twenty-seven percent of the time no transfer takes place. In these cases, the mind dies with the body."

Again a reluctant nod.

"Finally, I understand that in approximately fifteen percent of the remaining cases, the transfer is only partial. In the vast majority of these the mind is a pale reflection of what it once was, and the legal guardians will likely choose a virtual Do Not Resuscitate, that is, will not allow you to continue your suffering."

The third nod of the head, as if to say, "what choice do I have doctor?"

"I am sorry to be so blunt, Mr. Moscowitz, but you and your family need to know the risks. Our hospital is one of the leaders in the field, and we'll do our best to make this work. There is every likelihood that tomorrow you'll wake up in your new robo-body and feel better than you have in fifteen years."

Moscowitz slowly shakes his head once more as the neurologist leaves the room. Mrs. Moscowitz gives a quick peck on the forehead of husband, and then runs to catch the specialist as he walks briskly to his next patient.

"Doctor? Just a second, doctor."

Reluctantly, he slows his step; as usual on Tuesday rounds, he is already an hour and a half behind schedule.

"I know we met in your office and you explained everything to me. I went on all the sites and read everything I could, even the technical articles that I didn't understand. And my bridge partner, Bonnie Featherstone, I think you did her husband, anyway, she said it was all a big nothing. One day her husband was in the hospital and the next week he was out on the golf course with a 2 handicap. He eagled the third hole. His handicap in his original body was ..."

"Excuse me, Mrs. M., but I do need to get to my next patient. Did you have a quick question, or perhaps you'd like to schedule an appointment, in which case you can contact my avatar at the hospital's Web site."

"As a matter of fact, I do have a question doctor. You told us that if this works, mentally, it will be as if he never left us. The body will be different, I know that, and just between the two of us, I'm not complaining on that score. But I have just one question. Will the man that you give me tomorrow be the man I married? Doctor, tell me, will it be my Lenny?"

My Lenny. Two simple words, one a possessive and one a proper noun. With these two words, however, Mrs. Moscowitz has perhaps unwittingly stumbled upon what are arguably the two central questions in the Western intellectual canon. Furthermore, as will be demonstrated, without proper answers to these questions, progress in neural technology, the science and art of enhancing or otherwise modifying the human mind, can proceed only so far.

We can rephrase the dual aspects of Mrs. Moscowitz's question as follows:

1. After the transfer, to what extent will the new entity resemble the original?
2. After the transfer, to what extent will the entity be sentient, that is, to what extent will it have the same sensations as before?

The first question is ultimately a question about personal identity. What are the constituents of a personality, that is, why do people appear to external observers as unique individuals? Is it a matter of some undiluted essence, or is it a matter of having the right memories, or the right behavioral tendencies? The second question is ultimately about consciousness. What is it like to be a person, and under what conditions can mere matter be imbued with thoughts and feelings? Mrs. Moscowitz, of course, desires that the transfer be veridical in both senses. She wants her new husband to act in an identical fashion to her old one, and because she loves him, she wants him to feel as if he just fell asleep and awoke in a healthier and more fit body.

The second question is the more difficult (in fact, some skeptics believe that it may not have a proper scientific answer, as we shall see in Chapter 4), but let us take it up to begin with because without a positive answer, any answer to the first question regarding personal identity is moot. What Mrs. Moscowitz wants, and certainly what her husband desires, is that the new entity formed by the transfer not be a mere imposter, behaving and acting like the old Lenny, but "inside" feeling nothing at all. In short, the operation can only be considered a success if it avoids producing what is called, in the philosophical literature on consciousness, a zombie, a walking talking simulacrum of a human being that has no inner life whatsoever. It must produce a truly sentient being, with real pains, real joys, and real sensations.

For some insight on this concept, let us briefly consider the notion of blindsight. *Blindsight* results from damage to the primary visual cortex (area V1 in the occipital cortex of humans). Depending on the size of the lesion, the damage may cover the entire hemifield (left or right visual field), or may spare some portion of that field. This topic will taken up in full in Chapter 4, but for now we note the following properties:

a) Patients report having absolutely no visual sensation in the impaired field. Remarkably, however, if a light, for example, is flashed in the upper or lower part of the damaged field, and the patient is asked to guess in which part it appeared, upper or lower, they can answer with a fairly high success rate.

b) Experiments show that discrimination is possible for a wide range of visual attributes, including orientation, color, motion, and shape. In all of these cases, however, there is no conscious perception at all. Judgments are made but nothing is seen.

c) Although, as one might expect, blindsight does not have the same acuity as normal vision, careful experimentation also reveals that blindsight is not qualitatively similar to normal degraded vision. That is, it is not the same as when you can't recognize a face because of dim light. It truly is the ability to discriminate without the ability to consciously perceive.

A number of experiments also purport to show, under special conditions, blindsight in normal observers. For our purposes, perhaps the best example of this is the common experience of driving along lost in thought, when suddenly you realize that you haven't been paying attention to the road for the past thirty miles. The shocking part of this experience occurs when you realize that your conscious perception of the highway has been either weak or nonexistent. But it is easy to show that you have been processing visual stimuli, if not paying attention to them. If something unusual happens, say the car in front suddenly slows down, your foot will automatically move to the brake (full visual consciousness will probably return, also).

Now, let us generalize blindsight to processing in general. Suppose that in addition to a partially damaged hemifield, both visual fields are fully damaged. Further suppose that in addition to blindsight, our hypothetical being has "deafhearing," "dulltasting," "numbfeeling," etc. That is, he lacks an inner life entirely, but, and this is the key, he responds as if that life is fully present.[1] Further suppose that the transfer operation results in this sort of being. In this case, one certainly could not claim to have transferred anything at all from its original sentient form. In other words, not only would it not be Mrs. Moscowitz's Lenny, it wouldn't be anyone's Lenny, because no one would be "home".

Thus, it is not sufficient that the new Lenny act like the old Lenny. If he is not also conscious in the way that he was before, he would be worse than an imposter, because even an imposter has an inner life. The problem of consciousness,

[1]Whether this is possible with human beings as currently constituted is doubtful. For example, blindsight, as remarkable as it is, is quantitatively weaker than normal vision in almost all respects, and other qualitative differences are present, such as that between spontaneous and forced choice decisions. However, at the very least, such a being is a conceivable entity, especially when abstracted from realization by the human brain.

then, is crucial for the success of future neural technologies. It is a problem, however, that is a vexing one to say the least. Nevertheless, we will have the opportunity later in this book, with the help of a few reasonable assumptions, to present a framework indicating the necessary conditions for producing the real Lenny, i.e., the one that makes the new Lenny feel just like the old Lenny. Moreover, we will be able to show under what conditions

a) a neural-enhancing device will have no effect on conscious content
b) such a device will indirectly affect conscious content by altering the inputs to the parts of the brain that are implicated in consciousness
c) the workings of the neural device are directly included in the contents of consciousness

For now, we simply note that although condition c must, to some extent, be present if a full transfer of personhood from brain to machine is to take place, a and b are interesting in their own right, and certainly important in the development of technologies prior to that of full transference. For example, most of us would be delighted if the answer to an arithmetic problem would appear in our mind's eye merely by formulating the problem. For example, one would think "The square root of 1783" and 42.23 would appear in an internal head's-up display. This is an example of case b. Nor is case a) to be dismissed lightly. Many tasks carried out by the brain, especially those concerned with motor response, are entirely divorced from consciousness. For example, highly automated tasks, such as riding a bike, involve decisions outside of awareness (and if they did involve awareness, it is unlikely they could be made with such efficiency). Already, the technology exists to help paralyzed patients carry out motor tasks with the aid of external devices; this is reviewed in Chapter 6.

Let us now assume, for the sake of argument, that case c has been solved, i.e., we can provide an affirmative answer to the second question regarding the possibility of transfer of sensations from man to machine. There still remains the first question, namely, to what extent will the transfer of a particular personality be possible. After all, it is necessary but not sufficient that the new Lenny be a feeling entity. In addition to having just any sensations, he must have the sensation of continuity from the old entity to the new entity, and, of course, he must appear, to the person who knows him best, Mrs. Moscowitz, to act like the old Lenny.

Once again, we are in partially uncharted territory, and a full answer to the question of personal identity awaits future research. Nevertheless, unlike in the case of consciousness (in which there sometimes appear to be more theories that data points!), there is a more or less clear line of thought extending from Hume to Ryle to present philosopher-psychologists, such as Daniel Dennet, and one that is difficult to refute—having a personality, or being person x instead of y, is not a matter of being in possession of a special essence (Chapter 5 will present a fuller explanation). That is, it is not like the oft-appearing situation in science-fiction/horror movies in which the soul is imagined as a confinable, if ethereal substance that can be injected at will into another body, the body of the opposite sex, or even a dog. As will be argued, if you create a machine that is capable of sensation (i.e., one that is not a zombie), and if you imbue it with Leonard Moscowitz's memories and tendencies of thought, it will appear to that machine that it is Leonard Moscowitz. That is, the transfer of personal identity is a well-defined scientific problem, even if the technology to accomplish it as such is currently well beyond our reach.

This is at once an unsettling and promising proposition. It is unsettling first because there are a number of seeming paradoxes that arise, such as the possibility of keeping the old Lenny alive after the new Lenny is created. These will be treated in due course. Another unsettling aspect of this scenario that we will briefly touch on now is that this scenario runs directly counter to our notion of ego/body continuity. Ego continuity is intimately tied into bodily continuity for good reason. When we wake up from *sleep* (unconsciousness punctuated by brief periods of internally generated sensations called dreams) we are usually in the same place where we went to sleep, and we have more or less the same body. It is natural to entwine this bodily continuity with ego continuity. However, future technologies will allow us to divorce these two notions. This means, among other things, minds need not have a single embodiment, as previous stated, but also that minds will become more fluid objects, encompassing new forms of knowledge, new modes of being, or even blending with other virtual souls. These will be discussed more fully at the end of this book.

The most promising immediate consequence of this scenario is that the self will enjoy a continued existence above and beyond its original bodily form. Of course, capturing a lifetime of memories and behavioral tendencies (really just another form of memory, as we shall see) is by no means a trivial proposition. However difficult this may be, it is of a different and lesser order of difficulty entirely than trying to first extract a soul from a body and then placing it another body or machine. We do not have a science of souls, most likely because the notion of a soul is a reification error. It is an error based on elevating a group of contiguous spatio-temporal activities that are tied together because they emanate from a more or less constant form, a particular body, into a concrete essence that has its own physics and spatio-temporal boundaries. But it is unlikely that the soul has such an independent existence, and it will suffice to turn our attention, as we will see, to the collection of patterns of connectivity embedded within neural synapses in order to "lasso" it and subject it to technological manipulation.

Having now given tentative positive answers to the possibility of first creating a machine with sensations, and second, one that experiences a particular combination of sensations and actions that is ordinarily termed a personal identity, our initial conclusion is that the upload of a mind from its native brain-based instantiation to an alternative form is indeed possible. There is a third and final additional question, however, also implicit in Mrs. Moscowitz's. concern regarding her reconstituted husband. Assuming the new Lenny is, for all intents and purposes, the old Lenny with a more metallic sheen, then there still remains the issue of how long he will stay this way. Will his new existence prompt further changes in his behavior that ultimately will cause him to evolve a new personality, and one that his wife may not look upon with as much favor?

An alternative way of putting this is as follows: To what extent will coming neural technologies alter what it means to have a human or any sort of sentient existence? Here, intellectual honesty forces us to provide Mrs. Moscowitz with a somewhat more negative answer regarding the integrity of her postoperative husband. As we shall see, it is unlikely that a mind set free from its original bodily constraints, and furthermore possibly given new and better intellectual faculties, will remain as it was before for very long. To pursue this in more detail, let us take a brief excursion into the history of technology in general, and the way in which neural technologies are fundamentally different than these earlier developments.

All technological development from the invention of the wheel to the present may be placed in four broad (nonmutually exclusive) categories: transportation, transmutation, automation, and information. The first two technologies emerged first. Man's role in the evolutionary scheme is that of a generalist, and as such, we are neither fleet of foot nor is our endurance particularly noteworthy. From the earliest days, it was therefore necessary to tame animals that did possess these qualities in order for us to traverse long distances. Furthermore, while not lacking completely in protection from the elements, we fare rather poorly with exposure to these elements relative to almost any animal. This, along with associated demands from transportation, encouraged the transmutation of substances from their original form into those capable of serving as shelter. Beginning roughly at the turn of the nineteenth century, increasing mastery of the physical world permitted automation of the tasks necessary to produce these inventions, culminating in large semiautomated looms and ultimately in Henry Ford's mass-produced automobile. Yet none of these developments have fundamentally altered the nature of existence: a microwave oven may make life more convenient, but it does not alter the fundamental trajectory of the human condition from childhood to love and work and, ultimately, to death.

Projecting into the future does not necessarily alter this claim. *Star Trek* (the original) is a good example of this kind of linear projection, and in fact its creator, Gene Rodenberry, called it "*Wagon Train* in space" to emphasize that he was simply placing a conventional story in a somewhat unconventional context. With the possible exception of mind melds and the occasional bizarre energy configurations in deep space that confounded the ship's sensors, on the Enterprise it's life as we know it Jim. The ship may run at warp speed, and McCoy may order the onboard computer to mix a dry martini after a difficult day in sick bay, but the characters are at once recognizable, sympathetic, and most importantly for the current purposes, more or less static. *Star Trek* is the direct projection of the present into a future of molded tracksuits and smoky drinks on demand.

The one plausible exception to this view of technological development may be found in information technology (IT). To give the fast pace of current information technology change its due, let us imagine purchasing online not today, but on the enhanced Internet7 circa 2015, via the holographic web (most of us still have Internet1, the original Internet; Internet2 is a faster version of its predecessor which is slowly being introduced). To choose your clothes you navigate through a virtual store. As you point to items, they are displayed on a prestored version of your body; autoslimming and a variety of other enhancements can be set in the preferences. The key difference between this scenario and clothes purchased by a Mongolian on the steppes of central Asia in the twelfth century, and to a lesser extent, current methods of purchase, is not so much the fact that the clothes can be virtually viewed (although this is a nice feature). The main difference is the selection on offer. Prior to the Internet, if the local tailor does not carry or cannot manufacture a piece of clothing, you are out of luck. In contrast, the bandwidth of Internet7 (plus some creative programming on the server side) will permit one to purchase a pair of golf shoes, a one-piece suit that maintains constant body temperature in all climates, or indeed, a reproduction of a Mongolian warrior outfit. One can also envision a robo-tailor in each zip code that sews your choice in a matter of minutes, and ships the resulting product to your door via pneumatic tube. This would constitute a kind of bidirectional Internet, with information flowing one way, and products in the other direction.

Developments in IT such as these have a greater chance of fundamentally altering the nature of existence because they introduce a degree of flexibility in one's interaction with the material world in a way that is both impossible and also barely conceivable in their absence. However, the sorts of developments made possible by IT will pale in comparison to those offered by the coming neural revolution. The reason for this is that by altering the brain, and thereby mutating human consciousness, we are not merely changing the environment in which this consciousness dwells, but altering the very nature of identity. To say that the brain is the filter by which we view all things is a strong claim, but it is also an understatement. The brain is at once the generator and originator of those things. One does not have to be a philosophical idealist, that is, one who believes that there is nothing but thought and that the world is an illusion[2] to see that this is the case. We do not see wavelengths directly; we see colors, which are the brains interpretations not only of the wavelengths in particular regions of visual space but also the entire set of wavelengths in the visual field (see Chapter 2). For sounds, the situation is likewise indirect. Technically, what the ear is responding to is variations in pressure in the air, and the frequencies of those variations. But what we hear goes well beyond the detection of frequencies. We hear sound in chunks: words, notes and chords, melodic phrases, and so forth. This grouping function is one among many transformations that the brain performs that ultimately determines the contents of conscious thought.

The technical name for the idea that brain states determine mental states is *supervenience*. This principle is accepted implicitly by almost every current neuroscientist as well as explicitly by most philosophers of mind. What it means is that one cannot have two different thoughts without having two different brain states. In other words, mental states, or thoughts, are dependent on the brain for their realization. For our purposes, we are mostly concerned with an equivalent formulation of this principle. This states that in order to effect a change in mental states, we need to alter the physical substrate of those thoughts, the brain. Or, in engineering terms, the ability to tweak the brain into a given state, or to otherwise enhance brain-based processes, implies the ability to bend, create, or otherwise modify thought.

Here is a just a partial list of technologies, treated more fully in Part III of this book, that this ability will engender:

- Short-term and long-term mood alteration
- Auxiliary memory and computational devices
- Enhanced sensory input
- Accelerated and improved thought
- Direct mind-to-mind communication
- Direct acquisition of fields of knowledge without lengthy learning processes
- Alternative and more direct artistic experiences
- Survival of bodily death via uploading to alternative media
- Merging of identities once these have been captured in digital form

As is easily seen, this is not merely about making life more comfortable or more convenient, the goal of prior technologies. This is altering life to its core, by alter-

[2] Bishop Berkeley, perhaps the most prominent advocate of this view, famously asked whether a falling tree in the forest is heard with no one around. In the strictest interpretation of idealism, the answer would have to be no. Only thoughts are real, according to this philosophy, and if something is not contained in a thought, it did not happen.

ing human consciousness directly. However, before we broach such heady topics in full, this book will endeavor to provide both a firm understanding of first how the brain works. We will also touch on the topics of consciousness and personal identity, two key foundational areas for the development of neural technology, for reasons already mentioned We will then explore current neural technologies, how they build on our current knowledge of the brain, and how limitations in this piece of "wetware" present barriers to progress. We then look at the near future (5 to 20 years), in which many of these barriers will be overcome, and the start of a true science of neural manipulation will begin. Finally, we will see how these initial developments will culminate in the ultimate neural technology, identity uploads.

Let us now return to the case of Leonard Moscowitz. Having tentatively concluded that it will someday be possible to transfer his identity from the dying husk of his birth body to a more viable form, we were able to delight both his wife, and of course, Lenny himself. This download, however, is the start and not the end of the story. Let us begin modestly and assume that in his new state he chooses to lead a happier and more fulfilling life. The alteration of the workings of the brain, in its native state, is by no means a trivial affair. This can be seen, for example, in the current generation of psychiatric drugs. When they do work, they achieve their effects by altering large-scale processes in the brain via one or more major neurotransmitter systems. No doubt, future generations of these substances will be improvements on the current ones, but there is only so much one can do if one cannot control the fine-scale dynamics of the brain.

Now, however, having acquired the code for Leny-ness in the process of uploading him, so to speak, it is a far simpler matter to tweak his behavior and thoughts in more precise ways to achieve his emotional ends[3]. And why should we ask him (or anyone) to suffer more, especially after what he has been through? But the one thing we cannot do is effect this change and have Lenny remain the same. A less-cranky Lenny is not the same Lenny; he will act differently, he will think differently, and he will feel differently. It is hard to see him objecting to this, and Mrs. Moscowitz may also welcome the change, if at first it appears a bit unsettling. What if, though, Lenny begins to partake of some of the other possibilities that his new state has opened? The reason that he was originally drawn to accountancy, as opposed to nineteenth-century Romantic poetry is that he was both good at numbers and liked the order they imposed upon the world. Why though, should, he remain that way? If Byronic impulses are just another set of contingencies, they could easily be incorporated into his new persona.

[3] A more directed scenario for doing so will be presented later, but for now one can envision a "genetic" strategy of simply changing a few things at random and then seeing if his mood changes for the better.

In the initial months of his new life, he may be happy just to think as clearly as he did when he was twenty-five, and to possess the emotional exuberance of a younger man, but one day he is going to start flipping through the catalog of "courses" he can take as a virtual being: jazz drumming, the Hajj, dolphin intelligence, etc. These, though, are not lecture courses, but being courses. To pass such a course means, to a greater or lesser extent, to have incorporated aspects of these alternative identities into one's own. In essence, Leonard Moscowitz will be born again not once, but many times, and the final Lenny may exist only as a small kernel in the man that he will become. This may be a journey that his wife does not wish to follow, or, as is likely, cannot follow given her more limited embodiment.

Worrying about the marital status of disembodied superenhanced beings is something we can safely put off for the moment. Our first task in the treatment of neural technology is to: a) review both the functional anatomy of the brain as it is currently constituted, b) understand how computation is achieved in this remarkable device, c) show how personal identity arises in the brain, and d) discuss the theoretical and empirical results relevant to consciousness. These are the topics of the first third of the book, and once they have been considered, we can move to the second part, a description of the state-of-the-art in this field. Finally, in the third part, we will attempt to provide a scientific basis or some of the stronger, and no doubt counterintuitive claims made in this introduction.

Chapter 2 The Brain as Designed by Evolution: Functional Neuroanatomy

The more one examines the brain, the more one is impressed with evolution's accomplishment. Studying this organ presents an unprecedented challenge to scientists, as there are an almost unlimited number of phenomena to be explained, and many different levels at which they can be understood, from the molecular to the neural to the psychological. Mysteries pop into existence as fast as old ones dissolve, as if one is embedded in an elaborate intellectual video game set at a level just above that of one's current capability. None of this should be surprising; it is likely that the complexity of the brain far exceeds the complexity of the rest of the systems in the body combined. If it is also conceded that the body is the most intricate of any existing system, including man-made ones, we arrive by transitivity at the conclusion that the brain is the most complex system on earth.

This alone makes our task daunting, but as neuroengineers we must do more than pay homage to this remarkable design. As well as marveling at the elegance at which natural selection has crammed so much functionality into such a small space, we must also look at the brain as a work in progress, short of both its evolutionary end point, and more importantly, short of where we as engineers can carry this device. In short, neuroengineering demands of us both awe and a well-honed critical faculty, and the task before us is to a make a reasonable accommodation between these intellectual sentiments, which are normally at odds. This cannot be undertaken lightly, but it is by no means infeasible, and the first step in doing so is to understand the functional neuroanatomy of the brain in its current unmodified form.

We begin our survey with an overview of the gross anatomy of the brain. We then look at the elemental component of the brain, the neuron, and why it suffices to understand this component to understand the brain's behavior. We next examine some critical circuits and modules in the brain as a means of bridging the gap between anatomy and physiology. Finally, the chapter closes with a critical discussion of the limitations of the human brain. This last section would typically not appear in ordinary reviews of this topic, but it is one that is absolutely crucial for our purposes. As previously stated, the goal of this chapter, and indeed of this entire work, is to move beyond the notion that we must accept what Providence has given, and to think about what could be as well as what is.

THE BRAIN: AN EVOLUTIONARY OVERVIEW

There are many ways of viewing the organization of the brain. Perhaps the most natural, and the path most often taken, is to allow pedagogy to recapitulate phylogeny; that is, to describe the brain in terms of its evolutionary development, and that is what we will do here. Figure 2.1 presents three views of the brain; the first, an actual brain; the second, an abstraction, in which the lobes and some functional areas are labeled; and third, the most abstract, in which the brain is divided into three broad regions. The latter abstraction, the triune model of the brain first proposed by Paul Maclean in the 1960's (Maclean, 1990), illustrates how the brain has evolved in a succession of strata, rather than as an entirely new design at each phylogenetic step. The reptilian brain dates back approximately 300 million years and represents the most primitive aspect of the nervous system, responsible for regulating autonomic functions such as heart rate, temperature, and breathing. The mammalian brain emerged 100 million years later and is associated in humans with the limbic system, a set of tightly interconnected structures responsible for emotional and memory-related functionality. Finally, the neocortex ballooned to its current size with the emergence of Homo sapiens approximately 120,000 years ago; the neocortex occupies almost the entire visible brain at the top of Figure 2.1, with the exception of the cerebellum, which can be seen at the lower right.

Maclean conceived of the three divisions of the brain as operating quasi-independently and on equal terms with each other. Thus, the response to a given stimulus will divide into three components: a reflexive aspect generated by the reptilian brain, an emotive response by the limbic brain, and a reasoned response by the neocortex. In practice, usually only one of these responses will result in direct action, but it is important to note that the winner of the competition between the three will not necessarily be the neocortex. Moreover, when this higher layer is dominant, it is almost always under some degree of influence from activity from the two lower layers, which percolates upward and serves to

A

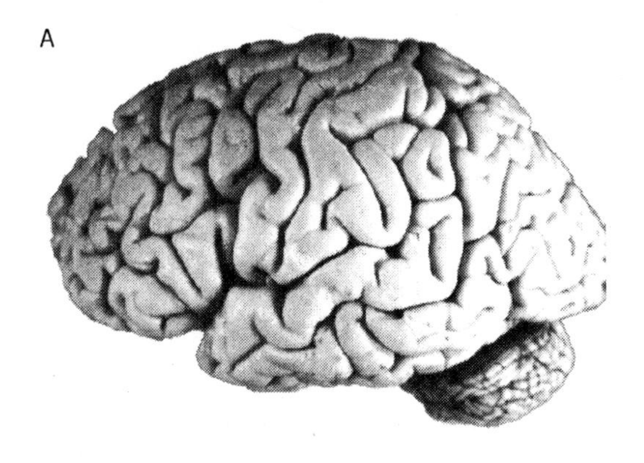

B

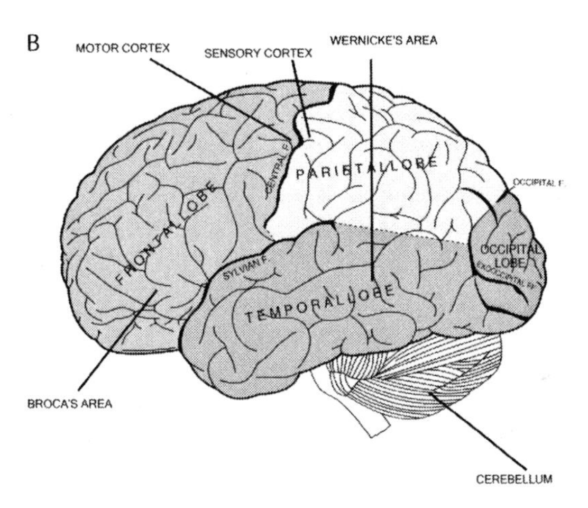

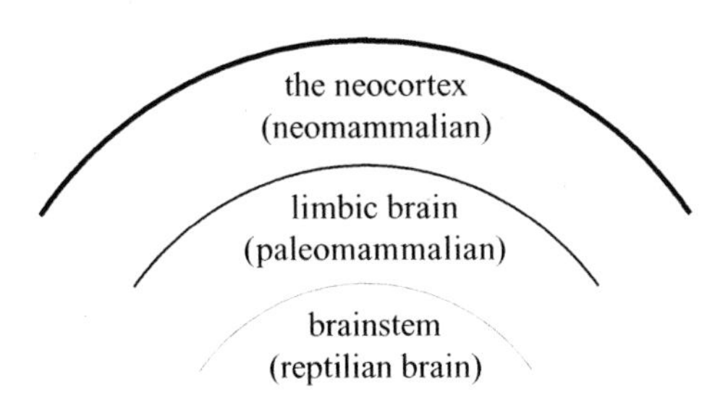

FIGURE 2.1 **Three views of the brain. An actual brain, a brain labeled by with lobes and major functional areas labelled, and a schematic of the triune brain.**
SOURCE: **A: http://www.healcentral.org/healapp/showMetadata?metadataId=40566. B: http://commons.wikimedia.org/wiki/Image:Gray728.svg.**

modulate action and thought. We will now consider each of the divisions in the triune brain and the major structures in each in detail.

The reptilian brain

The reptilian brain comprises the medulla, pons, cerebellum, and the mesencephalon. It may be thought of as that part of the brain that is reactive as opposed to contemplative, transforming an incoming stimulus directly into a motor action, assuming that this action is not inhibited by higher brain centers (there are no lizard philosophers, nor do they spend much time contemplating the meaning of life). This does not necessarily mean that the generated action sequence is not complex, however, and the human cerebellum in particular is a remarkable computational device; hence its name, Latin for "little brain." We will now describe the structures in the reptilian brain, including the cerebellum, which are shown in Figure 2.2.

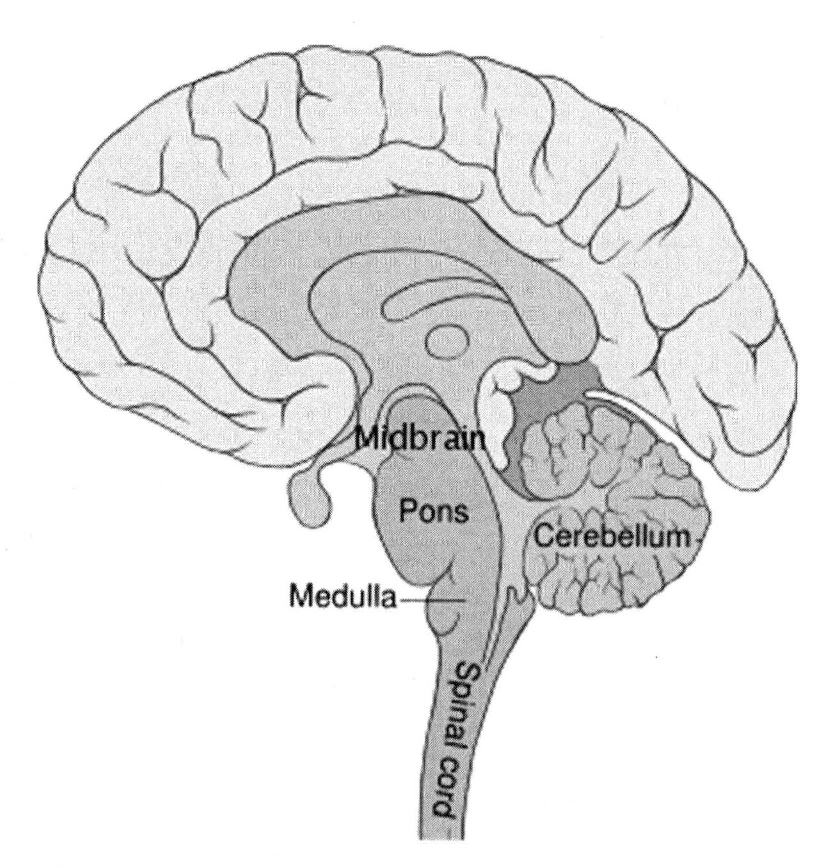

FIGURE 2.2 **A sagittal section showing the components of the reptilian brain.**
SOURCE: **http://commons.wikimedia.org/wiki/Image:Brain_bulbar_region.svg.**

- Medulla
  The medulla is responsible for the regulation of several autonomic functions including respiration, heart rate, and blood pressure.

- Pons
  The pons also plays a role in the regulation of respiration, and in addition acts a relay between the cerebellum and the neocortex. This structure may also help regulate arousal as part of the general ascending reticular activating system (see the fuller discussion of the arousal system, below), and may play a role in the production of dreams.

- Mesencephalon or midbrain
  The midbrain is just superior to these structures and contains the inferior and superior colliculi, the tegmentum, the crus cerebri, and the substantia nigra. The superior colliculus receives information from the retina and is involved in the generation of non-voluntary eye movements (saccades); in Chapter 4 we will discuss this structure as an alternative route for visual information to the cortex in the case of blindsight (visual recognition in the absence of consciousness). The inferior colliculus is an important structure in the processing of auditory information. The midbrain tegmentum is a network of cells involved in homeostatic regulation. The substantia nigra (Latin for "black substance") is responsible for dopamine production, a neurotransmitter implicated in reward and also the regulation of movement. Damage to this area results in Parkinson's disease and possibly affective symptoms; in Chapter 7 we will discuss deep brain stimulation as a means of compensating for these effects. The crus cerebri contains motor tracts from the brain to the spinal cord and back.

- Cerebellum
  The fact that up to 50% of the total number of neurons in the brain are contained in this structure is some indication of its significance. The cerebellum itself can be divided into three parts—depending on the stage in phylogenetic development. The archicerebellum (old cerebellum) regulates balance and eye movements. The paleocerebellum receives proprioceptive input and uses this information to help anticipate bodily movements. The neocerebellum is involved in the planning of movements, and recent functional magnetic resonance imaging (fMRI) evidence suggests that this structure may also play an as yet unknown role in non-motor-based cognition. Together, these sections combine to provide a feedback system to the motor cortex that produces routines to accomplish a given motor task; this is discussed in more detail in the section on the action system.

The limbic brain

Broca originally designated *le grande lobe limbique* as the area surrounding the corpus callosum and including the deeper structures within. He originally believed it subserved olfaction, although it is now known that only a small portion, the olfactory bulb, does so directly. It is now commonly held that the limbic system is linked to the four Fs: feeding, fighting, fleeing, and sexual reproduction. Structures in the hippocampus and amygdala are also implicated in memory, especially the laying down of long-term memories. It is not always an entirely trivial matter to determine where the limbic system ends and the rest of the brain begins[1], but Figure 2.3 gives an approximate overview of the major structures in this system and their topography.

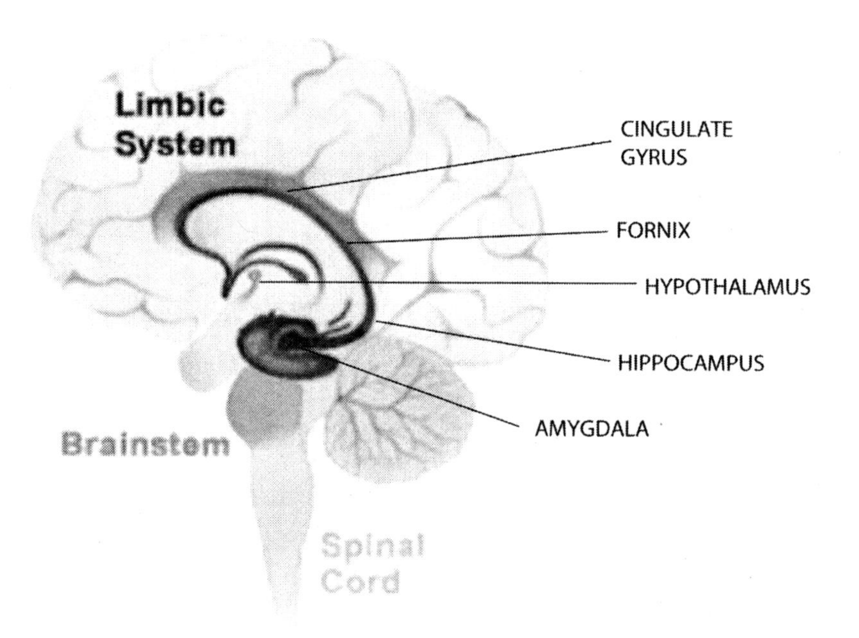

FIGURE 2.3 **An overview of the major structures in the limbic system.**
SOURCE: **http://commons.wikimedia.org/wiki/Image:Brain_limbicsystem.jpg.**

[1] Nor is it an absolutely essential matter; the brain from the broadest perspective is really just one big interacting circuit. Still, for pedagogical purposes as well as theoretical simplicity we will act as if the limbic system is well-differentiated from the rest of the brain.

The limbic system is a medial deep structure bordered below by the amygdala and hippocampus and above by the cingulate gyrus. It receives afferent connections from the prefrontal cortex as well as other neocortical structures, sends efferents back to the prefrontal cortex as well as other structures, including the basal ganglia, and also plays a regulatory role via its influence on the autonomic nervous and endocrine systems.

We will now review the role of the key structures in this system:

- Amygdala
  The amygdala (from the Latin for almond, because of its shape) has made its way into popular conception as the "bad boy" of the brain. Overactive amygdali (there is one on both sides of the brain) are supposedly to blame when we yell at a driver or when we lose our temper; not our inability to put things in perspective (a function, by the way, of descending prefrontal inhibition). There is a good deal of truth in this conception: when the amygdala is artificially stimulated, for example, animals will show characteristics of both fear and rage. However, it is also true that this is an important adaptive structure, and that a malfunctioning or lesioned amygdala will result in a wide range of abnormal behaviors. In so-called *Kluver-Bucy syndrome*, named for the researchers who discovered this disease, the afflicted animal shows a loss of fear, hypersexuality including indiscriminate mounting of both males and females by males, a reduction in maternal behavior in females, and an overall blunting of affect.

  The amygdala is also implicated in the emotional modulation of memory. Anecdotal evidence from our own lives strongly suggests that emotionally charged events are much more likely to be retained than neutral ones, and this has been demonstrated repeatedly in animal experiments. Animals with impaired amygdali show reduced ability to remember tasks that involve punishment or reward; conversely, drugs that stimulate this structure produce better performance on these memory-related tasks. One question not explicitly addressed by these experiments is the reason for the tight connection between emotions and memory. Most likely, the amygdali are encoding not a simple emotional charge but meaningfulness; given a finite memory capacity and a large amount of sensory input it makes sense to retain only those events that directly impinge on one's life, and to discard the rest.

- Hippocampus
  To a first approximation, the hippocampus may be thought of as the calmer less-emotional cousin of the amygdala. It too is concerned with consolidation of long-term memory, primarily of the episodic kind. Episodic memory is about events, and is normally contrasted with two other forms of

memory, declarative memory, that is memory about the world, such as the fact that honey is sweet, and procedural memory, that is memory of how to perform physical and other tasks, such as how to ride a bike. Probably the single most famous patient in the neurophysiological literature, H.M., suffered from bilateral ablation of the hippocampi (the result of treatment for temporal-lobe epilepsy) (Scoville & Milner, 1957). Like the protagonist in the movie *Memento*, he lived in a kind of continuous present, with each moment completely divorced from earlier events. The hippocampus has also been implicated in spatial memory. London taxi drivers have larger hippocampi than nondrivers, and the more time they spend behind the wheel, the greater this difference becomes. Hippocampal development in childhood is also thought to be affected by environmental influences; animal studies suggest that a rich environment encourages such development, whereas a stressful one inhibits the formation of postnatally developing cells (Xu et al., 1997), an effect that may have long-term consequences for the limbic system and the brain as a whole.

- Hypothalamus
  The hypothalamus, via its influence on the proximal pineal gland and its projections to a number of other brain structures, plays an important role in a number of regulatory behaviors including temperature, feeding behavior, thirst, circadian cycles, and anger; in the next section we will also examine its role in arousal. This structure is divided into a number (about 20) of nuclei each with their own relatively narrow role or set of roles. For example, the anterior hypothalamic nucleus (AH) is responsible for thermoregulation; the posterior nucleus (PN) is responsible for blood pressure increases. The various nuclei show considerable variation in size depending on gender, and in a well-publicized set of results, also on sexual orientation (LeVay, 1991). These results provide support for the claim that homosexuality is an innate characteristic, not a learned one. The hypothalamus has also been implicated in eating disorders, such as anorexia, via its role in satiation, although this is an excellent example of the danger of trying to tie a given function or disorder too closely to a single brain region. Anorexia has a complex set of determinants, including assessment at a higher cognitive level of the *meaning* of satiation signals that depends on cognitive information, such as ideal body image; this information is unlikely to be represented directly in the limbic system.

- Cingulate gyrus
  The cingulate gyrus is a large structure just superior to the corpus callosum (see Figure 2.3) and it is still not completely understood. Its anatomical position immediately suggests one possible role, however, that of

an interface between cortical regions and the other structures in the limbic system. In fact, evidence regarding the preponderance of so-called spindle cells in this region, unique to the higher primates, suggests that it may have more in common with higher cortical regions than with the limbic system (Allman et al., 2001).

The forward half of the cingulate, the anterior cingulate cortex (ACC), in particular has recently come under intense scrutiny as a possible gateway to the prefrontal cortex from the limbic system. Research originally found that high activity in the ACC is correlated with depression and, more recently, that inhibition of activity by deep brain stimulation in this region may relieve depressive symptoms. We will return to this important result in Chapter 7. The ACC has also been implicated as a possible mediator in the reward/reinforcement chain between the frontal cortex and the limbic system. In particular, it appears to be involved in error detection. When the error is affectively neutral, as in the Stroop test[2], then the dorsal (upper and therefore closer to the frontal cortex) ACC becomes active. Emotionally active stimuli, in contrast, tend to trigger activity in the ventral (lower and therefore closer to the limbic system) ACC. The importance of the ACC and the cingulate gyrus as a whole as a gateway to the limbic system from the neocortex will likely increase as neuroengineering attempts to modulate emotional states with various forms of stimulation. In later chapters we will consider a set of structures in the ventral striatum that are closely tied to those in the limbic system and are also important for the neuroengineering of affective states.

The neocortex

Recent evidence suggests that Homo sapien brain evolution was accompanied by a series of rapid and largely unprecedented genetic changes. This resulted in the mushrooming of the neocortex, and the accompanying spectacular growth in the frontal lobe in particular. It is unclear whether this growth was triggered by bipedalism and the resultant need to plan and control fine-grained hand movements, by the highly elaborate social structure of our species, or some combination of the two. In any case, it places man at a qualitative as well as quantitative reserve with respect to other species. We are the only species that has truly generative language skills, the only ones to use tools on a regular basis, and eventually, as will be argued, the only ones who will be self-modifying, by virtue of the achievements of neuroengineering.

---

[2] In the Stroop test the subject is asked to read words with characteristics incongruous to their meanings, such as LARGE and SMALL.

Before describing the structures that allow us to do this, let us first consider three important aspects of the neocortex. First, let us examine the structure of the cortex. It is approximately 3 mm thick; beneath this is the white matter of the brain. The gray matter consists primarily of the cell bodies of the neurons, and the white matter is formed by the myelinated sheaves that surround the axons of these cells. The cortex consists of six layers, each of which contains a characteristic population of neuron types and densities, although this varies throughout the brain as does the functional meaning of the processing done in each layer. Next, as can easily be seen at the top of Figure 2.1, the surface of the brain consists of numerous folds. The ravines created by these folds are known as sulci (singular sulcus) and the raised area between the sulci as gyri (singular gyrus). We will discuss the importance of folding when we examine the surface area covered by the cortex.

Finally, and most obviously, the neocortex is divided into two hemispheres. Severing the corpus callosum, the large fiber bundle that communicates between the hemispheres, illustrates that the two halves of the cerebrum serve different functions. Split-brain patients are able to name an object presented to their right hand or visual field, because this information projects to their language centers (Broca and Wernicke's areas, see Figure 2.1) in their left hemispheres. When the same information is provided to their right hemisphere, they lose the ability to describe the object. Conversely, an object presented to the left hand (right brain) can be recognized by touch with the same hand but not named. In popular accounts of hemispheric differences the left brain is analytic and linear and the right brain is holistic and creative, and the injunction is often added that we should spend more effort developing our neglected right brains. There is some truth to this account; for example, we will discuss in Chapter 9 that insight may be a function of right hemispheric activity. However, it is a vast oversimplification to state that creative activity as a whole is the province of the right brain; all mental activities involve the interaction between many neural modules, and also between hemispheres.

- Occipital lobe
  The occipital lobe, at the back of the brain, is responsible for primary and secondary visual processing in humans. This is described extensively in the section on the visual system.

- Temporal lobe
  The temporal lobe is the destination of the ventral stream of visual processing, which, as also discussed in the section on the visual system, is responsible for naming and recognizing objects in the visual field. It is also the home of primary auditory processing in the brain.

  Through its connections with the hippocampus, also located in this division, the temporal lobe is involved in the translation of short-term memories into long-term memories.

- Parietal lobe
  The parietal lobe is involved in integrating sensory information from the sensory modalities and visuo-spatial processing. It is the destination of the dorsal stream of visual processing, which computes the positions of objects in space relative to the body and thus is critical in determining motor paths. *Balint's syndrome* patients, with bilateral parietal lesions, cannot coordinate hand-eye movements as one might expect given the role of this region in motor action, but may also suffer from simultagnosia, or the ability to recognize more than one object in their visual fields at once. Other more severe attentional deficits are also possible due to parietal lesions. For example, right parietal damage can lead to neglect of the entire left side of the body, and in some cases, complete denial that this body half belongs to the patient. When asked who their left arm belongs to, patients will confabulate and say it belongs to the doctor, or that it is a fake arm.

- Frontal lobe
  The frontal lobe, the most anterior aspect of the brain, is separated from the parietal cortex by the central gyrus, and from the temporal cortex below by the lateral sulcus. The sulcus between the precentral and central gyri contains the motor cortex, and just anterior to this are the premotor and supplemental motor areas; these are described in the section on the motor system. The rest of the frontal cortex is not as well understood, but it is known that it plays a crucial role in so-called executive functions. These include tasks involving working memory, planning, and problem solving—any of these activities cause various parts of the frontal cortex to "light up" on fMRI and other imaging techniques designed to reflect cortical activity. The frontal cortex is also implicated in social functions and socially appropriate behavior, and damage to the frontal cortex may seriously impinge on these capabilities. Significantly, men with antisocial personality disorder show reduced volume in this lobe (Raine et al., 2000); it is also known that the white matter in the frontal cortex does not reach its full volume until the age of about 25 (although antisocial tendencies in late adolescence are likely to be multidetermined).

  This completes our review of neuroanatomy; we will now turn our attention to the functional unit of the brain, the neuron.

THE NEURON AND THE NEURON DOCTRINE

The human brain contains approximately 10^{11} neurons, and 10 to 50 times more glial (Greek for "glue") cells. The latter perform a number of different functions, including structural support, as their name implies, but also insulation serving to increase action potential speed and nutritive functions. But it is the neuron that is the primary functional unit of the brain, and as we will discuss next and in following chapters, it is the networks formed by the collection of these cells that are responsible for the transformation of sensory inputs into motor outputs, and also the thoughts and perceptions (both conscious and unconscious) that arise between input and output.

There are over 1,000 distinct neuron types in the brain; Figure 2.4 illustrates three common types. The cell on the left is a motor neuron from the spinal cord; here we see the three main components of most neurons, the dendrites, the cell body or soma, and the axon. The middle cell is taken from the hippocampus, and is called a *pyramidal cell* because of the shape of its cell body. Note the presence of two types of dendrites, apical (at the apex or tip), and basal, on the cell body itself. The rightmost cell is a *Purkinje cell* from the cerebellum. Note the extensive dendritic arborization, which serves to gather large amounts of input information into this cell.

The structure of a pyramidal neuron and its connections are depicted in more detail in Figure 2.5. The dendrites, both apical and basal are the sources of input to the cell. Other neurons communicate with the dendrites by sending neurotransmitters from the synaptic button (the triangle-shaped object) across the narrow synaptic cleft, approximately 20 nm wide. Receptors in the dendrites integrate this information and the resultant synaptic potential travels through the cell body and to the axon hillock. It is here that an action potential, the all-or-nothing signal that propagates down the axon, is generated. The myelin sheaf surrounding the axon insulates it and enables this signal to propagate at a

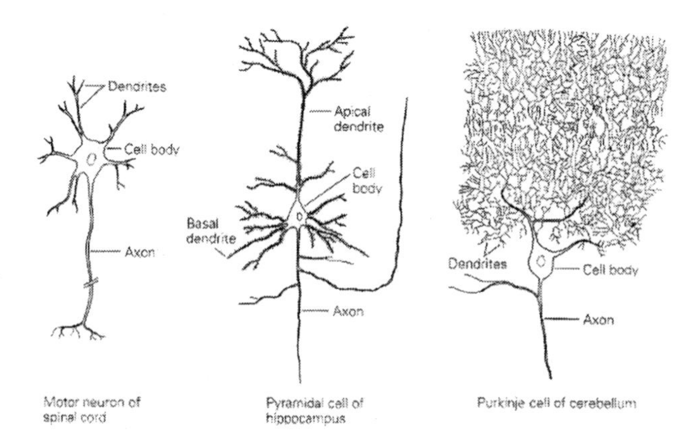

FIGURE 2.4 **Three common neuron types. The cell on the left is a motor neuron in the spinal cord, the middle cell is a pyramidal neuron, and the one on the right is a Purkinje cell.** *SOURCE:* **Principles of Neural Science, fig 2.4 bottom half. E. Kandel. (McG–H).**

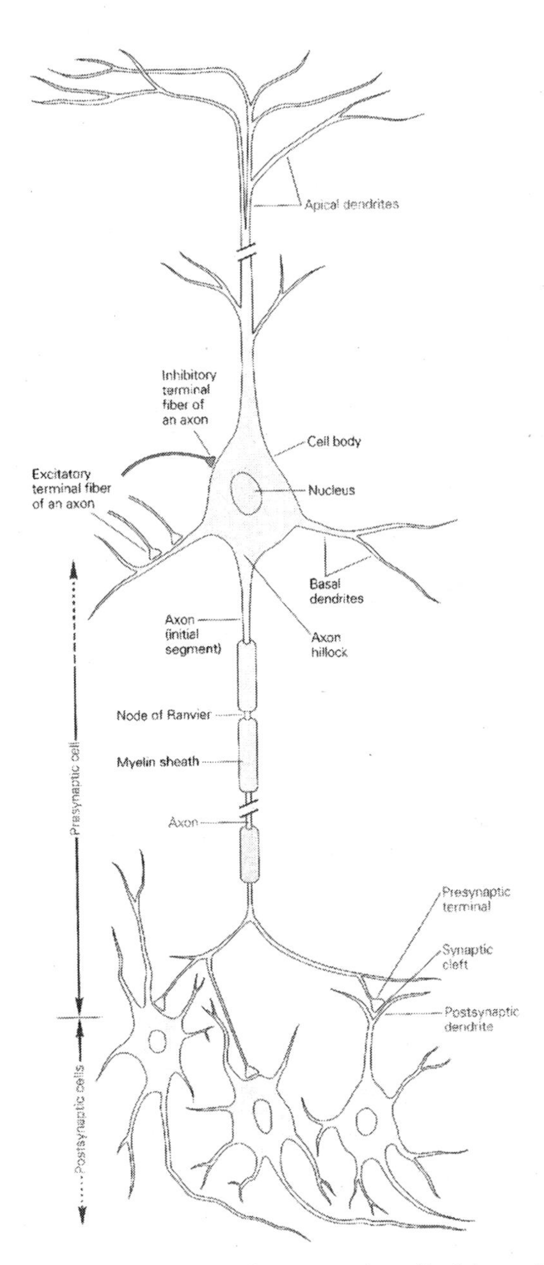

FIGURE 2.5 **The pyramidal cell in more detail. Input to the cell originates in the dendrites (apical and basal). The cell integrates these influences, and if a sufficiently large excitatory signal is detected, an action potential is generated at the axon hillock, which propagates the length of the axon and stimulates the release of neurotransmitters at the presynaptic terminals. Signal propagation is one-way only, that is, from dendrites to cell body to axon.**
SOURCE: **Principles of Neural Science, fig 2.2.**

higher speed than it would otherwise. When the signal reaches the presynaptic terminals, it initiates a release of chemical neurotransmitters, which then affect the behavior of the postsynaptic cells. Over 60 different molecules can serve in this capacity, including "classical" neurotransmitters such as serotonin, acetycholine, GABA, and dopamine, as well newly discovered ones including substance P, beta endorphins, and soluble gases such as nitrous oxide (hence the psychoactive properties of this substance).

Let us work through the typical chain of events that lead to the generation of an axon potential. Neurotransmitters released across the synaptic cleft are collected in special receptors on the dendrites. Depending on the neurotransmitter and how it is interpreted by the cell, the accumulation of these chemicals in the receptors will lead to either a depolarizing or hyperpolarizing influence on the synaptic potential. The former raises the membrane potential of the soma (the voltage difference between the inside and outside of this body) and the latter decreases the membrane potential. These respective excitatory and inhibitory signals are integrated at the body and propagate the short distance to the axon hillock. If this signal is sufficiently depolarizing, then an axon potential will be generated at this structure by a sudden influx of Na^+ ions through the cell membrane. This signal is then propagated down the axon, aided by the insulating properties of the myelin sheaf. At regular intervals there are breaks in the sheaf, called Nodes of Ranvier, where the signal is regenerated (motor neurons can have axons up to 3 meters in length).

Upon reaching the synaptic button at the presynaptic terminal, a neurotransmitter will then be released across the synaptic cleft. The amount of neurotransmitter will be proportional to the rate of firing. This is illustrated in Figure 2.6, which shows three different firing densities and the corresponding neurotransmitter concentration in the cleft. It is in this manner that the neuron converts an all-or-nothing signal, the action potential, into a graded response. It is also this fact that allows us to approximate discrete neural firing by a single real quantity representing the firing rate, as we will discuss in the next chapter.

There is one final point of interest of this process. The nervous system cannot possibly be hardwired for all contingencies at birth; some level of adaptation must take place. The neural mechanism that achieves this is known as long-term potentiation (LTP). LTP is achieved by a long chain of chemical events that are both incompletely understood and beyond the scope of this work. Nevertheless, we can summarize the process as follows. The same neurotransmitter that binds to a receptor to create depolarization binds to another receptor. If the neuron becomes sufficiently depolarized for sufficiently long, then the original receptor is made more receptive to the neurotransmitter by a chain of reactions set off by this second receptor. The net effect of this process is that high-frequency stimulation of a synapse accompanied by net depolarization of the cell makes this synapse more responsive to the neurotransmitter over many hours or days. Or as the

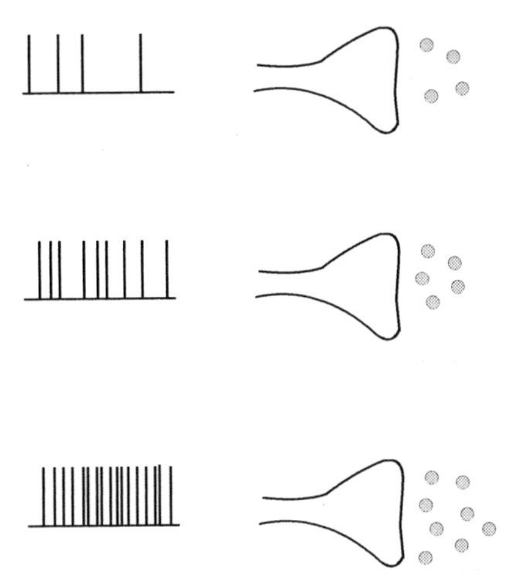

FIGURE 2.6 **The axon potentials are shown as all-or-none spikes. The more such spikes per unit of time, the more neurotransmitter released into the synapse.**

slogan goes, cells that fire together wire together. This can also work in reverse. Net hyperpolarization can result in long-term depression (LTD), in which the synapse is less responsive. Finally, presynaptic mechanisms may also be at play, in which the amount of neurotransmitter released is a function of coincidence of firing between the presynaptic and postsynaptic cells.

In summary, the neuron acts in an integrative capacity, receiving a potentially large number of excitatory or inhibitory signals, and on this basis "decides" to fire or not, thereby affecting the downstream cells it is connected to. The importance of the neuron for brain function has not always been obvious, however. Prior to the pioneering work of Ramón y Cajal the reticular view of the brain held sway, in which this organ was conceived of as a single, giant interconnected network. With staining techniques first developed by his scientific rival Golgi, however, Ramón y Cajal was able to see the outline of entire neurons, and thereby show that neurons were indeed distinct cells. This principle, that the neuron is a distinct anatomical and functional unit forms the foundation for the neuron doctrine. Other associated claims include the idea that information flow in the neuron is unidirectional, from dendrites to the soma to the axon, and that the neuron can only develop in the prenatal brain from an undifferentiated neuroblast. The most important of the extensions of the neuron doctrine for the current purposes, though, is the McCullogh and Pitts (1943) assertion that the neuron is the fundamental information processing unit in the brain. If this is the case, then to reach a given neuroengineering end, it suffices to manipulate a neural cell or cells in the appropriate way, *and nothing more*.

Why might we believe that neurons are the workhorses of the brain? The answer comes from a number of converging sources. First, and most obviously, when recording from single neurons, or using various neuroimaging techniques, such as fMRI and EEG (electro-encephalogram) (see Chapter 6), we can see that neural activity or the pattern of activity over many cells corresponds to an identifiable functional attribute. For example, in the section on vision we will discuss cells in the primary visual cortex that respond to the orientation of lines in their receptive fields. We will also see in Chapter 6, quite remarkably, that we can predict motor intentions by looking at the firing pattern in the motor cortex. Second, we know that stimulating neurons or group of neurons produces a regular effect. We can, for example, induce motor activity by simulating the motor cortex either with invasive electrodes or by magnetic means. There is also a large body of evidence that lesions of neural areas due to stroke and other diseases lead to a more or less regular pattern of pathology, depending on the area afflicted. We will examine in Chapter 7 an exciting recent finding indicating that deep brain stimulation may be able to alleviate the worst aspects of depressive illness. Finally, models of neural networks along the lines of what we will discuss in the next chapter have proved invaluable in understanding both the brain and intelligence in general. In the next chapter, we will refute the strong claim that intelligence is merely the appropriate pattern of neural activation, but this will not be a rejection of neuron doctrine per se, but rather a shift to a more abstract level of analysis that includes the neuron as its fundamental substrate.

Despite this impressive body of evidence, a number of recent findings have cast doubt on both the central tenet and associated claims of the neuron doctrine in their strongest forms. These include, *inter alia*, the existence of gap junctions or electrical synapses between neurons allowing a direct nonsynaptic connection, the discovery of neural generation after birth in the hippocampus and possibly elsewhere, a wide variety of modulatory mechanisms serving to alter the firing behavior over slower timescales than synaptic transmission, and a possible computational role for glia and other nonneural brain cells (Bullock et al., 2005). Judging just how serious these alterations to the neuron doctrine also alter the neuroengineering claims of this work is a matter of difficulty. In general, however, it is unlikely that any of the more important propositions to be made will be uprooted by these new discoveries alone. For example, the core claim of this work, that we can, like Lenny Moscowitz, outlive our corporeal existences depends on capturing enough Lenny-ness as it dwells within his synapses. If we must also measure the effective efficacy across his electrical junctions to record his memories, this simply implies that the means of liberating him may naturally have to be expanded somewhat, but not fundamentally altered. A changing view of the brain will necessitate a changing view of how to modify it, but barring a wholesale reappraisal of the workings of this organ, most of claims for the power of neuroengineering will carry through, *mutatis mutandis*.

SOME IMPORTANT SYSTEMS

We will now briefly review three important systems in the brain: the vision system responsible for visual pattern recognition and coordinating action, the motor system responsible for the generation of action, and the arousal system, regulating sleep and wakefulness. While each is important in its own right, the primary purpose of this section is to give the novice to this field a feel for how structures in the brain cooperate to achieve a given behavioral or cognitive/affective goal.

Vision

The visual system is the jewel in the crown of our sensory systems, and in many ways is the most advanced of all systems, sensory or otherwise, in the brain. The eye itself is remarkable for its acuity and sensitivity, and is able to detect the presence of a single photon (the rough equivalent of being able to see a candle on the horizon on a clear night). Our color sensitivity is exquisite; most of us can detect upward of 10 million unique colors. But it is the processing of the visual stimuli that is the most impressive. We can quickly and accurately survey a scene, and within no more than 500 ms, and typically less, we can identify all of the major objects in that scene. The visual system is the most well-studied brain module, although the consequent richness of data presents us with a dual-edged sword; we can with confidence describe in coarse outline how and why it works, but space prevents us from doing full justice to the current state of knowledge.

Figure 2.7.A shows the initial path of information from the retina to the visual cortex, which occupies the occipital part of the brain. Signals from the retina are routed through the optic disk in the eye and meet at the optic chiasma. There they are separated into two streams, with the right part of the retina from both eyes joining on the right, and likewise for the left retinal image. As with any lens, however, the eye reverses the image before it is projected onto the retina at the back of the eye. Thus, the right optic tract, (the optic nerve post-chiasma) actually contains signals from the left visual fields for both eyes, and the left optic tract contains signals from the right visual field. These proceed to the left and right lateral geniculate nucleus (LGN) of the thalamus, respectively, and from there to the respective parts of the visual cortex.

Figure 2.7.B schematically represents the processing of visual information from the LGN on. We will describe in detail each of these processing steps later

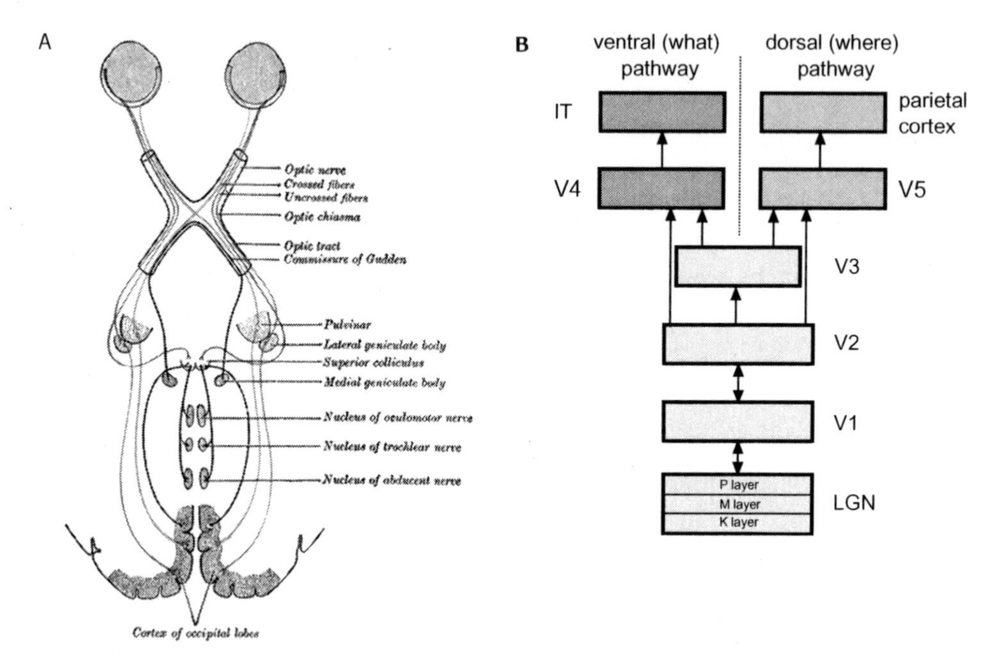

FIGURE 2.7 **The general flow of optical information is from the front to the back of the brain, as shown in A, and then in the reverse direction from the occipital cortex outwards and forwards, as shown schematically in B in the successive transformations from V1 onwards. Source: http://commons.wikimedia.org/wiki/Image:Gray722.png**

in the chapter. For now, we note that there are two main streams of information flow downstream of V3. The one illustrated on the left, the ventral (inferior) stream proceeds from the occipital cortex to the inferotemporal cortex (IT). Area IT is responsible for pattern recognition, that is, the labeling of objects in the attended part of the visual field. The dorsal path (right) proceeds from the occipital cortex to the parietal cortex. It was once believed that this path was responsible for the location of objects; hence the label "where path." However, for reasons to be described, it is probably better to think of this as an action stream that helps guide motor movement in actions such as maneuvering around and grasping objects.

We now turn to a brief description of the function of the various processing layers in the visual system. It is important to bear in mind during this discussion that while there is a considerable degree of modularity in visual processing, it is not necessarily the case that each anatomically distinct area is necessarily responsible for one and only one aspect of vision. Rather, it is the dynamic interaction between modules, including the reciprocal connections between them, that result in both the informational and perceptual properties of this sensory modality.

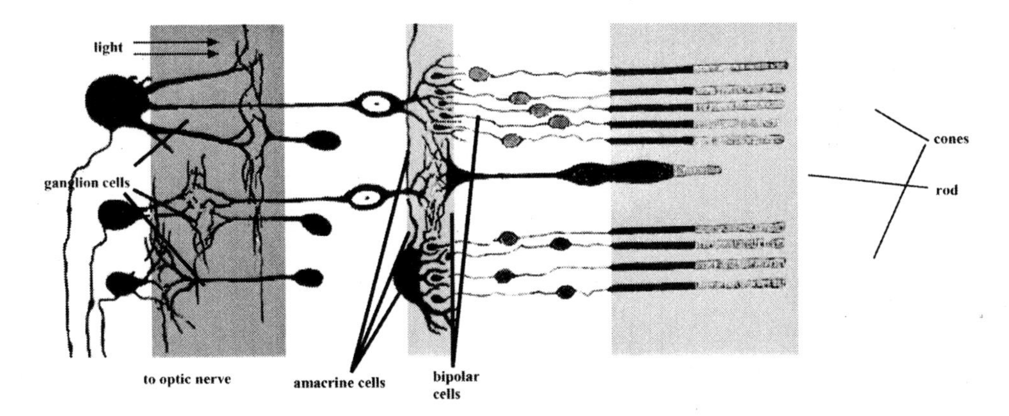

FIGURE 2.8 **A small section of the retina.**
Source: http://en.wikipedia.org/wiki/Retina

- The retina
  Light that enters the eye is focused onto the retina by the cornea and the lens. The retina contains two types of transducers, the *rods* and *cones*, named according to their approximate shapes (Figure 2.8). The cones, which predominate in the fovea (the most sensitive part of the retina containing the highest density of light detectors) are responsible for color vision; rods are responsible for detecting absolute light levels. There are three types of cones corresponding roughly to what we call red, green, and blue. Color perception is solely a function of the degree to which these cells are active, and this is why it is sufficient for monitors to contain only these hues.

  Considerable processing takes place in the retina, as evidenced by the fact that the rods and cones outnumber the ganglion cells, which feed into the optic nerve by a ratio of approximately 10 to 1. The bipolar cells are sensitive either to small patches of light with the surround off, or vice versa. The net effect of this is to highlight edges; thus, well before visual information reaches the brain proper, objects in the visual field have begun to be segregated from their neighbors and the background. The horizontal cells aid in this process by providing lateral inhibition between proximal detectors and effectively increasing contrast at points of difference and reducing it over constant color or lumination fields. Amacrine cells are thought to be responsible for detecting motion; as we will see, this aspect of vision is processed in a different stream than that for stationary objects.

As previously indicated, the layout of the retina is such that the center of the visual field, the fovea, is dominated by high-resolution cones; the rest of the field has relatively poor acuity. Among other things, this implies that the

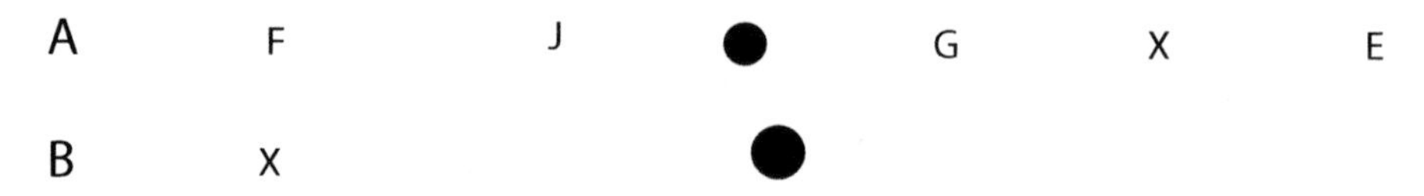

FIGURE 2.9 **A: A demonstration of the loss of resolution outside the fovea. B: A demonstration of the existence of the blind spot. See text for details.**

existence of a wide-angle clear field of view is largely an illusion constructed by a series of eye movements (saccades) that work with the visual brain to piece together a set of imagistic patches into a coherent whole. This can be seen by concentrating on a single region only. For example, close your left eye and focus on the circle at the center of Figure 2.9.A, holding the page approximately 6 inches from your face. You may see the letters surrounding the plus sign, but you will not be able to recognize the peripheral letters, with the possible exception of the "A," because of its size. Another curious feature of vision is due to the optic disk, the place where the ganglion cells leave the eye to form the optic nerve. This means that there is an absence of light detectors in this region, and therefore a blind spot in the visual field. Close your left eye and focus on the "X" in Figure 2.9.B with your right eye. Gradually bring your eye closer to the "X" and at a distance of about six inches the dot should disappear. A number of mechanisms in later visual processing allow the spot to be filled in under ordinary circumstances, and thus we are normally blinded to the existence of the blind spot.

- LGN
  The primary role of the LGN is to act as a selective relay station routing information from the eye to the visual cortex. When, for example, we are asleep this does not happen, and when dreaming images are endogenously generated. As indicated in Figure 2.7.B, there are three types of layers in the LGN that receive projections from the retina and then differentially project to V1. The P layers, responsible for transmitting the parvocellular pathway, originate in so-called midget ganglion cells, connect to layer 4CB in V1, and eventually terminate in areas V4 and IT. This relatively slow pathway is responsible for the processing of color and form. Another faster pathway, the magnocellular, transits through the M layer of the LGN, connects to layer 4Ca in V1, and eventually terminates in MT. This pathway is concerned with the processing of depth and motion. Finally, the K layer in the LGN is responsible for relaying information from the low-wavelength cones (i.e., blue-detecting) in the retina to the visual cortex. The LGN receives considerable feedback from V1, belying its role as a passive relay station, although the precise role of this feedback is still not well understood.

- Primary visual cortex (V1)
  As with the LGN, V1 is retinopically arranged, that is, proximal signals in the retina are also proximal in this module. Unlike LGN, V1 is responsible for early processing of the visual signal. For example, Hubel and Wiesel's pioneering work in the early sixties showed that a large number of cells in V1 are sensitive to the orientation of lines in their receptive fields (RFs).[3] The receptive field of a visual cell is the area in the visual field for which the cell is responsive. Orientation is a crucial aspect of shape, and the combination of orientations can define contours within a shape. In fact, lateral excitatory and inhibitory connections in V1 are also thought to contribute to the formation of longer lines composed of these smaller orientation detectors. Additional processing done in V1 includes binocular disparity, or computation of the difference between the way the same image falls on the different eyes, crucial for the perception of depth; latency between activity in neighboring cells used in the detection of motion; and other aspects of form and color processing.

- V2 and V3
  V2 and V3 carry on the initial work done by V1. While neither is primarily responsible for pattern recognition or motor detection as in the case of IT and MT respectively, both provide intermediate information important for these aspects of visual processing. One crucial feature of cells in these areas is that their RFs are wider and more encompassing than those in V1 or the retina. As one moves up the hierarchy of visual processing, information gained about features or whole objects comes at the expense of spatial resolution. This is of necessity. For example, in V1 it is possible to have orientation detectors at every point in the visual field, because there are relatively few of these detectors per region. However, it is not possible to have a lion, a giraffe, a pizza, and a falafel, etc., detector in every region of space, because the size of this set is extremely large.

- V4
  V4 is implicated in the processing of color, and damage to this area will lead to *color-blindness*. Patients with this affliction report that not only has their world been deprived of color, it also appears to be deprived of vibrancy, so it is possible that their vision is more impaired than if they were simply looking at the world as if it was an old black-and-white movie. Recent research (Walsh, 1999) also suggests that this area may be criti-

[3] It is also known that these detectors are not present at birth but arise through the appropriate stimulation in the environment. For example, kittens that are only exposed to a single orientation in the first few weeks of life never develop orientation detectors for other angles.

cal in two aspects of color vision, color induction and color constancy. In color induction a color will make an adjacent opponent color appear more saturated; for example, a yellow patch next to a blue patch makes the yellow more yellow and the blue bluer than it would be if they were standing alone. Color constancy means that we are largely able to ignore the effect of the illumination of a scene and perceive the colors as if they were in neutral lighting. In Chapter 4 we will see a dramatic example of this effect, and we will also discuss its implications for conscious perception.

- V5 (MT)
  One might guess that motion perception in the brain is simply the effect of noting change in a succession of static images, but in fact motion detection is too important in a hostile environment of predator and prey to leave to this mechanism. Instead, specialized cells compute both the direction and degree of change, and fire accordingly. The chief computational locus for this in the brain appears to be V5. One important piece of evidence for this fact is that damage to this area can lead to *akinetopsia* or motion blindness. A person with akinetopsia will perceive the world as a slow-moving series of images, with no connection between them. For example, they will find it impossible to pour water in a glass. Their first image is that of the glass empty, then they see a half-full glass, and then they seen water spilling over the top. V5 then appears to allow us to interpolate between successive states of visual awareness, and perceive the world as a smoothly flowing series of events.

- IT
  As has been previously noted, the ventral visual stream leading into IT is thought to be responsible for object recognition. In IT, for example, we find neurons that fire when an object is present, ignoring the position of the object in the visual field, its size, its orientation, and other factors. For example, a monkey cell in this area may fire whenever a banana is present, regardless of where it is in relation to the monkey, and even if it is half eaten. Invariant pattern recognition of this sort is an extremely important aspect of information processing, whether accomplished by machine or by organic computation, and enables an extremely large amount of visual information to be encapsulated for later use by the brain.

  In particular, one highly developed area in IT, the fusiform face area (FFA) is responsible for face recognition. One might hypothesize that this ability ought to be just a special case of object recognition, but apparently evolution has deemed it sufficiently important for social reasons as to endow us with a special module for this task. The existence of this finely tuned module is probably why faces seem so distinct to us; the FFA exaggerates the differences and conceals the considerable similarities from one visage to the next. We also know that impaired operation of the FFA leads to *prosopagnosia*, or the inability to recognize faces. Someone with a mild innate form of this ailment could pass a loved one on the street without noticing them. In severe cases, however, prosopagnosics are never able to recognize people from their faces, and use a labored combination of auditory and visual cues, such as gait, in order to approximate what for most of us is a trivial and unconscious task.

- Dorsal stream (parietal) processing
  Processing in the parietal cortex enables motor actions by assisting in the computation of coordinates of objects in the world relative to the position of the body. For example, when grasping for an object, it is not sufficient to know where the object is in an absolute coordinate space, but how the arms and hands must move in order to accomplish this task. Apart from anatomical considerations, there are two types of evidence indicating a dissociation of action processing in this stream from the form, color, and recognition processing in the ventral stream. The first, as usual, comes from looking at what happens when the relevant areas are lesioned. Patients with parietal lesions can recognize objects but cannot grasp or use them. Conversely, the ability to hold and manipulate objects can be preserved without the patient being able to identify an object or indicate its function when this cortical area is preserved but there is damage in the ventral pathway.

  The second bit of evidence can be seen without recourse to pathology. Figure 2.10 shows the well-known Müller-Lyer illusion, in which the shaft of the top figure appears considerably shorter than the horizontal line in the object below. Over the years, many different explanations have been generated for this effect, and consensus is still forthcoming, but what concerns us here is another phenomenon. When asked to grasp the objects in question, subjects will ignore perceptual evidence and treat

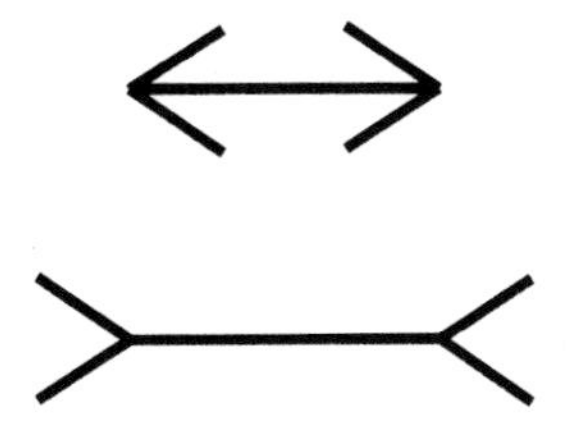

FIGURE 2.10 **The Müller-Lyer illusion. The horizontal line in the top figure appears to be much shorter than that of the bottom figure; they are, in fact, the same size. However, if asked to grasp the objects, subjects move their hands in identical fashions.**

> the lines as identical; this can be seen in their initial movements as well as their final ones, so this is not simply a last-second correction. Thus, not only are the dorsal and ventral streams separate, the former appears to be separate from conscious perception.

The previous discussion is by no means a complete picture of how the visual system works, but even if it were, we are left with one serious problem. If the modules from V1 on are involved in the processing of completely different aspects of vision, where (or how) do all of these features come together to produce a unified "picture" of the world that we perceive? One possible answer is that this information is integrated in some downstream area toward the front of the brain, such as the association cortex, or the frontal lobe. On the face of it, however, this is implausible, because of the loss of information that occurs at each successive stage of processing. For example, cells in IT can respond to a given object in a wide variety of locations and other alterations, such as rotation, but it can only do so because it is not responding to a precise configuration of pixels. But we consciously perceive the world at this level of detail, and moreover with the different features—shape, color, motion, etc.—"layered" on top of each other. Just how the mind forms an integrated image from these separate processing streams is know as the *binding problem*. The full solution of this important conundrum is not yet known, although we will suggest a partial answer in the chapter on consciousness in the next section of the book.

The Motor System

In many respects, the motor system is the simplest of the major circuits in the brain to understand, as it has the most direct set of consequences: the initiation and control of movement. Simple, however, is being used here in a strictly relativistic sense, and as with all aspects of the brain, many thousands of scientist-years have been already devoted to unraveling its mysteries, and many more will be spent in the future. An overview of the motor system is present in Figure 2.11.A. The primary motor cortex (area 4) lies immediately anterior to the central gyrus. The representation of the body in this area resembles the distorted homunculus (little man) shown in Figure 2.11.B. Note that areas that are highly innervated and over which we have relatively high control take up more of the cortical space. This includes, most obviously, the hand, but the face as well, the control of which is extremely important for social interaction. There is also a corresponding homunculus for the primary somatosensory cortex, on the other side of the central sulcus.

The primary motor cortex receives input not only from the prefrontal cortex, which is responsible for the planning and initiation of movement at the highest level of abstraction (i.e., that some action should be taken, but not how to achieve it), but also from area 6, which sits between the prefrontal cortex and area 4. Area 6 is divided into two zones, the premotor area, situated laterally, which integrates sensory information and controls muscles along the body's axis; and the supplemental motor area, situated medially, that controls complex movements, such as those involving the hands. The prefrontal cortex also receives information from the parietal cortex that contains a model of the position of the body in space, and therefore helps guide movement.

As shown in Figure 2.11.A, axons from the motor cortex project to the spinal cord via two pathways, the lateral and ventromedial systems. The lateral system contains two tracts, the largest and most important of which is the lateral corticospinal tract. This bundle extends through the midbrain to the medulla, where it is situated near the surface and resembles a pyramid; hence it is also known as the *pyramidal tract*. Before entering the spinal cord, the fibers originating in the left hemisphere cross over to the left and vice versa (a process known as *decussation*); hence, the left hemisphere controls the right of the body and the right controls the left.[4] The other, less significant tract in the lateral system originates in the red nucleus in the midbrain, which receives input from the frontal cortex. The ventromedial system (Figure 2.11.A) is composed of three main tracts. The reticulospinal tract is involved in the maintenance of posture in the face of gravity. The vestibulospinal tract transmits information from the inner ear and is involved in balance. Finally, the tectospinal tract receives input from the superior colliculus, which is involved in integrating sensory and visual information about the environment, and facilitates visual orientation.

Figure 2.11.A also shows that there are two control loops involved in motor action. The one shown on the right involves the cerebellum, and the left the basal ganglia. In both cases, they integrate a wide variety of cortical information and then send feedback to the motor cortex via the ventrolateral nucleus of the thalamus (VLo). The cerebellum, as has been discussed, is believed to contain procedural memory for the realization of particular motor sequences. It is where the "muscle memory" of the brain is stored, and the large number of neurons in the cerebellum is a good indication of the complexity of these motor routines. The other control loop centering on the basal ganglia is less understood. Some clue as to its utility may be found in Parkinson's patients, in which the functioning

[4] Why are both motor and sensory information processed on the contralateral (opposite) side? One possibility is that the evolution is actually being consistent, not between the world and the brain, but between the brain's view of the world and the rest of the brain's functionality. Recall that the lens of the eye, like any lens, inverts images. The left retinal field from both eyes (i.e., the right visual field) therefore projects to the left cortex and vice versa. Hence visual information effectively switches sides without a decussation. To be consistent with this important modality, all other sensory modalities and motor processing must therefore involve an explicit decussation somewhere in their processing chain.

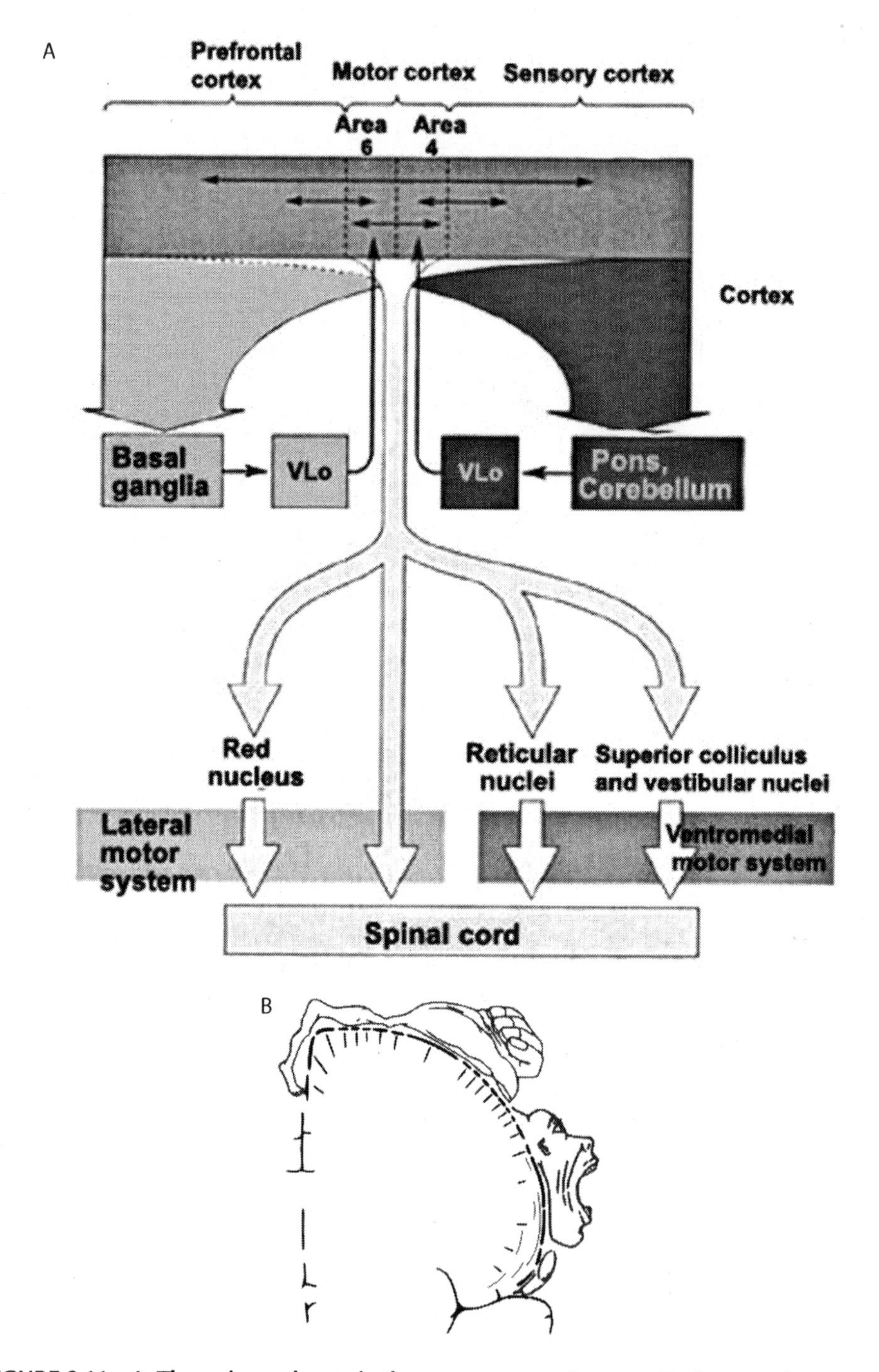

FIGURE 2.11 A: The major pathways in the motor system. Source: Brain top to bottom. B: The homuncular layout of the primary motor cortex. Source: http://commons.wikimedia.org/wiki/Image:Homunculus.png

of this loop is impaired by the lack of dopamine. One of the hallmarks of this disease is the inability to initiate action. Thus, it is likely that the basal ganglia are implicated in the decision to proceed with an action after the prefrontal cortex has decided that it may wish to embark on such a course, although it is also likely that these structures play a role in fine grained movement.

Let us examine the approximate course of events that takes place when you catch a tennis ball that has been hit in your direction. Sensory information to this effect arrives at the frontal cortex and you make the decision to catch the ball. This information is sent to the supplementary motor area (area 6), which combines information about the incoming ball with information generated by the parietal cortex regarding the position of the ball relative to the orientation of your body in space. A motor routine is generated by the cerebellum based on past ball catches and is adjusted to meet the demands of this circumstance. If the basal ganglia "agree" that this is an appropriate course of action, a sequence of motor neurons in area 4 will fire and signals will be sent via the corticospinal tracts to generate the muscle movements. If all goes well, the ball drops into your hand and you are off to your next motor routine.

One interesting aspect of this process is that only the initial desire to catch the ball enters consciousness. You are not generally aware of the computations needed in order to achieve this. Of course, you can feel your muscles move, and you may also see the ball coming and dropping into your moving arm, but this is almost like watching a movie, in which you just happen to be the actor. Furthermore, if you are too conscious of your movement (as you are, for example, when first learning the task), it interferes with your ability to perform it well. In Chapter 4 we will explore why this may be the case.

The arousal system

In this section, we will examine a large number of interacting structures that serve to regulate both arousal and the sleep-waking cycle. Let us begin in this case not with anatomy but with function, as it is simpler to start in this direction. There are three distinct states that we need to account for: a) the awake state, in which consciousness is present, and the brain is capable of responding to and processing sensory input; b) the nondreaming, non–rapid eye movement (REM) state, in which the brain is unconscious and unresponsive, and c) the REM state, in which the brain is almost completely unresponsive to external input[5], but is aware of its own dream content. States a and c both constitute a form of consciousness, so let us examine their joint difference with state b to start.

[5] As Freud noted, however, elements of the external world can be incorporated into one's dreams. For example, an ambulance sound may cause one to dream of being in the hospital.

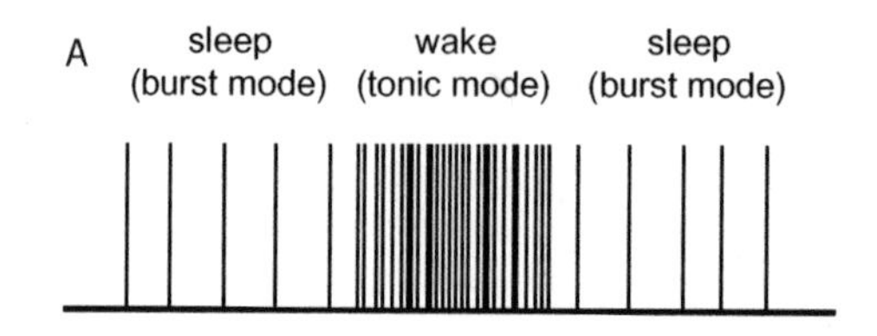

B

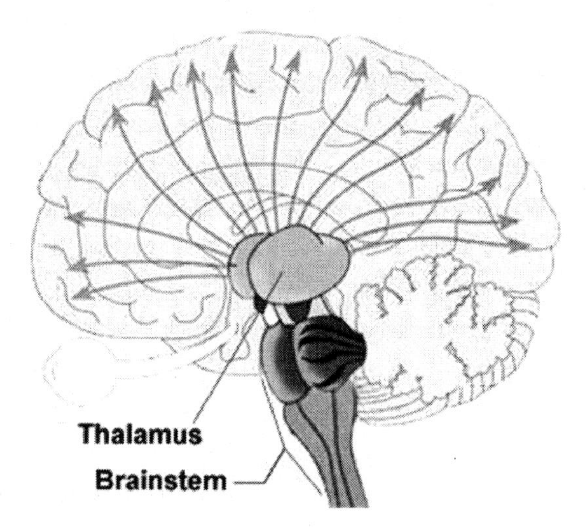

FIGURE 2.12 **A: The transition to and from the waking state as reflected in the activity of thalamocortical neurons. B: Diffuse projections from the thalamus to the cortex under control of the brainstem ascending the reticular activating system.**
SOURCE: **Brain top to bottom. http://thebrain.mcgill.ca/**

Non-REM sleep is characterized by synchronized cortical activity. This means that within a local area, the neurons tend to fire independent of each other (as we will see in Chapter 4, the reverse may be true *between* cortical areas; i.e., unconsciousness is characterized by lack of long-range synchronous communication). EEG, which measures the mass action of the synchronized dendritic potentials, shows a relatively high amplitude during this state; in addition, the EEG is characterized by a decrease in mean frequency from beta waves (>12 Hz) in the fully awake state to delta waves (<4 Hz) in deep sleep. The primary cause of this effect is the switching of thalamocortical neurons from burst firing mode to tonic firing mode under the influence of the ascending reticular activating system (RAS), originating in the reticular (network-like) formation in the brainstem. This is illustrated in Figure 2.12.A. The default state of these cells is to fire in burst mode, a cyclical and low-activity state. Under the depolarizing influence of the brainstem, they transition to tonic mode. This in turn provides tonic activation to the cortex via the diffuse thalamocortical projections (Figure 2.12.B) "waking" these neurons from standby mode to processing mode. Cortical computation, and presumably consciousness also, can only take place when cells are allowed to differentiably fire; when firing in unison, there is only a single state and no information can be represented.

What then of the difference between the two conscious states, the awake state and the dreaming state? As in ordinary wakefulness the cortex is relatively desynchronized during REM sleep. However, during dreaming the cortex is nearly completely cutoff from sensory input (this is also true of nondreaming deep sleep). It is believed that this sensory gating is the effect of pontine (i.e., located in the pons) nuclei on the thalamus. In addition to its role in regulating arousal, the thalamus acts as a sensory gateway to the cortex, as we have seen in the case of vision. Synchronization of relay neurons carrying sensory information to the cortex from the thalamus results in the loss of the ability to transmit sensory information, in the same way that synchronization of the cortex as a whole puts it "to sleep." Another distinguishing characteristic of the REM state as opposed to ordinary consciousness is paralysis; without this, we would reenact our dreams in the real world. Somewhat surprisingly, research shows that structures in the motor system show something like normal activity during REM sleep, so these are not inhibited directly. Rather, pontine structures are responsible for inhibiting motor neurons in the spinal cord, effectively disabling them from cortical commands. Finally, the dream state, unlike waking, is characterized by heightened emotionality and suspension of rationality. This is consistent with imaging data showing increased activity in the anterior cingulate, hippocampus, and amygdala (i.e., increased emotionality), and reduced prefrontal cortical activity (i.e., decreased executive thought), although the full purpose of the suspension of rationality is not yet known.

It is now believed that no single circuit alone is responsible for cortical arousal; rather, arousal is a function of a number of interacting structures, the most important of which are:

- The ascending reticular activating system, originating in the midbrain reticular formation. The discovery of this system in 1949 by Moruzzi and Margoun was important not only in itself, but also because it showed that arousal was not solely the result of passive sensory control. Rather, as we now know, active structures integrate sensory information as well as other internal variables to produce sleep and wakefulness.
- The mesopontine nuclei. As described previously, these structures are important in regulating the REM state.
- The nuclei of the locus correleus. These influence the thalamus, hippocampus, and cortex directly. The activity of these structures is proportional to the degree of wakefulness, and they are completely silent during REM sleep.

- The nuclei of the anterior raphe. These, in conjunction with the hypothalamus, appear to regulate the sleep-wake cycle, and induce drowsiness and eventually sleep when the waking state has continued for too long.

The importance of the arousal system for neuroengineering is twofold. First, and most obviously, the control of arousal could heighten both motivation and the ability to learn. A second, more long-term objective involves the implications of this system for consciousness, the understanding of which is crucial to both "ordinary" manipulation of the brain and eventually to mind uploading, as we shall see.

ANATOMICAL LIMITATIONS OF THE BRAIN

This section completes our brief survey of the major neuroanatomical landmarks in the brain. As previously suggested, however, to make progress in our primary goal we must be as severe in critiquing as we are generous in appreciating the brain, and view this marvelous contraption from the perspective of the most dour engineer. It is only then that its limitations will be revealed, and only then that we can begin imagine how it could be otherwise. Here, we will examine constraints imposed upon the brain by virtue of its nature as a physical device encased in a body that is attempting to realize multiple objectives at once, and in the following chapter we look at how these physical limitations place bounds on cognitive and other psychological powers.

Volume and weight

Hominid evolution shows a clear linear growth in brain size starting at about 400 cm^3 three million years ago to Homo habilis, possibly the first tool user at about 650 cm^3 two million years ago, to the current volume of approximately 1400 cm^3 and weight of 1500 grams in modern man. This growth was the direct development of a number of evolutionary factors, the most important of which was the transition to bipedalism. Walking upright freed the hands for other uses, including tool-making, but in order to be an effective tool user, one must have a more advanced brain. The motor movements alone involved in tool use encourage a more sophisticated action system, but it is the cognitive requirements that produced the greatest evolutionary spur. To use a tool one must plan ahead, and to plan ahead one must have a notion of the future, a representation of one's environment, and a means of manipulating these representations. In short, one must be more than a reactive system; one must be a thinking machine. It is thus no accident that the fossil record shows a clear correlation between brain volume and walking in an upright position.

However, bipedalism has a significant downside, and it is here we meet with our first and, in many ways, most serious constraint on brain capacity. Walking and running require a relatively small pelvis, but a small pelvis limits the size of the birth canal, which in turn limits the size of the brain at birth to about 400 cm^3, or a little less than $\frac{1}{3}$ of its final size. Given this constraint, evolution has come up with an interesting set of stopgap measures to maximize cortical volume. First, unlike most species, the size of the newborn's head is as large as the birth canal. It is only via expansion of the pelvis and an extremely painful (and dangerous) process that human birth is possible—giving birth has farcically been compared to pushing a bowling ball through a garden hose. Second, the human infant is born at an earlier physical and cognitive stage than most other species, to minimize birth size. One slightly perverse way of stating this is that every infant, even those carried to full term, represents a premature birth. Unlike most species, the human infant is an all-consuming ball of pure helplessness, completely dependent on its parents for all its needs, for at least the first few years of its life.

Evolution still has one final and difficult problem, which is that the brain needs to be well protected at all times, especially infancy. This means that a hard skull, and therefore a relatively rigid skull, must be in place. Here, evolution compromises somewhat and gives infants what are known as fontanels or soft spots consisting of fibrous connective tissues between the plates of the skull. This allows the cranium to expand as the brain develops and still affords a good deal of protection during this development. Clearly, all of these measures, though clever and successful, are at their straining point; this is a topic we will take up again at the end of this chapter when we see just how far natural selection can take the brain.

Of special interest, while we are on the topic of brain size, is the evolution of the prefrontal cortex to its current prominence in Homo sapiens. We cannot, of course, know precisely how large this soft tissue was from the fossil records alone, although we can estimate this trend by looking at existing species, and through other means. This is illustrated in Figure 2.13. The top of the figure shows the rapid growth of the prefrontal cortex as one progress from cats and dogs to primates to man. Indirect evidence for a burgeoning prefrontal cortex in the development of genus Homo is provided by the slope of the forehead, shown in the bottom of the figure. As discussed earlier, the prefrontal cortex is implicated in executive control, social interaction, abstract reasoning, and possibly consciousness; in short, the very requirements for intelligence as we know it. Given the previous discussion on the overall limits on brain size, however, evolution faces a difficult challenge in significantly enhancing this area. Moreover, there is some evidence that there is a trade-off between frontal and parietal volume (Allen et al., 2002); if you try to expand one area it must come at the expense of another.

In summary, it would appear that evolution has a difficult task before it with respect to increasing total brain size. All things being equal, a bigger brain is

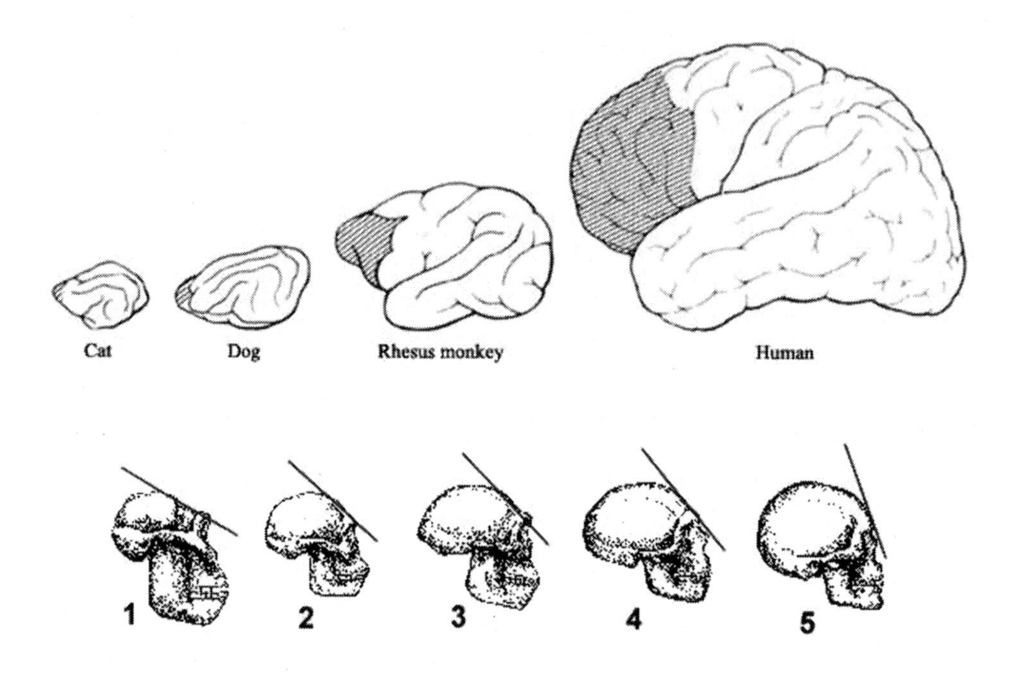

FIGURE 2.13 **The top of the figure shows the size of the prefrontal cortex in man and other mammals. The bottom shows the evolutionary development of the slope of the forehead (1 – Australopithecus, 2 – Homo habilis, 3 – Homo erectus, 4 – Neanderthal man, 5 – Homo sapiens).**
SOURCE: **Brain top to bottom.**

probably a better brain but this must be balanced against a number of other factors, including the need to walk upright and the consequent reduction in pelvic size, the need for infants to not be too helpless for too long, and the need to protect the infant brain from injury with a relatively strong enclosure. In following sections, we examine the same topic from two other perspectives, that of surface area, and that of connectivity.

Surface area

Brain size alone or brain size relative to body weight is a poor indicator of cognitive capacity. Men for example, have brains approximately 12% larger than women, and although there is an emerging body of evidence suggesting differences in cognitive strategies between the sexes, we know of no systematic differences in mean intelligence. Between species, the differences are even more dramatic. Man has a brain weight of 1500 grams, the elephant 7500 grams, and the mouse a mere .4 grams. Yet when we look at the ratio between brain weight to body weight it is the lowly mouse that wins hands down: his ratio is 3.2% compared to 2.1% for man and 0.15% for the elephant. Mice are clever little creatures as anyone who has tried to get rid of them knows, but they will never contribute toward the development of neuroengineering or any other scientific pursuit (except as unwitting subjects), there are no rodent symphonies, and they have yet to prove a theorem. It sounds faintly ridiculous to put it this way, but this fact needs to be emphasized, especially in light of recent attempts to blur the difference between man and other creatures.

If total size alone does not count, the next natural place to search for a correlate with intelligence is the size of the neocortex, the repository of the gray matter in the higher brain. Given that the cortex has a fixed depth of approximately 3 mm, this quantity will be directly proportional to the surface area of the brain. This is approximately 2500 cm^2 (about the size of a pillowcase) and comprises 20 billion neurons, or about 1/5 of the total in the human brain. If we approximate the brain as an ellipsoid with radii of 7, 8, and 4.5 cm (the mean brain width is 140 mm, the mean brain length is 160 cm, and the mean brain height is 90 cm), however, we obtain a surface area of only ~530 cm^2.[6]

What accounts for the nearly fivefold difference in these quantities? It is the previously discussed folds in the cerebrum that enable a greater surface area to be fit into a smaller space, and therefore more cell bodies in the same space. Humans have about 30 significant sulci, although their location and size may differ greatly from person to person. Significantly, these folds are not present in lower animals, although dolphins do have more folds than we have (cue dolphin intelligence theories). In one of many instances of ontogeny recapitulating phylogeny, the cortex of the fetus is almost completely smooth until the last trimester. In a process continuing into infancy, it then develops the major and minor sulci and gyri characteristic of the adult brain.

If folding provides space for more cortical neurons, why has evolution not taken this further? Why are there not 100 gyri, and why are the sulci not deeper? As with most engineering tasks, there are trade-offs. The more and deeper the folds, the greater the interference with the white matter and therefore the greater the effect on connectivity. Consider the fate of two neurons on either side of a sulcus. If the brain were smooth at that point instead, then the two would be able to communicate via axons just underneath the cortical surface. The presence

[6] Calculated using the approximation $S \approx 4\pi\left(\frac{a^p b^p + a^p c^p + b^p c^p}{3}\right)^{1/p}$ where a, b, and c, are the radii and p ≈ 1.605.

of the sulcus, however, means that these axons need to wend their way beneath this fissure, and then up again into the target cell, resulting in a longer path and therefore slower interaction.

Perhaps this is why Einstein's brain was found to have a shortened Sylvian fissure. The researchers who examined the physicist's brain found it normal in almost all respects except for this (Witelson et al., 1999). They speculate that the "defect" in Einstein's brain may have enabled greater communication between inferior parietal neurons, an area of the brain responsible for visuo-spatial representation. Einstein himself claimed to think in imagery, and he said little until he was nine years old. Much of this is purely speculative and there could very well be other reasons for his genius; for example, it was also found that Einstein had more glial or helper cells in the left inferior parietal cortex. But the fact remains that given a limited skull volume, itself the result of the live birth of mammals and bipedalism, there is only so much one can do to increase neural capacity. Similar reasoning applies to the next topic, the white matter of the brain.

White matter and conduction time

The gray matter of the brain increases roughly linearly with the volume of the brain; however, white matter, which contains the myelinated relatively long-range axonal connections, obeys a 4/3 power law as a function of volume (Zhang & Sejnowski, 2000). Hence, in the pygmy marmoset white matter occupies 9% of the brain; in humans, approximately 34%. What accounts for this increase? Apparently, the brain attempts to maintain a constant level of infiltration, or connectivity, to other neurons regardless of the size of the brain. This invariance applies to both connections within a specialized area (some 30 to 50 in the human brain) and between areas. However, as brain volume increases, so do the number of neurons, and therefore the total number of synapses increases. In order to maintain this degree of arborization, the size of the axonal trunk must increase accordingly, and therefore the overall white matter volume (the sum of the trunks) must also grow.

But this growth presents a new and serious problem. Given a fixed axonal conduction time, increasing white matter volume necessarily implies greater conduction time between neurons; a given area will, of necessity, be farther from any other area. Slower conduction time, in turn, implies slower overall response, and the overall performance of the brain decreases. To some extent this factor can be mitigated by increasing the number of areas of specialization, and having most communication take place within an area rather than between areas. But there is still the need to distribute the products of these computations to other areas; unless integrated with other knowledge, they will not be very useful.

One solution to *this* problem is to let the modules deposit the results of their computation in a central store, and let all the other brain modules have access to this area. This will reduce the number of interarea fibers because only one bidirectional cable is needed to connect to this hub. This is a common network solution, and one that package carriers use, for example. If you send a carton from Jersey City to Tampa it will not go on a plane between these two cities; it would be too costly to have a specialized route for the relatively few packages that go between these cities. Instead it is sent through a hub, like Atlanta, and then all the other packages from the East Coast en route to Tampa arrive there together. It may be that evolution has also hit upon this idea, and later in this work we will discuss the idea that the central store can also be identified with the contents of consciousness. But this is not a panacea. Unlike in the package routing analogy, the central hub of the brain has a limited capacity—only so many items can be stored in short-term memory because this form of memory is based on active neural participation. Therefore, the greater the number of neural modules, the less chance that any given module will have that its products are distributed to the rest of the brain. We will revisit this topic in the next chapter when we discuss limitations on the content of short-term memory and the associated bottleneck of consciousness.

Barriers to the evolution of the brain

Our final question concerns the time course of the evolution of the brain in light of these results. Is it reasonable to invoke, for example, a modified form of Moore's law[7] to cerebral development stating that cerebral capacity will continue its linear ascent as it has since the introduction of genus Homo? Will mankind one million years from now not have evolved into a race of megacephalics, giant heads perched on slender bodies engaged in all manner of arcane philosophizing and scientific exploration? And if this will occur, would it be better to wait, let nature take its course, and forego possibly dangerous neural modifications?

There are strong reasons, however, to believe that nature, unaided, will not be able to accomplish much more than what it has already achieved. As we have seen, the brain's *bauplan*, the master blueprint for the layout of this organ, imposes a number of limits on its future development. There are constraints as to how big the brain can be, how much surface area it can have, and even if these can be relaxed, on how fast signals can propagate to distal areas. Cochrane et al. (1998) estimate, that given the trade-offs between conduction speed, synaptic processing time, and neuron density that the brain is only 20% to 30% below its optimal performance. There is only so much that can be done with neurons as they are currently constructed and given a relatively fixed cranial capacity.

[7] Moore's law states that computing capacity doubles every 18 months or so.

What about the possibility of subverting the bauplan? There are three ways that this could be done. First, we could conceive of a fundamentally different neural unit than the neuron, perhaps one that is much smaller and more efficient. Second, we can envision a new fourth brain layer on top of the existing three, which we might want to dub the sageocortex, for its hypertrophied intellectual capabilities. This part of the brain would be similar to the neocortex, but without its predecessor's clumsy reasoning abilities. Finally, and most radically, an entirely new form of superior intelligence could arise. For example, evolution could take a page from insect interaction and endow us with a superior collective intelligence. We cannot rule out any of these (or perhaps other possibilities that we cannot conceive with our current limited cortical machinery), and evolution does have a knack for surprising us with its ingenuity. But if any of these designs do arise, they will do so many millions of years from now. In other words, we should not expect that evolution will carry us upward in an unabated fashion; rather we will shortly reach our intellectual limits, and it will only be after a radical change in course triggered by a fortuitous set of mutations that mankind will be able to truly go beyond these barriers.

Given these sober facts, we are left with three choices: a) be satisfied with what we have, b) wait an indefinite period and hope that evolution hits upon another solution, or c) begin to tinker with the brain ourselves. This book will suggest that a is unsatisfactory and complacent, that b can only be a collective hope, of course, not an individual one, and a very long-range one at that, and therefore only option c is open to us. Before studying how to engage in this enterprise, however, we need to make the transition from hardware to software, and look at the brain and its limitations from a computational point of view; this we will do in the following chapter.

REFERENCES

Allen, J., Damasio, H., & Grabowski, T. (2002). Normal neuroanatomical variation in the human brain: An MRI-volumetric study. *American Journal of Physical Anthropology, 118*, 341–358.

Allman, J., Hakeem, A., Erwin, J., Nimchimsky, E., & Hof, P. (2001). The anterior cingulate cortex : The evolution of an interface between emotion and cognition. *Annals of the New York Academy of Sciences, 935*, 107–117.

Bullock, T., Bennett, M., Johnston, D., Josephson, R. M., & Fields, R. (2005). The Neuron Doctrine, Redux. *Science, 310*, 791–793.

Cochrane, P., Winter, C., & Hardwick, A. (1998). *Biological limits to information processing in the human brain.* British Telecommunications Publications.

Hubel, D., & Wiesel, T. (1959). Receptive fields of single neurones in the cat's striate cortex. *Journal of Physiology, 148*, 574–591.

LeVay, S. (1991). A difference in hypothalamic structure between heterosexual and homosexual men. *Science, 253*, 1034–1037.

Maclean, P. D. (1990). *Triune Brain in Evolution: Role in Paleocerebral Functions*. Norwell, MA: Kluwer Academic.

McCullough, W., & Pitts, W. (1943). A logical calculus of the ideas immanent in nervous activity. *Bulletin of Mathematical Biophysics*, 5, 115–133.

Moruzzi, G., & Magoun, H. (1949). Brain stem reticular formation and activation of the EEG. *Electroencephalography & Clinical Neurophysiology, 1*, 455–473.

Raine, A., Lencz, T., Bihrle, S., LaCasse, L., & Colleti, P. (2000). Reduced prefrontal gray matter volume and reduced autonomic activity in antisocial personality disorder. *Archives of General Psychiatry*, 57, 119–127.

Scoville, W., & Milner, B. (1957). Loss of recent memory after bilateral hippocampal lesions. *Journal of Neurology, Neurosurgery, and Psychiatry, 20*, 11–21.

Walsh, V. (1999). How does the cortex construct color? *Proceedings of the National Academy of Sciences, USA, 96*, 13594–13596.

Witelson, S., Kilgar, D., & Harvey, T. (1999). The exceptional brain of Albert Einstein. *Lancet, 353*, 2149–2153.

Xu, L., Anwyl, R., and Rowan, M. (1997). Behavioral stress facilitates the development of long-term depression in the hippocampus. *Nature, 29*, 497–500.

Zhang, K., & Sejnowski, T. (2000). A universal scaling law between gray matter and white matter of cerebral cortex. *Proceedings of the National Academy of Sciences USA*, 97, 5621–5626.

Chapter 3 The Brain as Designed by Evolution: Low- and High-Level Cognition

We now enter the second phase of our inquiry regarding the nature of the brain, its capacities and capabilities, and its limitations with a view as to how it may be enhanced by future neurotechnologies. In this chapter, we concentrate not on anatomy and physiology but on functionality. Specifically, we ask the following question. What sort of computational device is the brain, and how does it achieve its computational ends? Surprisingly, there are two entirely different sets of answers to this query, one that concentrates on low-level computations (sensory processing, pattern recognition, and the like) and one that emphasizes high-level cognition (reasoning, model-building, creativity, and the higher powers). The latter is often given short shrift in treatises on neurophysiology and neural technologies, but we do not have that option if we are to ultimately consider devices that, among other things, interact with the brain in its capacity as a symbol manipulation device. Accordingly, both sides of this divide are given equal weight. The chapter concludes with a discussion of the constraints imposed by the brain's design on the brain's ability to think; understanding these will prove crucial to further discussion in which overcoming these limitations are discussed.

NEURAL NETWORKS

The working assumption of this section is the *Neuron doctrine*, which, as stated in the previous chapter, claims that computation in the brain is the result of a network of discrete cells known as neurons. It may well be that this doctrine needs to be cleaned up around the edges; e.g., it is increasingly accepted that direct electrical (i.e., nonsynaptic) interaction is possible. However, the fundamental model of the brain as a collection of interacting neurons need not be altered, and unless glial cells turn out to have as yet undiscovered computational properties, we remain in fairly safe theoretical waters.

We will also require recourse to another principle that we will call the *Auxiliary Neural doctrine*, which states that it suffices to consider only two aspects of the operation of individual neurons when attempting to predict or understand the behavior of a network of such cells: a) the means by which the cell integrates the various influences of its afferent cells (i.e., the totality of synaptic influence at its dendrites), and b) the means by which it transforms this signal into a volley of axonal firings. Nothing else, from the point of view of modeling the network, matters (e.g., the chemical nature of the implicated neurotransmitters, including the details of ionic transport, the influence on conduction by the myelin sheath, etc.) except and in so far as these influence the calculations for a and b. This makes sense if we are only interested in the behavior of the network as a whole. Whatever happens internally in the neuron is not of interest as long as we have a good approximation to the input-output behavior of these cells.

There are two types of approximations that we can make. The first, and more detailed, approximation is the so-called integrate-and-fire model. There exists many different models of this type of varying degrees of complexity and accuracy but here we consider the most basic. As described in the previous chapter, neurons are characterized by all-or-none spiking behavior. Let us also assume that we can reduce the degree of influence on one model neuron, or unit, to the next with a single real quantity proportional to the synaptic efficacy, which we will call a weight. This is illustrated in Figure 3.1.A, in which three neurons are connected to the target unit; excitatory weights are positive and inhibitory

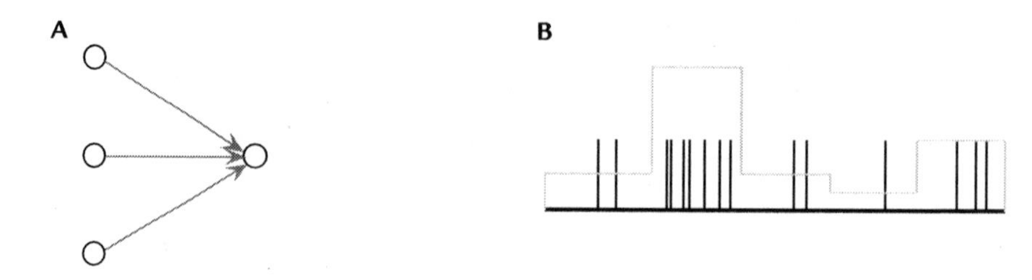

FIGURE 3.1 **A: A simple neural network. Three neural units impinge on a single unit; the weights are represented by real numbers. B: Neuronal outputs of an integrate-and-fire and rate-coding model. The former produces individual spikes (like a real neuron); the latter approximates this with a single firing rate per time quantum (gray lines).**

negative. The target unit works by integrating the influence of these incoming signals over time, in proportion to the indicated weights and the firing of their respective units, to produce a running sum (an inhibitory or negative weight means that this total is subtracted from rather than added to). That is, the more these presynaptic units fire, and the larger the absolute value of the weight, the greater their effect on this sum. In addition, we will regularly subtract a small quantity from the sum, to allow for the fact that the unit is integrating over a fixed time period; otherwise, we could cause a unit to eventually fire by very slowly impinging it with small excitatory signals, contrary to empirical results. If at any time the sum exceeds a predetermined threshold, then the target unit itself produces a spike, and we reset the sum and other parameters to zero and begin again.

Generally speaking, integrate-and-fire units are only used when we are interested in fine-grained temporal behavior (see, the section on synchrony). Most models approximate this behavior by using a single quantity representing the number of spikes per unit time, or what is also called the firing rate. Figure 3.1B shows how his works. Superimposed on the spiking pattern is the firing rate for this unit (gray lines). The firing rate ignores the precise temporal positions of the spikes and simply reports how many there are for each time quantum.

With simplification in place, we can characterize the output behavior of a single neural unit by the following equation:

$$r_i = F_{nl}\left[\sum w_{ij} r_j - \theta_i\right],$$

where r_i is the firing rate of the unit in question, the r_j are the firing rates of the incoming units, w_{ij} is the weight between r_j and r_i, θ_i is the firing threshold for unit i, and F_{nl} is a nonlinear output function that takes this weighted sum and threshold and turns it into an output. Two kinds of output functions are commonly used, a hard threshold and a sigmoidal function. In the former, the unit fires only if the weighted sum exceeds the threshold, and will continue to do so as long as this is the case. With a sigmoidal function, the output is an S-shaped function of the input quantity (the weighted sum minus the threshold). This allows for nondiscrete levels of activity, in keeping with the fact that although the axonal signal is close to an impulse or spike, the rate of firing in a given time period is a continuous value (see Figure 3.1.B).

Pattern recognition and computation

So far, we have considered the operation of neural-like units in relative isolation. As in the brain itself, the real power of these models comes about when we connect these units into a larger network configured in the appropriate way. The first and most natural application of neural networks, also known as connectionist systems, is to pattern recognition. Most networks of this type make heavy use of Hubel and Wiesel's original (1959) finding in the cat visual cortex, since confirmed in countless studies and on numerous animals, that as one proceeds higher up the chain of processing the receptive field of cells grows and the features that they are responsive to become more abstract. At the highest level of the hierarchy, for example, the corresponding area IT in humans cells are responsive to faces and other objects, regardless of their position in the visual field, as we discussed in the previous chapter.

As an example of pattern recognition that conforms to this principle, consider the three-layer network in Figure 3.2. The bottom layer consists of a collection of orientation detectors similar to those that are prominent in the primary visual cortex. Only shown are those that are triggered by the input, a letter "A"; in reality, there will be multiple orientations at every region in space. These feed three line detectors in the intermediate layer, corresponding to the three lines in the letter, and the joint presence of these lines in the appropriate position will activate the "A" detector in the topmost layer.

One important property of such a model, consistent with human pattern recognition, is that it is resistant to noise. For example, if one of the line detectors in the bottom layer was not fully activated, either because the segment was slightly bent relative to the optimal triggering angle, or because of noise internal to the system, then the model would not fail catastrophically. Rather, the activation in the corresponding unit in the middle layer would be slightly reduced, and this would also reduce the activity of the final output, also. Despite this property, however, the model as it stands is not as flexible as a human pattern recognizer. For example, it fails to account for the ability to recognize objects regardless of

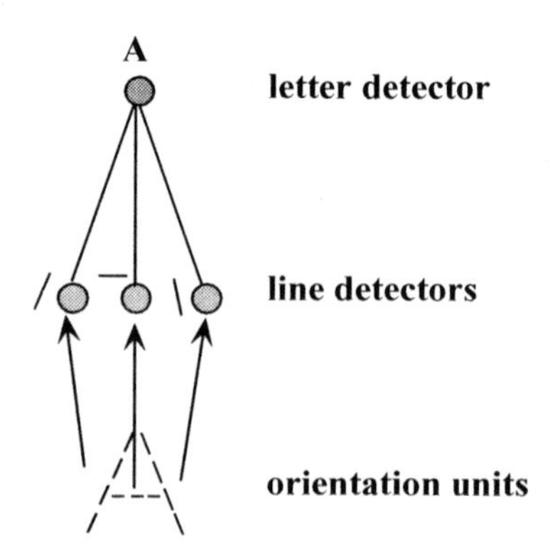

FIGURE 3.2 **A pattern recognition example. Orientation detectors in the bottom layer trigger line detectors in the subsequent processing layer, which in turn activate the unit for the letter "A."**

position, rotation, size, and a host of other distortions, nonlinear and otherwise. It would also not do well without considerable modification, for example, on a CAPTCHA, the distorted letter sequence used to prevent automatic response to Web site forms.

Nevertheless, connectionist systems with the appropriate modifications have proved extremely successful in explaining human pattern recognition. One important and ubiquitous aspect of human information processing that falls almost immediately out of the model is the integration of bottom-up (i.e., sensory) and top-down (i.e., preexisting) information. Figure 3.3 shows an example of this phenomenon similar to that in the Rumelhart et al. (1986) book that was largely responsible for launching the modern era of neural network models. Shown are the successive states of a hypothetical network that recognizes the word "CAT." In this instance, the middle letter stands halfway between an "H" as an "A"; however, we have no trouble seeing it as the latter. The relaxation of the network shown in the diagram explains why this is the case. Initially, bottom-up information weakly triggers both the "A" and the "H" in the middle layer (left). This, however, in combination with the more highly activated "C" and the "T" units is sufficient to select the word "CAT" in the word layer. Top-down priming then descends to the "A," increasing its activation relative to the "H." Eventually, a resonance loop of excitatory feedback forms between the word and letter layers and provides the "A" with almost as much activation as if it had been unambiguously present in the stimulus (there is no word "CHT," and thus the same thing does not happen for the "H").

One important property of neural networks not illustrated in the previous networks is lateral inhibition. Inhibitory cells are ubiquitous throughout the brain, and inhibition serves to keep excitation localized and within a limited range. In neural network theory, lateral inhibition, that is, inhibition between units at the same layer of processing, serves a number of purposes. Figure 3.4.A

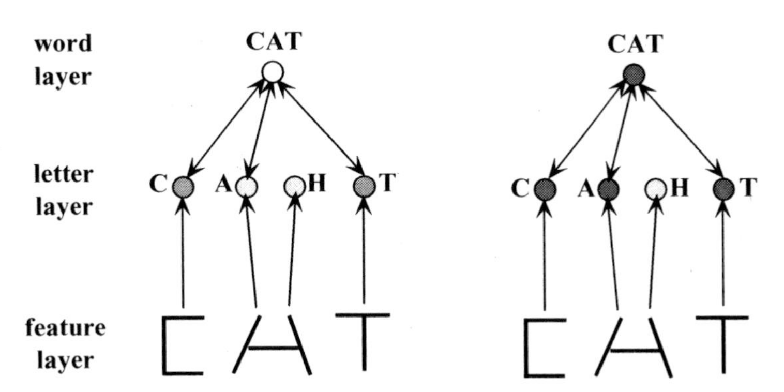

FIGURE 3.3 **Bottom-up and top-down activation serve to disambiguate the middle letter of the stimulus pattern, which could be either an "A" or "H."**

shows an example of this form of connectivity applied to pattern recognition. The high degree of similarity between block letters "A" and "O" implies that the one will also be strongly activated when the other is present. But this is not the desired result; the recognition layer should ambiguously state which is present. We do not see a partial "O" when we look at the "A"; we see only the latter. One way of achieving this is to have excitatory self-connections (the loops from the units to themselves), and inhibitory connections between the letters. Initially, the correct letter will receive slightly more activation (left). On the next processing cycle, it will send more activity to itself, and will also inhibit its neighbor more. This process will continue, until the correct letter achieves full activation, and the other is driven to a minimal value (right).

Figure 3.4.B shows another application of lateral inhibition. One remarkable property of the human visual system is color constancy, or the ability to see approximately the same colors regardless of the light illuminating the scene. The color of one's car, for example, appears the same whether it is in daylight, dusk, or an indoor parking lot. If you took a photograph of your car, however, and then used Photoshop or a similar program to find the actual color levels, you would see considerable changes based on these conditions. The human visual system is able to discount the illuminant, that is, subtract the impinging light on a scene, leaving the "true" color behind.

It has long been a mystery as to how the brain achieves this feat, although it has been suggested that a more elaborated version of the relatively simple network shown in Figure 3.4.B provides a possible explanation (Wray and Edelman, 1996). Shown are the detectors for a single color, for example, blue for a small subset of the visual field. Each detector is connected by a small inhibitory weight to its neighbors, and this process continues across the visual field. If there is an excess of blue in this field (left), this inhibition will lower the perceived level of blue by lowering the activation levels of these detectors (right). Thus, the system

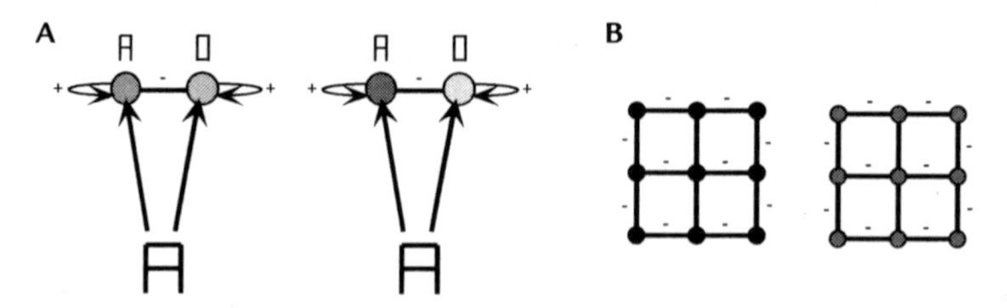

FIGURE 3.4 **Two applications of lateral inhibition. A: A winner-take-all network at the top of the pattern recognition hierarchy. This subnetwork increases the original activity difference. B: Lateral inhibition between identical color detectors will lower the overall activation of that color (only connections to the immediate neighbors are shown).**

guesses that the reason there is so much blue across a wide part of the visual field must be due to the illuminant, and not due to the scene itself. This is a reasonable approximation, because nature rarely presents a uniformly colored field to the eye. What happens if there really is a large blue patch? The model predicts that it will appear less blue to us than a smaller blue patch embedded in a larger field of colors. But this is exactly what we find from psychophysical measurements. Lateral inhibition may also play a role in color contrast, in which the blueness of a patch is exaggerated by adjacent patches of yellow (Katz, 1999).

Representation

A critical question that we have been avoiding concerns the nature of representation in neural networks. So far, we have been pretending that single units, or so-called "grandmother" cells are capable of encoding a given concept. But this is unsatisfactory for a number of reasons. If the unit (or neuron in the brain) were damaged, then this concept would be lost. We lose somewhere between 100 and 1000 neurons per alcoholic drink; if one of these happened to encode for grandma, we would lose the concept of her entirely. In addition, single-unit encoding is inefficient. Suppose, for example, we had ten units, and each stood for a unique concept, then we could represent, at most, ten separate concepts. Alternatively, suppose that we encoded each concept with a unique set of 2 units. Then we could represent up to 45 concepts (10 choose 2).

In general, then, population coding, also known as coarse coding, will be superior to grandmother or single-cell coding. Of special interest is the coarse coding used by the visual system; Figure 3.5 shows two instances of such. The first, on the left, shows a representation of position in the visual cortex consistent with the principle that the receptive field of cells grows as one ascends the processing hierarchy. These fields are represented by circles in the diagram. The position

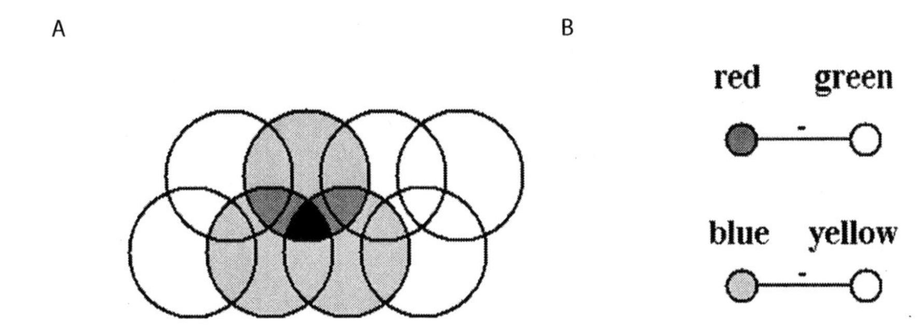

FIGURE 3.5 **Examples of coarse coding. A: Spatial position can be represented as the intersection of activated detectors with overlapping receptive fields. B: Color can be represented by the relative values of two of the four cells in the color opponent process.**

of an object can be represented by the intersection of the activated receptive fields, as in the diagram. Note that this mechanism provides a means of indicating fine-grained positions with relatively few detectors with large receptive fields.

The representation on the right shows how color is believed to be encoded in area V4 of the secondary visual cortex. Color is mediated by an opponent process, with blue and yellow, and red and green detectors antagonistic to each other. Illustrated is the representation of maroon, approximately three-quarters red, and one-quarter blue. This example also shows that coarse coding can be made even more efficient by the use of continuous values. An extremely large number of units would be required to represent each shade, depending on how finely one wanted to cover the space of all colors. Instead, the entire range of perceptible hues and shades can be reduced to four real valued numbers, implemented in the brain by relative rates of firing of the opponent colors (there is a also a white-black opponent system, not shown, that represents the brightness of the color).

So far we have been speaking as if we can assign a label to each processing unit that corresponds to a well-defined concept. One interesting aspect of neural networks, and especially ones that learn (see the next section), is that it is not necessary that the behavior of these units be semantically transparent. That is, the dynamics of a given unit do not need to bear any obvious relationship to the input, to the output, or indeed to what we believe to be the mechanism that transforms the input to the output. For this reason, connectionist systems are often called *subsymbolic processors* (Clark, 1991), in contrast to good old-fashioned AI (GOFAI) systems, where the symbols correspond to meaningful concepts. The implications for neuroengineering are direct: a) when reading from a cell, it may be difficult to assign a simple interpretation to that cell's firing behavior, and b) when attempting to stimulate a cell, the consequences of this stimulation may be difficult to predict without understanding the entire network in which it is embedded. As we shall see, however, these problems may be at least partially overcome with an adaptive mechanism that attempts to learn how best to interpret or manipulate the neurons in question.

In summary, representation in neural networks can be: a) distributed, b) nondisjoint, c) continuous, and d) subsymbolic. There is one further possibility, relating to our original distinction between rate encoding and integrate-and-fire networks. A number of empirical and theoretical considerations have suggested that fine-grained firing patterns are responsible for the binding of elements of a stimulus into a unified whole. Engel and Singer (2001), for example, have presented evidence that when the components of a stimulus are part of the same object, synchrony is found between the spiking patterns among cells subserving these components. This result suggests that the precise temporal firing pattern may be useful to the brain, and if this is the case, it will be necessary to use an integrate-and-fire model to capture this behavior.

In Chapter 4 we will present an additional reason for including the timing of spikes in our models when we look at the way in which the brain produces conscious content.

Learning

One of the most important and remarkable properties of neural networks is that learning can be reduced to a single operation, the alteration of synaptic efficacy, or, in model-theoretic terms, the weights between units. The simplest learning rule of this type is also the oldest. As far back as 1949, the psychologist Donald Hebb suggested the following in his influential book, *The Organization of Behavior*:

> *When an axon of cell A is near enough to excite cell B and repeatedly or persistently takes part in firing it, some growth process or metabolic change takes place in one or both cells such that A's efficiency, as one of the cells firing B, is increased.*

More formally, we can express this associative principle with the following rule:

$$\Delta w_{ij} = \lambda a_i a_j$$

where Δw_{ij} is the change in weight between units i and j, a_i and a_j are the firing rates of these units respectively, and λ is a parameter that controls the rate of learning. Thus, when the two units tend to fire together, the strength of the connection between them increases. Although extremely simple, this rule can account for numerous psychological results, including the fact that the frequent conjunction of two stimuli will cause the one to act as a prime for the other, and the fact that frequent presentation of a pattern allows the brain to recognize and respond to that pattern faster. This may also be the basis for the well-known familiarity effect, whereby an aesthetic stimulus, such as a song, both seems more familiar and also gives greater pleasure after it has been experienced a few times (see Katz, 1999). Finally, Hebbian learning, in conjunction with a mechanism to learn inhibitory as well as excitatory connections, can explain how parts of frequently presented stimulus cohere together, and how parts that don't belong are inhibited.

One drawback of Hebbian associative learning is that it does not allow for feedback from the environment. There is no teacher telling the system whether or not it is doing the right thing in altering the connection strengths; these emerge solely based on the coincidence of firing. In the late fifties, Rosenblatt (1958) proposed a different rule, Perceptron learning, that overcomes this difficulty with the use of a teaching signal. Like all supervised learning techniques,

Perceptron learning entails the presentation of numerous examples to the network. The role of the teaching signal is to tell the network whether the example is a member of a category or not. For example, if the category to be learned was dogs, examples of various dogs and their features would be presented to the network, and these would be marked as positive examples. Cats, mice, and tables and chairs would also be presented, and these negative examples would be accompanied by a signal that they are not members of the dog family.

Formally, Perceptron learning can be implemented by the rule

$$\Delta w_{ij} = \lambda(T - a_j)a_i$$

where, as before, Δw_{ij} is the change in weight between units i and j, a_i and a_j are the firing rates of these units, and λ is a parameter that controls the rate of learning. Here, unit j takes on a special meaning, that of classifying the examples; it is 1 if the net input to this unit exceeds a threshold, and 0 otherwise. The teaching signal T is also set to 1 if the example is positive, and 0 if the example is negative. It is easily seen that if the example is correctly classified (i.e., the teaching signal T matches a_j), no learning takes place. Alternatively, if the teaching signal is 1 and the classification signal a_j is 0, then the connection between the feature$_i$ and this unit is strengthened, and if the teaching signal is 0 and the classification is 1, then this weight is weakened. In this manner, the network gradually learns to connect the features of a category with the classifying unit, and to weaken the connection between features that do not belong to this category and the classifying unit.

Rosenblatt had high hopes for his Perceptron as a general adaptive mechanism, hopes that were shattered, however, by the seminal 1969 publication of Minsky of Papert titled, simply, *Perceptrons.* These researchers proved that Perceptrons not only could not learn the class of nonlinearly separable concepts, they could not even represent them. That is, one could not hardwire a Perceptron to be able to distinguish between concepts that are not cleanly divided by a plane in the feature space, with all of the positive examples on one side and the negative ones on the other.

We will study this in more detail in Chapter 6, but for now we will simply point out that an extra or hidden layer is required in the network to surmount this difficulty. That is, we need not just inputs and outputs but a third layer in between that performs an intermediate transformation. If we tried to put this layer into a Perceptron, we would have no way of training it, because the teacher can only tell us what the output should be, not what this hidden transformation looks like. Fortunately, there is a way of propagating the teaching signal back from the outputs to the middle layer, and this algorithm to do so is not coincidentally known as backpropagation (1988), or backprop. Backprop, and its variants, have been used to great advantage in modeling both putative human knowledge acquisition methods and as a general classification algorithm for large data sets, such as stock market prediction. It should be noted, however, that the critical element that makes backprop work, the propagation of the teaching signal, has not been found in the human or any other animal nervous system.

This brief survey by no means exhausts the field of neural network learning. It is possible, however, to make an absolutely essential point with the limited knowledge that we now possess—that there is a serious problem with the foregoing mechanisms as accounts of the *psychological* learning process, as opposed to simple categorical or associative learning. Consider, for example, what is going on in your brain as you read this book. You are learning new ideas and your horizon is being broadened (hopefully!), but can we really assimilate this process to any of our learning schemes? This seems a stretch to say the least; you are not merely forming new associations, nor are you being provided with categorized examples in a straightforward manner. The alteration of synaptic efficacy, while certainly taking place—otherwise, you would have the same brain before reading this book as after—does not capture the essential aspect of this type of knowledge acquisition. What appears to be going on instead is a reorganization of abstract knowledge structures to meet the demands of the novel incoming information. In other words, neural network learning is at the wrong level of granularity to explain the kind of learning that most of us do every day in a wide variety of contexts. The next section examines this problem in more detail, and attempts to resolve it by looking at learning and cognition in an entirely different way.

HIGH-LEVEL COGNITION

Taken at face value, connectionism states that the brain is simply a very large set of interacting networks, and in principle, patience and hard work ought to be able to unlock its secrets. To see just how misleading this seemingly innocuous way of thinking is, let us look at an analogy. Suppose you wanted to understand the current behavior of your computer. The following is a true statement: Given a full understanding of solid-state physics, the configuration of the CPU and other chips in your machine, and current state of all the electrons in these chips, you could figure out exactly what your machine was doing by figuring out the paths these electrons will take. The problem is that this is also a vacuous truth. The complexity of the device in question makes these calculations risibly difficult; you would need a machine many orders of magnitude more powerful than the one you are studying to simulate all the events going on at the subatomic level.

Suppose, however, that you knew that your computer was running program x, and that you had access to this program. Then you could make the same predictions with computing power on the same order as your machine. Moreover, in an important sense program x describes what the computer is "really" doing. Suppose the program was waiting for input from the keyboard, for example. We can see this immediately by studying the code. On the other hand, this fact is not at all evident by looking at the state of the electrons—at a minimum, describing this in solid-state terms would involve a complex conditional stating that if there was an influx of charge at one of many regions, then a sequence of actions would initiate to do such and such to the screen (to display the character) depending on another set of charges (indicating where the cursor was), etc. Note that our abstract code-informed decision ignores most of these details. We can simply say that whatever key is struck, it will be displayed at the current cursor position. We could also run the same program on a different machine of the same type or even on a different platform and still make the identical predictions. We would need to redo all our calculations, however, if we were trying to understand the behavior of the new machine.

What does this have to do with human thought? The answer, as with all models, brain-related or otherwise, is that one must approach the problem at the right level of granularity if one is to end up with a workable description. Just as it makes little sense to describe computer behavior at the level of the electron, it may be inappropriate to understand human cognition at the neural level. It may be far more difficult to do so, it may not be as predictive, and it may mask the essential commonalities between different cognizers, i.e., human beings.

The importance of this conclusion for neuroengineering cannot be overestimated. It means that ultimately, neuroengineering must piggyback onto cognitive engineering if it is to achieve its full ends. However, before examining this conclusion in detail, let us attempt to determine just what is missing from the picture of cognition as being a purely neural-driven enterprise.

The tale of the monkfish

The monkfish is bottom-dwelling creature that is both spectacularly ugly to look at and surprisingly delicious to eat (it is sometimes called the "poor man's lobster"). What concerns us here, however, is how the monkfish catches *its* food. As Figure 3.6 shows, as in all anglerfish, a number of spines protrude from the head of the monkfish. The longest and first filament, the illicium, ends in an irregular growth, which attracts fish that swim past when the monkfish wiggles it. When this filament is triggered, the monkfish brain tells its body that is has a catch on the line, and with its large mouth and sharp teeth it is able to quickly kill and swallow its unsuspecting prey.

The question we now wish to pose is the following: Does the monkfish know that it is actually an angler, or is this more in the way of an automatic reflex? We have absolutely no reason to believe the former. The neurophysiology

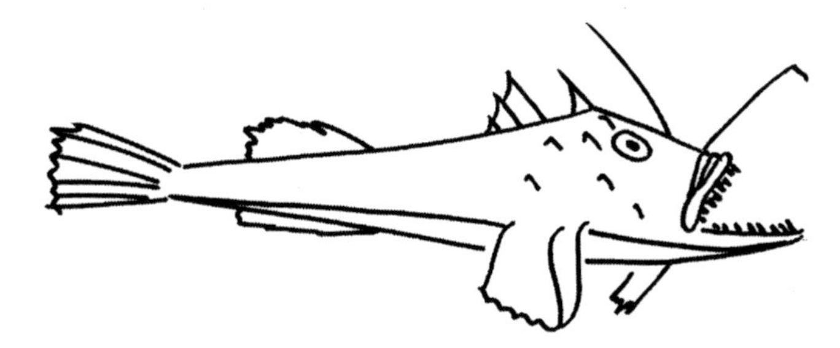

FIGURE 3.6 **The monkfish, with the illicium (natural fishing rod) shown as the foremost extension.**

of the monkfish is not well studied, but it extremely likely that this is little more than a reflex action, akin to the knee-jerk, albeit in this case one routed through the central nervous system. When the fish detects a certain type of tug on its "line," this initiates a motor sequence causing it to leap up and bite. There is no need to posit anything more complex. This is not say that evolution has not endowed this species with a remarkable gift, and one that is fine-tuned to work extremely well in its environmental niche. It simply means that we will never spot a monkfish waving a piece of seaweed in front of other fish with its mouth, because it never really knows it is fishing in the first place, and has no concept of what fishing is.

How does this differ from human fishing? In humans, there is an explicit representation of the fishing procedure, and one that can be manipulated as circumstances dictate. We know this because human beings, as a species, engage in a wide variety of fishing activities. Moreover, not only are we aware that we are fishing, we are aware of why we are fishing. This is a crucial distinction, and one that separates Homo sapiens from the monkfish, and all other species, with the possible exception of the higher primates. Knowing that simply means that one can catalog one's experience. Knowing why means that one can adjust this experience by concentrating on only what is essential in the relevant chain of causation.

Let us make this a little more concrete. Let us say that human beings are in possession of something like the following rule, which we will label GenFish for future reference:

GenFish

To catch prey X:

a) present X with something X wants
b) wait
c) confine X, hook X, or otherwise stop it from moving when it arrives

This looks like no more than commonsense, but in fact, the rule is anything but common in the animal kingdom. GenFish may seem vague, but in fact its power derives from its nonspecificity. The actual prey is not indicated in advance, but is left open as a variable. It can then be filled in the appropriate manner to allow deep-sea fishing, by putting bait on the line, to trout fish, by dangling a fly or something that looks like a fly on the water, or to catch sharks by throwing chum on the surface. If we like, we can also extend the concept further without too much difficulty to traps on land, to landing a client or account, or even to the phishing form of fishing, that is, fishing online for user information.

What this analysis points to is a crucial difference in the nature of generalization between so-called symbol crunchers, and reflex systems, regardless of the complexity of the reflex. A *symbol cruncher* may be defined as a system that is able to store and use expressions, such as a, which invoke variables with unspecified values (X in parts a and c) of GenFish. The monkfish, it is true, does respond to a variety of circumstances in its angling behavior; otherwise, it would only be able to catch one type of fish that stimulates its illicium in a single way. But this generality may be thought of as a single halo of possibilities in the space of possible prey. It cannot, without something like the above rule, consider other entirely different sorts of possibilities.

The astute reader may object: if we know that neural networks can learn, and we also know that the monkfish brain is a kind of neural network, then why is it that it could not learn to do other kinds of fishing? There are two responses to this question. First, the monkfish would have to be exposed to these other circumstances. It would have to be put in a stream to watch the behavior of trout, and it would need a broadband connection on the ocean floor, at a minimum, to understand phishing. Note that a symbol cruncher in possession of GenFish could grasp these concepts, at least in outline, without the direct presentation of representative examples. For example, if you are simply told that trout like flies, you would come up with some rudimentary system for catching these fish. You would not have the same success as if you had read the *Compleat Angler* and had been on the river for twenty years, but you would have a chance of feeding yourself.

Second, every learning experience would be completely novel for the monkfish, because they would not share any superficial features in common. Dangling your illicium and tying a fly and casting it, are two entirely different experiences, from the perspective of pure action (conceptually, of course, they are related, but the monkfish by hypothesis has no concepts of this sort). This means that either the monkfish, or evolution acting the monkfish species, would have to undergo an extensive training phase before it could learn each type of fishing.

Instead, GenFish gives us instant access to a host of fishing experiences, not all of which need involve water or eating. It is important to note that evolution *has* endowed the monkfish with an exquisite ability to catch fish; we would not fare so well on the ocean floor in camouflaging ourselves, in our dangling ability, or our snapping action. Instead, however, we have been given an entirely different sort of ability, and in many ways, a more powerful one. This is the ability to represent the world symbolically, and to manipulate these symbols to produce inferences that bear no resemblance to each other, except at the conceptual level.

In summary, it appears that man is at least partially a symbol cruncher rather than simply a very large and dynamic neural network. Of course, if the Neuron Doctrine is correct, then this symbolic machine must be implemented by a neural network, and symbol manipulation is, in essence, an abstraction that sits on top of this underlying neural dynamics. However, it is an abstraction that is ignored only at one's intellectual peril, in the same way that one cannot ignore the program currently running on a computer to understand what the machine is doing. If someone sends you on a snipe hunt and tells you that these creatures are attracted to peanut butter spread on balloons, high-pitched doo-wop singing, and dirty socks, they know exactly why you are making a fool of yourself in the woods,[1] without any neuroimaging tools whatsoever at their disposal, nor would these help if they had them. Their understanding follows directly from the knowledge that you are implementing GenFish in the hopes of catching these elusive creatures. The next section explores in more detail how a symbolic representation may be realized, and more general properties of this model of thought.

Symbolic cognition

If we are to understand and ultimately interact with man in his capacity as a symbol cruncher, we must understand how symbolic knowledge is represented. One natural conjecture is that the brain runs a program, in much the same way that computers do. However, programs as representations have long since fell out of favor in both AI and Cognitive Science, as they are too complex to work with, and do not adequately reflect the commonalities between cognitive systems. This is not to say that cognitive scientists never invoke an algorithm to explain a given effect, but that they are careful to separate the data that the algorithm runs on from the code itself. It is this data that we need to represent.

Symbolic logic presents the next most obvious means of doing so, and indeed, it is one of the most commonly invoked forms of representation. A complete summary of logic is beyond the scope of this book; here we simply summarize its use as a representational tool. The first-order predicate calculus, the form of interest

[1] A snipe hunt is a hunt for a non-existent creature, invented by camp counselors to distract their wards for hours on end.

here, consists of predicates, variables, connectives, and quantifiers. Propositions are constructed by quantifying over expressions comprising the predicates, variables, and connectives between them. For example, assuming that GenFish can be reduced to simply presenting the prey with something it wants, part a of the rule, we could represent this concept as:

$$(\forall x)(\forall y)(\exists z)\ \text{Present}(x,y,z)\ \&\ \text{Likes}(y,z) \Rightarrow \text{CanCapture}(x,y)$$

This translates roughly to "for all x and for all y, if there exists a z such that y likes z and if x presents z to y, then x will be able to capture y." Slightly awkward, to be sure, but the important thing about this rule is that y and z do not need to be specified in advance. When you are told that snipes like balloons then y becomes bound to snipe and z to balloon and you (x) can therefore conclude that you need to display a balloon to attract the snipe. As previously argued, the nonspecificity of this representation, far from making it inexpressive, is the property that allows it to be harnessed for a wide variety of situations.

Logic itself is a very general representational scheme, but it is not without considerable difficulties in practical use. One problem is that things in the real world are usually neither true nor false, but somewhere in between. This has led to various relaxations of logic, including *default logic* in which one believes something to be true unless there is direct evidence to the contrary, and *fuzzy logic*, that permits explicit degrees of truth for both premises and conclusions.

An equally influential view, and one that is slightly more ambitious with respect to the scope of explanation, invokes the notion of a *mental model* (MM) (Johnson-Laird, 1986), or a set of causally interacting parts. We have numerous mental models from those pertaining to our bodies, to cars, to the workings of other people's minds and even our own. A closely allied notion is that of man-the-scientist. In this view, man is constantly building models of the reality around him, albeit of a lesser complexity than most "real" scientific models. However, like a scientist, he adjusts his views as new data comes in, and on occasion may also undergo a paradigmatic shift in his views, as in formal science.

It is important to realize that an MM need not be either correct or well specified to qualify as a bona fide member of this category. For example, children below the age of six usually have some form of flat earth theory. What makes this a true MM, apart from the fact that it is not a completely inaccurate view of the world, from the local perspective of a child (i.e., without access to GoogleEarth™), is that it is productive and consistent. If you ask the child what would happen if you come to the edge of the earth, they typically say that you fall off into space. In this instance, of course, no one gets hurt in this way, but other models are not so innocuous. The bad blood model of disease, the prevalent notion for many centuries, encouraged leeching in a wide variety of circumstances, including those in which the patient needed their strength most.

It is not the truth of the MM that is at stake, that is, the correspondence of the model to the actual world, but the degree to which it enables an external observer to understand what is going on in the head of the modeler. This will be the case for any representational scheme. Thus, the notion that symbolic cognition matters is equivalent to saying that we can understand the progression of thought at the level of this representation significantly better than if we looked at neural firing patterns alone.

To approach this idea more systematically, we need to turn to the so-called Language of Thought Hypothesis (LOTH) (Fodor, 1975). The LOTH argues that computation in the mind bears a crucial resemblance to natural language in that it has a compositional structure. This means that the language of thought (often called Mentalese) is built out of constituent parts, or semantic tokens, to produce complex structures, or sentences, on with meanings that are a function of these tokens and the syntax in which they are embedded. For example, when we think that "John loves Mary," or the Mentalese equivalent of this sentence, what we are thinking is that John has a certain relation to Mary, namely that of loving her.[2]

We can see the importance of the LOTH to the current discussion by referencing Fodor and Pylyshn's influential (1988) article defending the so-called classical model of cognition from the onslaught of connectionism. Fodor and Pylyshn's claim that cognition has four necessary properties:

a) Productivity
We can entertain and generate a virtually unlimited number of thoughts. For example, here are a few: "The monkfish was hungry," "The monkfish was hungry at night," "The monkfish was hungry at night because it had a hard day at work," and "The monkfish was hungry at night because it had a hard day at work attempting to convince other monkfish philosophers that its fishing was more than just a reflex action." You probably have never seen any of these sentences, and especially not the last one, but you have no trouble understanding them. This follows directly from the fact that the semantic content of the sentences can be constructed from the meanings of the individual words and the syntax of these sentences. It is hard to see how this generativity would occur without such a structure.

[2]Exactly what Mentalese looks like and how it behaves is still a matter of debate. However, it need not correspond exactly to the way we think what we think. For example, it may or may not be close to what is known as folk psychology, the commonsense view that our actions our explained by intentional states, such as beliefs, desires, and the like. The key to the LOTH is that thought, however it is realized, has a combinational structure.

b) Systematicity

Systematicity means that the ability to understand an utterance is syntactically proscribed. That is, once a syntactic form is mastered, one can systematically understand other items with the same form. It is not possible, for example, to understand "John loves Mary" without also being able to understand "Mary loves John." Here, the contrast with connectionism is stark. There is no reason to believe that a network trained to understand the first sentence would be able to make the leap to grasping the second, *unless the network was implicitly implementing a LOTH*. What is preserved between the two is the structural relations between being a lover, the act of loving, and being a lovee, and unless this is captured, systematicity will be lost.

c) Compositionality

Compositionality pertains to the fact that structural similarity implies semantic similarity. Thus, to say that "John loves Mary" and that the "Prime Minister of France loves his new wife" is to say something with a very similar meaning, despite the fact that they share only one word. That word, however, is the relation on which the semantic content of the sentence is built. This is perfectly understandable from the compositional standpoint, but much less from the connectionist one, in which generalization is made on the basis of the similarity of the language tokens. Again, we could tell the connectionist system that the love relation is crucial to understanding the sentence, but this is tantamount to making it into a LOTH-type processing system.

d) Inferential coherence

Inferential coherence means that we can draw similar conclusions from similar rules. For example, it would be hard to conceive of an entity that truly understood GenFish, had applied it to trout fishing and snipe hunting, but was utterly baffled as to how to catch a mouse, after being told that they like cheese. All are just instantiations of the more general rule in the first part of GenFish, namely, that you can attract an animal with something it likes to eat. In other words, the syntactic structure of Mentalese produces a systematicity of inference as well as a systematicity of thought.

Fodor and Pylyshn then make the following claim. "Given that these properties are truly necessary, then connectionist models will be missing some or all of them, and therefore fail to adequately model cognition, or, in the case that they do possess them, they will therefore be acting in a mere implementational capacity". We should no more call a model of mind connectionist than we would call a model of cognition LISPist if it happened to be written in this language; LISP is just the way the model was implemented, and it could also have been programmed in Java, or even Fortran.

Whether cognition has, essentially, a connectionist or classical flavor is still a matter of hot debate, and this is not something that we can settle here. It should also be noted that some sort of hybrid of the two is possible, with the neural network model playing a greater role in at least some aspects of cognition than as a mere implementational substrate. For our purposes, however, if cognition retains a classical feel to *any* extent, we must seriously reconsider the nature of the neuroengineering enterprise. If it is to be fully successful in the human domain (as opposed to the monkfish or other animal domain), it must also be a form of cognitive engineering. This is a topic we will explore next.

Implications for neuroengineering

Let us assume, then, that the human brain is best understood, at least partially, as a device that implements a language of thought, rather than simply a large dimensional dynamical system—as connectionism would suggest. Certainly there is abundant evidence that this is the case. Human beings are the sole users of language, itself a combinatorial system strongly resembling Mentalese. Humans are able to apply principles such as GenFish in a wide range of environments; this is why the globe is populated with members of our species in all but the harshest environments. Finally, it would appear humans build models of their world, and this applies as much to a four-year-old as to a working scientist. Models, like other combinatorial structures, are constructed from simpler atomic parts.

Taken together, this evidence engenders a great deal of confidence that the brain is more than merely a dense ball of fibers that perform a series of transformations on sensory inputs to convert these into behavioral outputs; it is a semantic engine. Yet, one can look and high and low in the neuroengineering literature and never come across a mere scintilla of reference to this fact, which ought to be obvious to anyone with a glancing familiarity with human cognition. To a large extent, this is a byproduct of the unfortunate increasing Balkanization of academic life in which to be successful, one must be ruthless about ignoring developments at the periphery of one's immediate interest. Engineers do not talk enough to philosophers, psychologists simply do not have time to engage with physicists on an everyday basis, and everyone ignores economists (including other economists).

But there is a larger, nonsociological reason for this state of affairs. Neuroengineering is a field in its infancy, and accordingly, the goals of its practitioners are exceedingly modest in the larger scheme of things. As we shall see, the primary thrust is to understand motor intentions, with the goal of helping the paralyzed achieve some degree of control over their environments, and these intentions do seem to be adequately coded at the neural level. No one doubts, for example, that given enough electrodes situated over the motor and parietal cortices, and with the appropriate signal-processing technology, we will *eventually* be able to create a brain machine interface (BMI) that allows the completely locked-in patient (i.e., with no muscle movement whatsoever) to exert control over either a robot or exoskeleton to accomplish anything a normal person can. The other primary area of interest is in sensory prostheses. Here success has been more rapidly forthcoming, and engineers are currently able to construct a passable cochlear implant, and retinal implants are on their way (see Chapter 7).

It does make sense to begin with both inputs (prostheses) and outputs (motor intention detectors). At either end of the processing stream the brain resembles a pure connectionist processor, as it must; a language of thought requires intensive computational resources not present at the periphery. But what happens when we start wanting to influence the contents of thought more directly? For this we will need to access the computations of the brain, at the level at which the real cognitive work is accomplished, and this may entail an entirely different strategy the one currently on offer.

Once again, we can see this most clearly by an analogy with machine computation. Suppose we wished not only to understand what a computer was doing, but also influence these processes. To do so by inserting a wire into the machine, and pushing signals onto a chip pin (or even worse, into the chip itself) would be a fool's errand. It is unlikely that even the computer engineer who designed this machine would attempt such a task. We interact with a computer either by running premade programs, or by writing new programs, *because this is the level of granularity that computation in these machines takes place.* Everything else is implementational detail, which we can ignore, unless something goes wrong, like a rare bus error or chip error instigated by cosmic radiation.

Likewise, to the extent that the LOTH is valid, it will not suffice to insert wires into the brain at random and hope to direct its cognitive computations. This is not to say they will not have an effect, it is just that we will not be able to produce the effects we desire by these means. In summary, the methods that neuroengineering has been built on to this date have served it well because they dovetail nicely with its current goals; it is not at all clear they will scale-up to other sorts of manipulations. Most likely, we will need to produce an interface into the brain, in its capacity as a semantic engine, to both read its contents and to affect those contents. We will reprise this issue in later chapters, where we also suggest some methods for doing so.

COMPUTATIONAL LIMITATIONS OF HOMO SAPIENS

The symbol-crunching ability of Homo sapiens, especially when combined with an exquisite and fast visual processing system, has resulted in a truly remarkable species. It is also noteworthy that evolution has only been able to manage this feat once, unlike say, flight (both insect and bird populations) or camouflage (numerous species throughout the animal kingdom). Genus Homo may be characterized as the result of a lengthy evolutionary arms race, but a very peculiar one, in which the goal is not strength or speed or a specific wile (as in the monkfish) but a more general ability along the lines we have argued, most probably tied, in anatomical terms, to a burgeoning frontal cortex.

However, before we pat ourselves on our collective backs too much, we need to appreciate that the physical restrictions on the brain of the sort discussed at the close of the previous chapter are, of necessity, accompanied by a number of computational limitations. To put it in the bluntest possible terms, if there is intelligent life on another planet, then there is a high probability that we will seem no more intelligent to them than the lowly flatworm does to us. This follows from the fact that our cultural and physical evolution has lasted a mere hundred K years, a drop in the bucket in a 13-billion-year-old universe.

The following, non—mutually exclusive, and nonexhaustive list provides a more detailed examination of our cognitive limitations:

i) Short-term memory
Miller's (1956) so-called magical quantity, 7 ± 2, is now firmly part of psychological (and popular) lore. This means that our short-term memories are limited to between 5 and 9 items, depending on circumstance. An important aspect of this formula is that it pertains to chunks and not to raw information. For example, if you are given 15 colors to remember, you will be able to recall roughly 7 of them. But if the first three are red, white, and blue, then the capacity can be increased, because they can be put into a single chunk—the colors of the flag. A related complication is that various encoding strategies can also increase capacity, such as imagining the items in a visual context. Our visual capacities are, in general, greater than our linguistic ones, in keeping with the large part of our brains dedicated to this modality. However, it remains the case that short-term memory capacity is severely limited with respect to even the most primitive of computational devices.

ii) Long-term memory
We are under the impression that long-term memory is virtually infinite, and in many cases, it acts as if it has no bounds. For example, we can

remember up to 10,000 faces or more, which is astonishing considering how indistinct they are from the point of view of feature similarity[3]. But any finite device has finite memory, and this is especially true of a three-and-a-half-pound device that uses relatively large objects, such as neurons, for storage. Estimates based on neural capacity tend to be misleading, however, because we do not know the degree of redundancy in the brain. This makes the work of Landauer (1986) especially inviting, because it is based on an actual performance estimate. Landauer looked at how much people were able to absorb in a wide range of contexts, including text, pictures, and music. He found that this rate was approximately 2 bits per second. Extrapolated over a course of a lifetime, and assuming minimal forgetting, this yields 10^9 bits, or about 100 megabytes. This is the average hard disk capacity of a desktop from 1988! In this respect, then, it appears we are no better then the very first personal computers.

iii) Processing speed

A key factor in assessing any computational device is speed, and this is no less true of the brain. The evaluation of the brain's speed, a large-scale parallel processor, is a much more complex matter than that of a computer, however. Much depends on the nature of the task. Humans, for example, have an extremely well-designed visual processing system that still exceeds, in many respects, that of the fastest machine. On the other hand, rational thought processes are plodding, inefficient, and often wrong (see the discussion of irrationality that follows). If you like, we can say that the human processing system, in its capacity as a connectionist device, is a well-oiled well-honed machine, but in its capacity as a symbol cruncher it lumbers along in the hundreds of millisecond (or worse) range. However, both aspects of processing are open to performance improvements, the former by increasing the degree of parallelism, and the latter by also increasing the degree of parallelism, by increasing the speed at which individual inferences are generated, or both.

iv) Processing performance

Earlier it was claimed the language of thought hypothesis allows us to explain the generativity of thought, and how it is possible to both create and understand sentences that we have not seen before. But there is a limit to the degree of recursivity in this generative process; this is the

[3]Of course, different faces look very distinct to us; this is probably why they are so easily remembered. As a social animal, it is to our benefit to extract small difference in appearance and remember these salient details better than the numerous similarities.

origin of Chomsky's (1969) famous competence/performance limitation. English grammar may make "The horse that raced past the barn fell" a perfectly valid sentence, but it is difficult to parse, and the sentence "The horse the Czechs bought on their trip to Kentucky raced past the red barn with the Amish Hex that looks like a Mandala fell" is more awkward still. Short-term memory limitations, with other shortcomings of a more linguistically specific origin, combine to limit actual linguistic and therefore representational performance. This is especially acute in spoken language, where the inability to rescan a sentence turns most conversations into a series of terse and easily understood propositions.

v) Bounded rationality

The time-honored definition of man as a rational biped has come under severe attack in recent years, and for good reason. One of the most obvious among many flaws in human reasoning is irrational exuberance and its opposite, or the tendency to believe that current trends will continue indefinitely. This, for example, drives the housing and stock markets above their "natural" values during a bullish cycle and below during a bearish one. It may be objected that these are rational *investment* strategies (as opposed to rational evaluations of value), but careful study in the laboratory reveals a number of provable departures from ideal reasoning do indeed occur.

One of Tversky and Kahneman's (1973) famous examples will suffice to make this point. Subjects are told that 31-year-old Linda was a philosophy major in college and heavily active in antiwar and antinuclear protest movements as a student, and asked which is more likely: a) that Linda is a bank teller, or b) that Linda is a bank teller and a feminist. Approximately 85% choose b, but this answer is a subset of that given by a, and therefore *objectively* less likely. Answer a includes the case in b as well as the possibility that Linda is a bank teller and a stay-at-home-mom, a designer of conservative t-shirts, etc. What appears to be happening is that we are associative as well as reasoning agents, and the strong connection between being a philosopher/activist and being a feminist blinds us from making a rational evaluation of probabilities. More disturbingly, Tversky and Kahneman present a number of examples in which doctors and other professionals, responsible for life and death situations, regularly depart from rationality when making judgments in their own fields.

vi) Creativity

As a tentative, working definition of creativity, we can say that it involves the ability to produce items that are both novel and useful (we will have much more to say about this subject in Chapter 9). Given this view, it must be concluded that most of us do not possess this trait in abundance. It is relatively easy to produce something novel, but to create something that is also of great utility is an entirely different matter. To take an extreme view, but one that is not without its adherents, we could argue that each generation contains only a few dozen truly creative geniuses. Everyone else, to be sure, engages in minor acts of creativity (where to seat the cousins from New York so that they don't argue with the relatives from Georgia, and at the same time keep Uncle Frank away from both) on a daily basis, but solving these conundrums is not on par with the creation of General Relativity, the invention of the lightbulb, or what may be the most underrated invention of our times, the humble key drive. It would appear then, that creativity is thinly distributed, and moreover, among those that are blessed with this trait, it remains annoyingly and bafflingly capricious.

vii) The bottleneck of consciousness
When we look out at the world, it appears to us like a wide-screen television, in high definition, and also in 3D. But this is largely an illusion. In fact, we really piece together a picture out of successive surveys of the scene, made by frequent and rapid eye movements (saccades). To see this, place two objects in front of you, moderately distant, say about 6 inches. Then try to concentrate on both. You will find that you can really only pay attention to one at a time, usually the one that strikes the fovea, or central area of vision. As with vision, so with the rest of consciousness. In fact, some have made the argument that Miller's magical quantity as applied to contents of consciousness should be reduced to 1 ± 0! That is, we only can attend to one chunk at a time, although as with short-term memory, this chunk can have constituent parts, as long as they can be taken in as a single Gestalt. The stream of consciousness is just that—a single line of successive thoughts, rather than multiple tributaries—and to attend to one implies the obliteration of another.

viii) The narrowness of thought
Finally, let us take a step back and look at the brain's output—thoughts—with a wider lens. The space of possible thoughts may be defined as the set of all thoughts that could possibly be had by any thinker, be they man, machine, terrestrial, or alien. Do we have any confidence that the collection of human thoughts occupies anything more than a sliver of this larger space? Furthermore, of the thoughts that we can think, how easy is it for the average person to reach these? The mind may

be compared to a physical landscape, with the flow of the water over this tableau likened to the flow of thoughts, and valleys and troughs controlling the course of those thoughts. As one ages, these indentations become deeper through erosion, and thoughts that could once flow in all directions are now forced to flow down familiar pathways. This is perhaps the most serious limitation on computation—because of a lack of will, or because of more ingrained limitations, we simply cannot explore more than a small fraction of "thought space" in our limited lifetimes. We will have much more to say about this neglected topic later in this book, when we discuss the possibility of incorporating other cultural perspectives and other minds into our own.

The implication of this list for neuroengineering is clear and direct: we can and should do better. There is no reason why we need to be addled with a thinking device that nature has yet to perfect. Other organs no doubt operate closer to optimal performance, because evolution has had much more time to work on them. The fact that heart transplants are still preferred over artificial hearts is a testament as much to the economy of nature's design as to the lack of our engineering prowess. But the brain, unlike the heart, does not have a fixed agenda—it is not limited to pumping out a certain number of thoughts per minute. There is no reason that we could not improve on both the quantity and quality of these thoughts, other than a timidity in confronting what nature has been so far able to produce. It is noteworthy that we feel no such timidity in the physical realm, and if we did, we would not have bulldozers, skyscrapers, plasma televisions, or computers.

Before exploring how mental enhancement could work, however, we need to provide a wider intellectual foundation commensurate with the wider goals of this book. This is to show that not only can we produce better brains, but that we may also be able to transfer our psyches from their original organic homes to entirely nonorganic and computationally powerful substrates. Accordingly, in the next few chapters, we explore the notions of consciousness and identity, with a view to showing how this could be accomplished.

REFERENCES

Chomsky, N. (1969). *Aspects of the Theory of Syntax.* Cambridge: MIT Press.

Clark, A. (1991). *Microcognition: Philosophy, Cognitive Science, and Parallel Distributed Processing.* Cambridge: MIT Press.

Engel, A., & Singer, W. (2001). Temporal binding and the neural correlates of awareness. *Trends in Cognitive Science*, 5, 16–25.

Fodor, J. (1975). *The Language of Thought*. Cambridge: Harvard University Press.

Fodor, J., & Pylyshn, Z. (1988). Connectionism and cognitive architecture: A critical analysis. *Cognition, 28*, 3–71.

Hebb, D. (1949). *The Organization of Behavior*. New York: John Wiley &Sons.

Hubel, D., and Wiesel, T. (1959). Receptive fields of single neurones in the cat's striate cortex. *Journal of Physiology, 148*, 574–591.

Johnson-Laird, P. (1989). *Mental Models*. Cambridge: MIT Press.

Katz, B. (1999). An Ear for Melody. In N. Griffith, and P. Todd (Eds.), *Musical Networks: Parallel Distributed Perception and Performance*. Cambridge: MIT Press.

Katz, B. (1999). Color contrast and color preference. *Empirical Studies of the Arts, 17*, 1–24.

Landauer, T. (1986). How much do people remember? Some estimates of the quantity of learned information. *Cognitive Science, 10*, 477–493.

Miller, G. (1956). The Magical Number Seven, Plus or Minus Two: Some Limits on Our Capacity for Processing Information. *63*, 81–97.

Minsky, M., & Papert, S. (1969). *Perceptrons*. Cambridge: MIT Press.

Rosenblatt, F. (1958). The perceptron: a probabilistic model for information storage and retrieval in the brain. *Psychological Review, 65*, 368–408.

Rumelhart, D., McClelland, J., and Group, P. R. (1986). *Parallel Distributed Processing* (Vol. I). Cambridge: MIT Press.

Tversky, A., & Kahneman, D. (1973). Availability: A heuristic for judging frequency and probability. *Cognitive Psychology*, 207–232.

Walton, I. (2006). *The Compleat Angler*. Lenox, MA: Hard Press.

Werbos, P. (1988). Backpropagation: Past and future. *Neural Networks*, 343–353.

Wray, J., & Edelman, G. (1996). A model of color vision based on cortical reentry. *Cerebral Cortex, 6*, 1996.

Chapter 4 A Framework for Consciousness and its Role in Neuroengineering

The so-called mind/body problem has occupied a central and influential role in Western philosophy since the inception of intellectual inquiry. Traditionally, the goal has been to show that a particular conception of the relation of the mind to the body is both possible and coherent. Our concerns, though drawing as needed on the voluminous literature, are largely tangential to that of the philosopher, and at the same time more ambitious. As scientists, we wish to know just what it is about the brain that allows conscious content to occur, and as technologists we want to know under what circumstances this aspect of the brain can be reproduced. It is not enough that our resulting conception be consistent, either internally, or with our intuitions about the world.[1] It must also be fertile, in that it has concrete predictive value, and it must be well-specified, so as to serve as the intellectual base for practical manipulations of consciousness in possibly nonorganic media.

This is, admittedly, a task of high order, and the sheer lack of empirical content of the sort to be discussed would have made it next to impossible just a few decades ago. It is true that these results still do not unambiguously point to one or theory or another, but then, empirical findings at the start of a scientific endeavor rarely do so. Indeed, we will attempt to demonstrate, among other things in this chapter, that most explanations for consciousness are beset by serious flaws, well

---

[1] Not only is correspondence with intuition not sufficient for the scientific enterprise, it is probably not necessary either. If we find that a good explanatory account conflicts with intuition, it is often best to jettison the intuition and keep the account.

before the evidence is taken into account. In particular, these theories do not possess the virtue of simplicity, the hallmark of a good scientific account.

This fact has been largely obscured by the driving methodology of the philosophical enterprise, that of analysis, which by its nature favors consistency over beauty of form. With the aid of a simplicity heuristic, and some additional facilitating considerations, we will be able to provide something less than a complete theory of the relation of the mental to the physical, but more than a conjecture. In addition, as will be argued, this proto-theory will be sufficiently detailed so as to provide a framework in which to discuss the future of neural technologies, as seen in Part III of this book, as well as provide an understanding of the limitations of current technologies, as discussed in Part II.

Before reaching these topics, however, much work needs to be done. We begin by discussing two foundational topics, that of *qualia*, or specific conscious contents, and the concept of supervenience. We then introduce the growing body of empirical research on the nature of consciousness, and consider topics such as blindsight, binocular rivalry, and masking. We then examine the major possible explanations for consciousness. These are in two camps, physical and process based. It is first discussed why physical theories, which state that consciousness is a function of some material aspect of the brain, have had a declining impact on the scientific community. Most believe instead that consciousness is a function of the way the brain processes its input and converts it to the appropriate outputs. These process-based theories may be divided into three types: computational functionalism, reduced functionalism, and state-based functionalism. It is argued that only the second of these is tenable, simple, and, as it turns out, compatible with the known data. For what it is worth, we will hang our technological hat on this possibility, although many of the conclusions reached in later chapters will be compatible, *mutatis mutandis*, with more general notions of functionalism. The chapter concludes with the implications of our musings on consciousness for neuroengineering, a topic that will be will be explored in greater detail in later chapters.

INTRODUCTORY CONCEPTS

We begin our sojourn by looking at two very important conceptual questions. The first asks just what it is that we are studying when we study consciousness. Science always requires, if not a perfectly well-ordered set of data, then at the least some loosely organized body of evidence that it can explain. In the case of consciousness, it is not entirely clear what that body would consist of, and thus some initial conceptual clarification is in order before actual data are approached. The second question pertains to the relation between the physical and the mental. We know that there must be some relation—when you hit your head, you may not see cartoon stars, but your stream of consciousness is definitely affected. Can we describe the relation between the brain and conscious content with sufficient clarity so as to encourage the formation of reasonable hypotheses without precluding in advance viable possibilities? It will turn out that we can, but before doing so, let us turn to the first difficulty, the vexing problem of what are known as "qualia."

Qualia as the data of consciousness

We could facetiously define qualia as that which if subtracted from the universe it might just be explicable. Qualia present the hardest challenge to scientists of the mind today, and many believe the hardest challenge in all of the sciences. In intuitive terms, the word "qualia" refers to the so-called raw feels of sentient existence. Qualia (single "quale") encompass the experience of looking at waves rolling in to the shore, the smell of fresh-baked bread, and the feel of a newly ironed shirt on the skin. Although qualia may lead to a behavioral response, or to an inner contemplative reaction, they need not produce a direct reaction above and beyond the experience itself.

There is some debate as what counts as qualia, but most would include at least the following:

- Visual experiences of all sorts
- Auditory experiences, including that of speech and music
- Olfactory experiences
- Tastes
- Kinesthetic feedback
- Tactile feedback
- Heat and cold sensations
- Complex combinations not reducible to either component, such as the mix of taste and smell when eating
- Pleasures and pains
- Emotions such as happiness, sadness, anger, and rage
- Itches, irritations, and the like

The following are more controversial, and in some cases may be merely an admixture of the previous list, but this is still unclear:

- The internal stream of thought
- Aesthetic enjoyment
- Understanding
- Confusion
- Craving

- Longing
- Love and hate
- Nausea
- The Aha or Eureka feeling
- Religious experience, epiphanies, orgasms, and other ecstatic experiences

Regardless of the nature of the feeling, the central scientific problem concerning qualia, and the one that this chapter attempts to address, is how the brain produces, generates or otherwise causes qualia to arise. The central problem for neural technologies that attempt to manipulate qualia is what aspects of the brain need to be modulated to achieve these effects. Satisfactorily solving the second technological problem is strongly dependent on at least a partial solution to the first scientific problem. While it may be possible to provide some degree of technological advance without a good scientific understanding of qualia, it is unlikely to result in the complete solutions that we are after.

To see that this is the case, consider the following analogy. It has been known for many centuries that exposure to a mild form of a disease can vaccinate one against more virulent strains of this disease. For some time it was thought that Edward Jenner originated the practice in England and Cotton Mather in the American colonies, but in fact it goes back countless centuries before this. Jenner learned of the procedure from the writings of Lady Mary Wortley, who observed it in Turkey while her husband served as the British ambassador to the Ottoman Empire, and Mather learned of it from his slave Onesimus. Thus, we do not have a good picture of the origins of vaccination, but most probably it arose through the simple observation that those exposed to a strain of the disease, no matter how weakened, do not get it again. This is why it was developed in at least two distinct cultures, neither of which had any real understanding of the immune system. Vaccination, then, as a purely empirical fact of nature, is certainly not without value, and has saved an untold number of lives. However, without an understanding of *why* it works, as opposed to the fact that it merely does, it is not possible to develop drugs that bolster the immune system as a whole, or to develop drugs that target diseases after they strike, let alone treat an autoimmune deficiency such as AIDS.

Likewise, it may be possible through trial and error to discover special cases in which mental life could be manipulated[2], but without a theory of the generation of qualia we cannot attempt more general manipulations of the sort proposed in Part III of this text. One problem in the search for a scientific

[2] A fascinating example of this predating modern neural technologies can be found in the experimentation with the direct stimulation of the pleasure centers by Robert Heath of Tulane University, reviewed in Chapter 10.

solution, and unique to qualia in their capacity as scientific data, may be termed the qualia-as-data (QAD) problem. The scientific endeavor can be characterized as the attempt to formulate theories that predict the outcome of experiments. These outcomes are presumed to be objectively available; that is, anyone with the proper equipment can observe the data, and with sufficient expertise can determine how well these data match the predictions of the relevant theory. With qualia, however, we are dealing with ephemeral effects that usually "reside" only in the mind of their bearer. On the surface, therefore, it appears that qualia do not meet a critical condition for scientific data, namely that they are available to any scientist who wishes to inspect them. Ordinary cognitive and other psychological sciences are beset with enough difficulties as it is; the QAD problem throws an extra wrinkle into the process that could easily lead to the abandonment of the entire enterprise were it not the case that drawing consciousness into the scientific realm will enable us to solve one of the most fundamental puzzles of the universe.

A partial solution to the QAD problem can be seen with respect to the phenomenon of binocular rivalry, to be discussed in more detail in the following text. In this effect, two different images are presented to the two eyes, and subjects report to the examining scientist which image they saw, that presented to the left eye or that presented to the right eye. It turns out that in most cases only one of the images makes it into consciousness. Assuming that the subject's report can be taken at face value, it can be transmitted to the scientist. Once the scientist has this data, he can then broadcast this to the larger community, and they in turn can get to work to determine the invariants in the brain associated with the seeing of one image rather than another. Thus, the private data can, in principle, be made public via reporting (or other behavioral signs in nonhumans).

The reason that this is a partial solution rather than a complete one is that the subject's reporting of his states presents a potentially weak link in the chain. If one wished to be skeptical, one could say, for example, that the act of reporting may alter the original qualia, or that reporting is a secondary process, subject to information loss due to memory retrieval problems or the like. More extreme forms of skepticism are also possible. For example, one could claim that the image in the so-called unattended eye is really seen, but not by the same consciousness that is connected to the linguistic area of the subject's brain. Although this is unlikely, there is no way or refuting this without a solid theory of consciousness saying why this is not the case. But such a theory can only be forthcoming if the data itself has a solid base, or at least one that is not constantly slipping.

Chicken- and- egg situations such as these are not necessarily impossible to resolve, however. The initial data of consciousness, it is true, must be viewed through the fog of subjectivity, but just enough may slip through to form a tentative theory. This theory, in turn, can help in delimiting and interpreting the data, which in turn may then result in a stronger hypothesis. This bootstrapping method will also prove useful in this chapter. We begin, however, by examining another key premise of this enterprise.

Supervenience

The notion of supervenience is one of the most commonly used analytic tools in contemporary philosophical treatments of the mind-body problem, and will prove equally useful in our scientific survey of theories of consciousness. *Supervenience* is a general concept that may be applied to relate two arbitrary sets of phenomena, but here we define it with respect to the physical and the mental:

> The Supervenience Principle
> *The mental is supervenient on the physical if, and only if, different mental states imply different physical states.*

In this definition, the mental supervenes on the physical, or, if we like, the physical acts as a kind of substrate (in a manner to be clarified shortly) for the mental. By mental we mean all aspects of the psychological, including perception, reasoning, emotion, etc., but most importantly for our purpose the complete set of possible qualia that can be experienced. The physical includes, for the current purposes, all the physical elements of the brain, such as the neurons, and possibly their helper cells and/or the functional workings of this organ.

The supervenience principle dictates the possible types of mappings from the mental to the physical. Figure 4.1 shows these relations. The graph on the left illustrates a possible mapping consistent with supervenience. Here, four distinct physical states result in three possible mental states (the instances 1 and 3

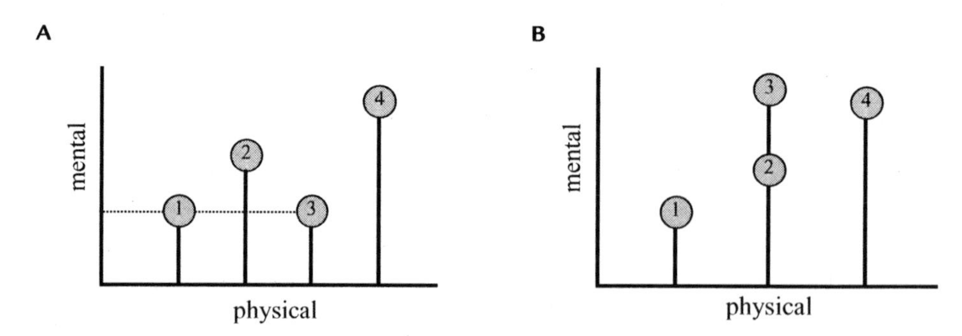

FIGURE 4.1 **A : If the mental supervenes on the physical, then each mental state corresponds to a unique physical state (although the differing physical states may produce the same mental state). B: This situation is not allowed by supervenience, because the same physical state produces two different mental states.**

represent identical mental states). Note that this kind of identity is permitted by the principle; all that is required is that if the mental states are different, then the physical states also are. Instances such as 1 and 3, which are the same mental states, may or may not have identical physical substrates. What is not permitted is the situation represented by the graph in Figure 4.1.B. In this case, differing mental states (2 and 3) have the same physical realization.

An alternative and equivalent way of expressing the same principle can be achieved by invoking the mathematical notion of a function. A function is a possible many-to-one mapping. That is, for a given argument to a function, there will be a unique result. On the other hand, different arguments may result in identical results. For example, $y = x^2$ is a function because for each x a unique y is produced. However, the converse need not be true: when $y = 25$, x can be either 5 or -5. To say that the mental supervenes on the physical, then, is to say just that the mental is a function of the physical. In Figure 4.1.B, mental events 2 and 3 must be a function of something else other than the physical, because they are not uniquely determined by the physical.

One of the reasons that supervenience proves so useful to the philosophy of mind is that it is a minimal claim. What it does not say is as important as what it does, and it is extremely liberal in the kinds of theories of consciousness it allows. For example, it says nothing about *why* the physical acts as a substrate for the mental. It could be something to do with some aspect of the physical, in this case the brain, or it could have something to with the operation of the brain. It is even possible, although less likely, that a nonphysical substance that moves in lockstep with its physical counterpart could be "causing" the mental; this would still result in the mapping constraints implied by this doctrine.

However, despite its minimal logical scope, it's just powerful enough to provide the intellectual justification for the construction of artificial consciousness. For if the mental truly supervenes on the physical, and if this is true for all aspects of the mental, including consciousness, then to produce a conscious machine it will suffice to construct a machine with the correct physical substances and/or operational characteristics. Of course, it will not tell us how to do so, only that it is possible in principle. This is the job of the rest of the chapter.

What sort of evidence is there for supervenience? We first mention the admittedly indirect evidence that its broad scope has allowed it to be one of the few principles in this often tortuously difficult field that is almost universally accepted in one form or another. For example, much has been made in recent years of the so-called hard problem of consciousness. This states that in contrast to (in principle) easy problems, such as pattern recognition or linguistic comprehension, the problem of how qualia arise from mere matter is one that we do not even know how to begin to solve. To bolster this intuition, the philosopher David Chalmers (1996) offers the following scenario. We make an atom-by-atom copy of you. You are still conscious after this operation (we hope!) but it is at least conceivable that the copy is not—it could still behave in an identical fashion to you without "giving off" the qualia that you routinely experience. There is no entailment from the physical to the mental in the same way in which once one knows all the physical properties of snow, one also must conclude that it is white, it is slippery to the touch, and so forth. On the other hand, knowing all the physical properties of the brain does not pin down the mental properties; they could be otherwise, or entirely absent. Thus, your copy need not, by necessity, be conscious in the way that you are.

Yet, as Chalmers also notes, a zombie copy of this sort is not a nomological possibility, only a metaphysical one. That is, given the laws of our universe, including the laws that govern the transformation from the physical to the mental, the copy would indeed have the same qualia under the same circumstances. The import of the argument is not to show that consciousness does not supervene on the brain—as previously stated, this is a widely held principle—but rather to show that consciousness presents a unique challenge, in that if we are lucky enough to produce a reasonable theory of this phenomenon, we may still not know why it makes sense. In the following treatment, we hope to finesse this issue to some extent by relying on the additional buttressing granted by the principle of parsimony.

Regarding the scientific evidence for supervenience, we can first provide evidence consistent with Figure 4.1.A. As discussed in the following, binocular rivalry produces activity in the area of the primary visual cortex for the both the attended *and* at least some for the unattended eye. But the only image that is seen is for the attended eye. Therefore, if we were to either remove the activity or alter the activity associated with the unattended eye, via Transcranial Magnetic Stimulation (see Chapter 7) or some other means, there would be no effect on conscious content; we would be eliminating a stimulus that does not make it to consciousness anyway. Thus, two different physical states can give rise to identical mental states. This much is uncontroversial, and it is generally accepted that not all brain activity gives rise to consciousness.

Showing that Figure 4.1.B is *not* possible is a more difficult proposition. However, we can obtain a high degree of confidence, if not absolute certainty, by noting that that this state of affairs has never been observed. Both large-scale and small-scale changes in consciousness, as far as we know, are always characterized by changes in the brain. Take, for example, the comparison between the awake state and the nondreaming sleeping state. As will also be discussed, sleep is invariably characterized by a set of slower and more synchronized set of brain waves, probably due to modulation of cortical activity by the thalamus. Or consider the changes associated with the ingestion of alcohol. This is associated with a set of well-known changes in brain functioning. It may be possible to feel drunk without drinking under certain pathological circumstances, but these are all presumably accompanied by the appropriate changes in the brain. We do not get tipsy for no reason, that is, for no physical reason. Changes in visual perception are also reflected in neural events. As was discussed in Chapter 2, color contrast and color induction effects, for example, are produced by a plausible neural mechanism. Finally, although the specific cause is often not known, no one expects that a "causeless" itch could arise. The itch might not have an immediate peripheral origin, such as a mosquito bite, but a suitably powerful neuroimaging device will eventually be able to register the itch as a central nervous system event. In summary, whenever they have been measured, changes in consciousness are always accompanied by changes in brain states, and we have no reason to believe that this principle will be violated in the future.

If supervenience is correct, as is likely, a number of broad classes of technological advances in neural engineering are, in principle, realizable:

a) First, it will eventually be possible to build, as the itch example suggests, a means of inferring conscious content from brain states. This is already true in a few limited examples, but supervenience offers hope that this can be done on a larger scale. For example, it should be possible to build a device to hear the thoughts of a thinker without those thoughts being uttered.
b) If the physical provides the substrate for the mental, then the mental is *determined* by the physical. This means that arbitrary modifications in the mental are possible via the appropriate modifications of the physical. Thus, devices which alter raw qualia, such as colors, or complex qualitative mental states, such as the degree of sobriety, are in principle possible. The wag may reply that we already have such devices; they are called TV and beer. The reply to this is that we will have much greater flexibility in what we can produce with consciousness modification devices. For example, it may be possible to induce a state in which you are much more likely to have a eureka moment, something currently not possible (see Chapter 9).
c) Finally, supervenience opens up the possibility of artificial consciousness, or consciousness residing in a nonhuman substrate. This principle as it stands is not chauvinistic, i.e., it does not state that that which determines mental content must be unique to the brain. Unless, for some reason, the aspect of the physical that generates consciousness can only reside in humans or perhaps some higher animals (and we have no reason to believe that it could), the laws governing the transformation of the mental to the physical will be in force for at least some alternative realizations.

Supervenience by itself is far too weak to provide specific mechanisms for these achievements, however. What is needed above and beyond this principle is a firm grasp of just what aspects of brain activity create and sustain conscious processes. However, the truth of supervenience gives us confidence that in the minimal case, we may understand the laws that transform the physical into the mental, even if we fail to understand why those laws should hold. We will first review in more detail the recent empirical data in this area, and then in the subsequent section offer a more specific suggestion regarding the form these laws may take.

EXPERIMENTAL DATA

The past ten to fifteen years have seen a flowering of both interest in consciousness as an important theoretical construct and as an object worthy of empirical study. Before describing the reasons for this minor renaissance, it is worth taking a peek into the attitudes that prevailed prior to this revolution. The study of the mind began in earnest in the second half of the nineteenth century, by pioneers such as Wilhelm Wundt in Germany and William James in America. As so often happens at the start of a new field, these men and their students had a wide range of interests including instinctive behavior, associations, and learning, but also comprising such diverse "internal" topics such as aesthetics, free will and volition, and consciousness itself.

The generation following these stalwarts chose to see things differently. Internal states of the mind were thought to have both a dubious ontological status as well as to be scientifically unproductive. The task of psychology, as advocated by staunch behaviorists such as B.F. Skinner, but also more or less by the rest of academic psychology, was to properly describe the input-output relations of the mind. Needless to say, such a psychology was an abject failure (although it took about fifty years to realize this). Any organism above the level of a paramecium not only transforms sensory inputs to behavioral outputs, it maintains a model of its world independent of behavioral contingencies. For example, it is not hard to show that the rat (Skinner's behavioral system par excellence) holds a mental map of its environment, and can use this map to anticipate and prepare for future actions.

Starting in the, '60s, onto this narrowly paved view of psychological processes strode the white horse of Cognitive Psychology. This approach marked a return to the vision of the founding fathers, in that mental representations were seen as the key to understanding the mind. There is no doubt that cognitive psychology was a major advance over behaviorism, but one major problem still remained. Although it was never stated by its proponents so boldly, cognitive psychology entertained the idea of internal states primarily as a means of better explaining behavior, and not as worthy subjects in their own right. The c-word, consciousness, was rarely uttered in polite psychological circles, and only then, *sotto voce*.

Why has this changed in the past decade or so? One reason is that many, such as the author, who were raised comfortably within the swaddling cloth of the cognitive paradigm, simply wanted more than their intellectual elders. The feeling was that if consciousness is real, and it appears to be, then we ought to be able to do more than armchair philosophizing about this most critical of problems: we ought to be able to manipulate it in the laboratory, and then provide scientific explanations of our results. This section highlights some of the important results that have come to the fore in this area. In the interest of space, no attempt will be made to be comprehensive—for example, two important sets of results, inattentional blindness, and the temporal relation between neural events and awareness, are not treated here. Rather, the hope is that a reader new to this area will find a representative class of results sufficient to demonstrate that the scientific approach is worthy, and that they may also form the basis for a future engineering of mental content.

Blindsight

One of the most dramatic examples of the separation between visual awareness and response to visual stimuli is provided by the phenomenon known as blindsight, a term coined by Lawrence Weiskrantz. As was discovered in the early 1970s, patients with significant damage to V1 (or the primary visual cortex, see Figures 2.7a,b), most often due to stroke, report complete loss of vision in the corresponding visual field (Weiskrantz, 1996). Typically this is either in the left or right visual field only, and in many cases, so-called macular sparing occurs. That is, they maintain vision in the unaffected field, and within the lesioned field there is visual awareness in the fovea, or central area of vision.

In a typical blindsight experiment, a light is flashed in the impaired visual field of the patient. They report seeing nothing whatsoever. If asked, however, to guess in which part of the field the light appeared, say top or bottom, they do considerably better than chance. The counterintuitive nature of these results caused early researchers to question whether this was a genuine effect. It was hypothesized that either: a) stray light in the retina may allow sufficient information to reach a nonimpaired cortex to allow a correct decision, or b) that islands of intact cortex within the lesion may be responsible for this latent ability (Gazzaniga et al., 1994). However, the first concern was largely allayed when it was shown that when light is flashed onto the natural blind spot (where the optic nerve meets the retina) in nonimpaired subjects, blindsight is not possible (Weiskrantz, 1986). If light were scattered to other parts of the retina, one would expect a similar result to that of impaired patients. The concern about intact islands has also been addressed by showing that blindsight is possible in patients in which the entire primary visual cortex has been damaged, as revealed by neuroimaging.

Furthermore, if blindsight were simply a form of degraded vision, one would expect it to produce similar qualitative characteristics. However, this has been shown not to be the case (Persaud and Cowey, 2007). With blindsight, there is the complete lack of ability to identify whether a stimulus has occurred; it is only within a forced-choice paradigm that this ability manifests itself. A blindsight patient cannot say, for example, whether a circle appeared in a given time interval, but if asked to guess which of two time intervals it did appear, they will do so with remarkable acuity. In contrast, the abilities of a normal subject to answer yes-no questions regarding the existence of the stimulus and forced-choice questions regarding intervals degrade in parallel as the signal degrades (this can be done, for example, by briefly presenting the stimulus and lowering the contrast between it and the background). Paradoxically, then, blindsight patients have access to information about visual signals but cannot make statements about the existence of such signals with any confidence; in normal conscious vision, these two aspects are inextricably linked.

Finally, additional evidence for the existence of blindsight comes from experimentation with similarly lesioned monkeys (Cowey & Stoerig, 1995). Assessing awareness in animals is problematic in that they cannot make verbal reports regarding their conscious states, and depending on the animal, there may be doubt as to whether consciousness exists at all. Nevertheless, experiments have revealed similar results to humans. Monkeys with lesions of the primary visual cortex could learn to point to stimuli in the lesioned field. However, they could not distinguish a trial in which a visual stimulus appeared from one in which it was absent.

One of the remarkable qualities of blindsight, besides its mere existence, is the sheer number of visual abilities that are possible without awareness. Some shape detection is possible, and when asked to grasp at an (unseen) object, patients do so much better than chance would suggest. Color discrimination is possible, as well as flicker and orientation of the main axis of the stimulus. In addition, stimuli presented in the blind field can influence behavior in the seen field at the semantic level. For example, if the word "money" was presented in the blind field, when asked to define the word "bank" appearing in the intact field, subjects were more likely to choose the meaning relating to a bank as a depository of funds rather than a river bank (Marcel, 1998). The same study showed that information can flow from the intact field to the blind field. Figure 4.2 shows the drawings of two blindsight patients when asked to reproduce a pattern. When shown a half-circle in the blind field, they cannot reproduce it. However, when this half-circle is part of a larger circle, the appearance of the left half in their intact field, in conjunction with latent information in the blind field, appears to allow completion.

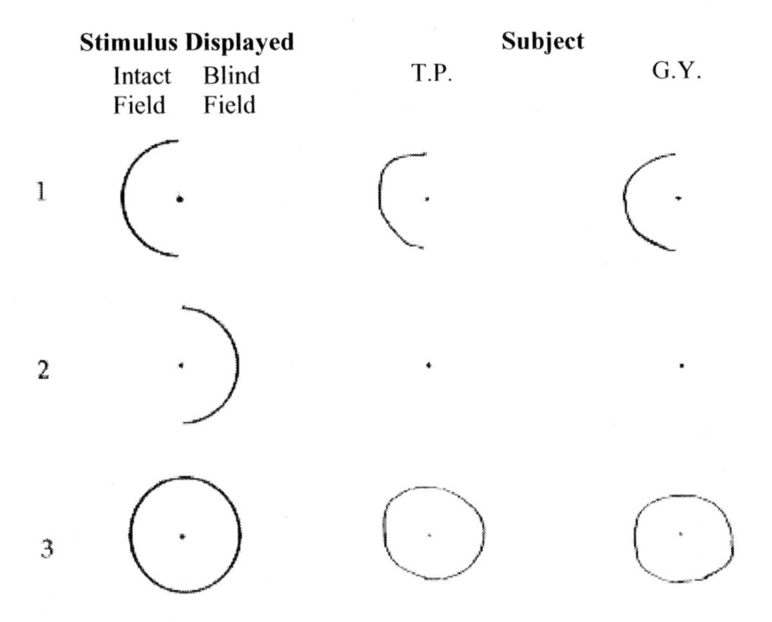

FIGURE 4.2 **Drawings of blindsight patients of shaped flashes of light. Note that in the second case, the shape in the blind field is not seen, but in the third case, in which the semicircle is part of a larger circle, it does enter awareness (from Marcel, 1998).**
SOURCE: **From Marcel, A. J. Blindsight and shape perception: deficit of consciousness or of visual function? Brain (1998) 121, 1565–1588.**

Interestingly, in cases such as these, subjects reported actually seeing the presence of the object in the impaired field, indicating that the label "blind" field is somewhat of a misnomer; blindsight and actual sight appears to be possible in blindsighted patients in the impaired region. It is worth noting that a similar filling-in effect occurs with nonimpaired people. As previously mentioned, there is a blind spot where the optic nerve meets the retina. How is it that our vision does not contain a black hole in this area? As it turns out, the visual cortex makes intelligent guesses about what should be there on the basis of the surroundings, and this is what we see.

Blindsight may not be as anomalous as it appears; a similar effect can be found in normal subjects (i.e., with intact primary visual cortices). When dots are shown in a small portion of a screen that are moving in a direction orthogonal to the dots in the rest of the screen, there is often no awareness of this contrary motion (Kolb & Braun, 1995). As with blindsight, observers exhibit little or no confidence that they can correctly identify the portion of the display in which the contrary motion occurred. However, they show a remarkable ability to guess the quadrant in which the anomalous motion took place (their guesses are correct approximately 70% of the time; 25% is chance). Blindsight in normals has been difficult to reproduce with this paradigm, and thus has proved somewhat controversial.

However, there are at least two incidences where the effect is fairly stable. In the first, to be described more fully in the section on masking, a briefly presented stimulus followed shortly by another pattern, the so-called mask, obliterates the original stimulus from awareness, but the brain is able to still use latent information about the stimulus in making discriminations. The second kind of blindsight involves the induction of temporary scotoma, with Transcranial Magnetic Stimulation (TMS) (Ro et al., 2004). Magnetic pulses were delivered to the visual cortex in normal volunteers while they saw either vertical or horizontal lines. They reported seeing no lines whatsoever, but were able to guess the orientation of the line with a significantly higher frequency than chance would predict.

What can account for blindsight abilities, either in patients with lesions in the primary visual cortex, or otherwise? The main visual pathway from the LGN to the primary visual cortex and then to extrastriate visual cortex (areas V2,V4, MT, and IT) is just one source of information by which the rest of the brain obtains visual information. The retina also projects to the superior colliculus, which in turn projects to these secondary visual areas, as well as at least six other neural structures. These pathways are not as well understood as that projecting from the LGN, but it must be that sufficient information is carried to make the appropriate discriminative judgments.

What is not generated in these alternative pathways is visual awareness, however. The reason that blindsight is so important to consciousness researchers is that it provides a clear example of processing of visual stimuli (albeit in an altered fashion) without the awareness of such. Apparently, something in V1, not present in these alternative pathways, is critical for visual consciousness, and the identification of this aspect could provide an important clue to the nature of consciousness as a whole.

Unfortunately, taken alone, this piece of the puzzle is weaker than it appears. Strictly speaking, blindsight only tells us that V1 appears to be *necessary* for visual awareness under ordinary circumstances, but this does not imply that V1 "creates" awareness, or has any causal role in awareness whatsoever. Compare this situation with the following: it is necessary to open one's eyes to have visual awareness (ignoring dream-based imagery and hallucinations). Yet no one would claim that the eyelids are the body structure responsible for generating consciousness. There have been many theories of consciousness proffered in the past few years—probably too many—but this is curiously not among them. On the other hand, the basic fact of blindsight, and experimental results associated with blindsight, are not without evidentiary value. Later in this chapter we will discuss why it is probably unproductive to try to pinpoint a specific area in the brain, V1 or otherwise, as the locus of consciousness, and suggest an alternative framework.

Binocular rivalry

Binocular rivalry represents an important piece in the consciousness puzzle complementary to the results regarding blindsight. Ordinarily, the two eyes receive nearly identical images of the world, but offset slightly with respect to each other. The visual cortex is able to fuse these offset images into a single *perceptual* image, and in the process, the illusion of depth is created (this illusion is also fostered by other factors, such as relative size, shadows and blocking, and perspective). What happens, however, when two entirely different images strike the retina? The answer is not intuitive—one might expect that a blend of the two images appears in consciousness, or perhaps parts of one image will alternate with parts of the other. The latter effect does happen on occasion, but most often, as discovered by Porta in 1593, usually only one image or the other is perceived, and these two images alternate in the mind's eye. Porta found, for example, that he could read two books in succession if he placed one before one eye and the other before the other eye.

Binocular rivalry has of late been investigated intensely because of the light it can throw on the nature of consciousness. With this phenomenon, there is a well-defined perceptual change in the absence of stimulus change. If, by neuroimaging or other means, we could identify the neural alterations that accompany the shift from one image to the other, we would have valuable clues as to the neural correlate of consciousness. To take an extreme example, suppose it turns out that image A normally produces imaging signal A' in a given brain area and image B produces signal B' in this area. We then present the two images to different eyes and we note that whenever the images alternate in consciousness, then we see the A' signal become the B' signal or vice versa, and in addition, *we do not see any other changes in the brain at this time*. We would then have reason to believe that the brain area in which the changes did occur was strongly implicated in the generation of consciousness; it is the only structure changing as consciousness changes. Of course, the matter turns out to be not so simple, although the results are still instructive. We summarize here three major findings of interest:

i) Activation and rivalry
There was an initial flurry of excitement around the findings that single-cell recording studies in monkeys (Logothetis & Schall, 1989) indicate that a greater proportion of neurons in later visual areas correlate with the changes in perceptual dominance (the shift from one image to the next in consciousness). This result is consistent with the notion that consciousness is a function of higher levels of processing in the brain. However, since this time, a number of cell recording experiments have concluded that the involvement of the primary visual cortex may have

been underestimated (Polonsky, et al., 2000). In addition, in functional magnetic resonance imaging (fMRI) study in which the images were tagged by using different levels of contrast, Polonsky et al found that the correlation between change in V1 and ocular dominance were comparable to those in V2, V3, and V4. Thus, it would appear that all levels of visual processing are equally implicated in the maintenance of the dominant, consciously perceived image.

ii) Pattern rivalry or eye rivalry?
Is the competition in the brain between the differing patterns, or does it involve a competition between the separate representations of monocular information, that is, between the eyes? Competition, presumably mediated by lateral inhibition (see Chapter 2), is a near-universal property of the brain, and allows this organ to focus on the dominant and, ideally, most important set of incoming stimuli. If the competition were between patterns, it would implicate secondary or higher areas of visual processing because these are responsible for pattern classification. If, however, the competition were between the eyes, earlier areas responsible for representing purely monocular information would be implicated. Blake (2001) has argued that except under special circumstances, the rivalry is eye based. In a clever experiment, he showed that if the patterns shown in the dominant and suppressed eyes (the eye in which the image is and is not seen, respectively) are suddenly reversed, the pattern that will be perceived is the one that was formerly projected onto the suppressed eye. In other words, the dominant eye continues its dominance even when the input to it is changed. If the rivalry was pattern based, one would expect no change in the perceived image; at all times this image is being registered by at least one eye.

iii) Synchronicity and dominance
In what will turn out to be a crucial piece of evidence for the theory to be proposed, Fries et al (1997) found that the degree of synchronization between neurons in areas 17 and 18 (primary and secondary visual cortex) of the cat increased for the dominant eye, and decreased for the suppressed eye. They also found no significant differences in firing rates between these two cases, suggesting that temporal synchrony rather than total activity per se is critical for visual experience, in contradiction to some of the earlier studies suggesting differing levels of activity correlate with dominance.

Backward masking

Masking phenomena have a long history in psychology, reaching back to the nineteenth century (the interested reader is referred to Breitmeyer and Oğmen, 2006 for a review). The backward masking effect occurs when a stimulus presented a short time *after* (usually between 50 ms and 500 ms) an original stimulus prevents this first stimulus from reaching consciousness. This is a perplexing effect: How can a future event prevent the occurrence of a past event? This perplexity, as well as the theoretical implications of such, has generated a voluminous amount of data over the past century or so, and we cannot begin to summarize this content. Here we concentrate instead on the four very basic results depicted in Figure 4.3.

In the first, and standard, masking experiment at the top of the figure, the target (solid square) is presented before the center-aligned mask (notched square). The rectangles indicate when and for how long each of these objects is presented. Depending on a number of conditions, including the time difference between the onsets, or stimulus onset asynchrony (SOA, 100 ms in the diagram), luminance, distance between mask and target, as well as individual differences, the target may suppress the appearance of an earlier target. A standard explanation for this effect is that in order for an object to register in consciousness, it must reverberate for sufficient time in memory or higher visual areas; the presence of the incoming mask inhibits the target, and prevents this reverberation (Turvey, 1973). This theory correctly predicts that when the SOA is either too short or too long masking will not occur. If too long, then the target will have a chance to enter consciousness; if too short both target and mask will be perceived together and both will enter into consciousness in tandem. Thus, this model correctly predicts that when the SOA is 0, as in Figure 4.3.B, both are perceived.

So far, so good. However, if we lengthen the display of the mask, as in Figure 4.3.C, then the target is again suppressed. Apparently, SOA is not the only variable. One can rescue the inhibition theory by claiming that the extra presentation time afforded the mask makes its activity higher, and therefore makes it more likely to suppress the target. But then we run into trouble again with the final example at the bottom of the figure. In this case, in which the mask appears before the target, the mask is again given more presentation time, but this time it does not wipe out the target.

Additional assumptions could be made in order to rescue the theory once again, but at this point we should also be asking if there is a simpler account for all four of these instances. Indeed, we will show that the model of consciousness introduced in this chapter does provide such an account.

Context-driven perception

Perception is never of the pure stimulus, but is modulated both by the context of the stimulus and the way in which the brain chooses to interpret the stimulus. There is a trivial sense in which this is true: qualia do not exist in nature, and therefore they must be constructed by the brain. For example, "blueness" is a psychological property, not a property of the reflection of light on surfaces.

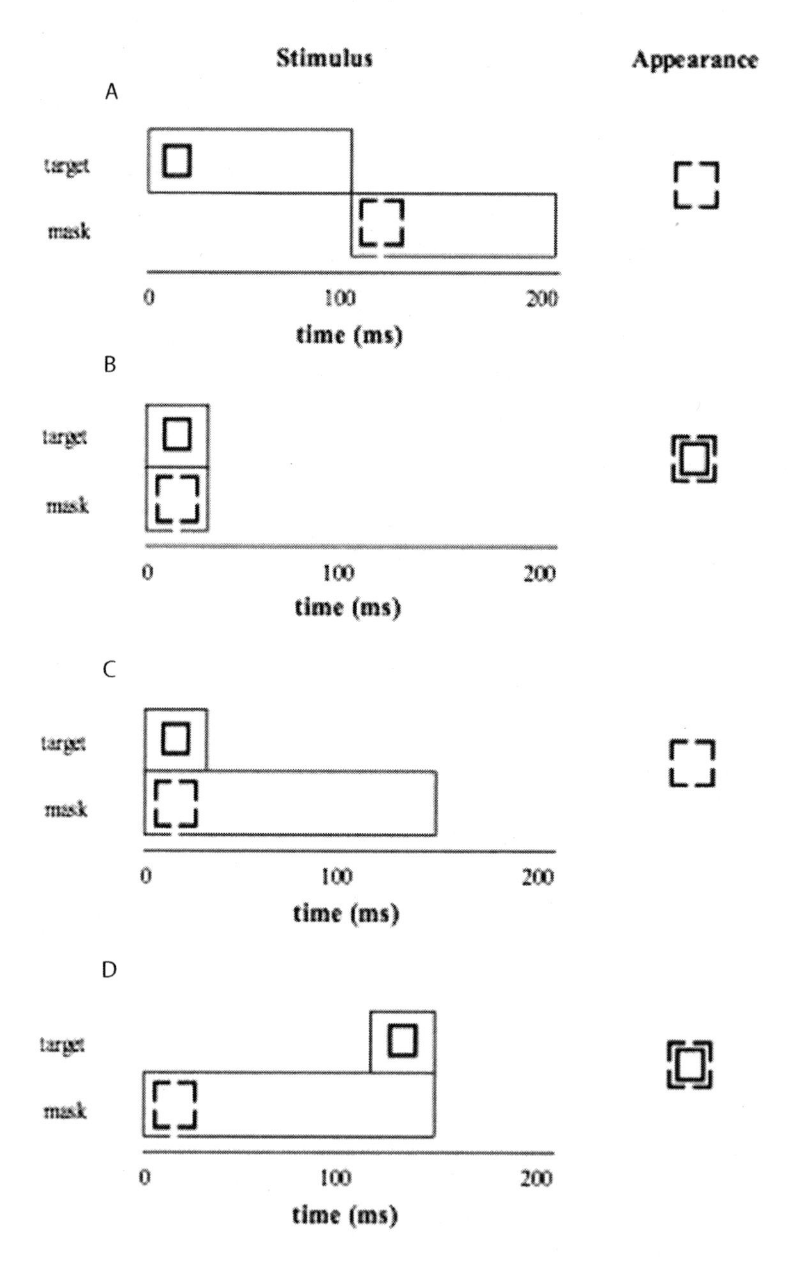

FIGURE 4.3 Target and mask presentation patters and the perceived image. A: The target is presented for 100 ms, followed by the mask for the same time. B: The target and mask are briefly presented together. C: The target and mask onset is the same, but the mask is allowed to persist. D: The mask is presented for 150 ms, at the end of which the target is briefly presented.

SOURCE: From Katz, B. F. (2008). Fixing Functionalism Journal of Consciousness Studies.

But, for the purposes of this section, the sense in which this will be important is that perception will invariably be altered by the surrounding in which it is embedded.

The case of color will again suffice to demonstrate this. In Chapter 2, we discussed a model for color and shading contrast. Figure 4.4, from Adelson (http://web.mit.edu/persci/people/adelson/checkershadow_illusion.html), provides a particularly dramatic example of this effect. In the board on the left, squares A and B appear to be radically different shades of gray. In fact, as the bars of constant gray demonstrate in the diagram on the right, both are identical. The shadow cast by the cylinder makes everything in its path darker, but color and shading is always perceived relative to surroundings. Thus, square B appears light because it is surrounded by much darker squares, and A appears dark because it is surrounded by light squares.

One could also appeal to numerous auditory examples to make the same point. For example, spoken word recognition (and perception) is a function of syntactic and semantic expectations generated in the course of listening to the entire auditory stream, and not merely the immediate word. Likewise, the qualitative feel of a note depends on the melodic, harmonic, and rhythmic context in which it is embedded. Interestingly, in both cases perception is influenced, to some extent, by events that take place after the processing of the immediate stimulus, as well as those before it.

What we will eventually need to explain is that we are only aware of the contextually modified stimulus, never the original form, and definitely not the two in succession. For example, in Figure 4.4, at no point do we perceive the squares as having an identical shade of gray, and we would scarcely believe it if it was not demonstrated directly as in the diagram on the right. It is thus important to realize that this is not merely a fact about pattern recognition, but a central fact about consciousness, and hence our interest in this phenomenon.

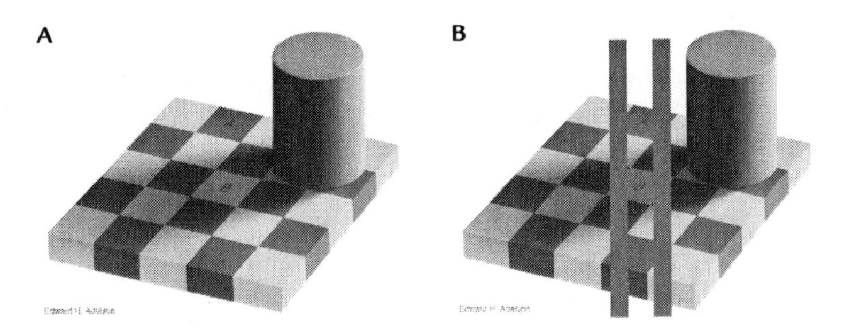

FIGURE 4.4 **A: Squares A and B appear to be different shades of gray. B: In fact they are identical, as the bars demonstrate.**

SOURCE: **A: Checker/shadow illusion; B: http://web.mit.edu/persci/people/adelson/checkershadow_illusion.html.**

Sleep and unconsciousness

So far we have been ignoring what is probably the most basic fact about consciousness. This is that every night, for every one of us, it fades away and then ceases to be present. If there were an invariant, anatomical, functional, or otherwise, present in the awake brain that is not present in the nonawake brain, then this could serve as the neural basis for consciousness. As usual, the simplicity of this empirical program is beset by a number of theoretical complications, the most serious of which is the possibility that any discovered invariant may be serving a facilitatory role in the generation of awareness but not acting as the true causal basis for such. With this caveat in mind, let us review the basic neurophysiological facts regarding the awake and sleeping states.

a) The extended reticular activating system (ERTAS) is responsible for regulating the nonaware states (non-REM sleep) and the aware states (dreaming and waking). The reticular activating system, described in chapter 2, is a diffuse collection of fibers originating in the midbrain reticular formation (MRF), extending through the nucleus reticularis (nRt) of the thalamus, and from there to the cortex. Lesions within the MRF, for example, can result in *coma*, that is, extended periods of non-awareness with little or no dreaming.
b) ERTAS achieves its regulatory effect via thalamo-cortical modulation. More specifically, tonic thalamic activity results in desynchronized, low-voltage, and relatively high-frequency electro encephalogram (EEG) signals, correlated with conscious states, and burst or phasic thalamic activity results in synchronized, high-voltage, and relatively low-frequency EEG signals associated with drowsy or unconscious states.
c) Anesthesia typically reduces overall brain metabolism, the amount varying between patients and between agents, although reduced brain metabolism is not a necessary concomitant of anesthesia.
d) A relatively recent set of results by the anesthesiologist E.R. John (2002) has suggested that coherence both between different parts of the brain and between hemispheres is correlated with the aware state, especially if measured by EEG gamma activity.

Independent of the preceeding evidence, the most natural first guess as to the neural correlate of unconsciousness is that the brain fully or partially shuts down. Furthermore, barring views of the mind that allow it survive bodily death, we know that if brain activity is sufficiently suppressed, then consciousness must cease. However, as c states, it is possible that overall metabolic activity may only slightly reduce in the unconscious state. Just what else could be going on we postpone until the introduction of a more general account of consciousness.

SUPERVENIENT THEORIES: A SURVEY

In our earlier discussion of supervenience it was mentioned that while it is likely that the mental supervenes on the physical, there is no necessary reason why it must do so. In other words, if it looks like a duck, walks like a duck and quacks like a duck, it still might not be a full-fledged duck with an inner life—we must invoke additional laws above those responsible for producing duck behavior to get duck qualia. Kripke (1972) has stated the case as follows: After God created the universe, with matter and the laws that govern material objects, he still had extra work to do, namely, the creation of laws that govern the way in which the physical is transformed into the mental.

The task of this section is to evaluate possible forms of these laws, with respect to congruence to the evidence of the sort examined in the prior chapter, but also with respect to their theoretical elegance. Einstein captured both of these criteria, empirical consistency, and theoretical elegance, with his famous dictum that a theory should be as simple as possible, but no simpler. We will term this principle Einstein's razor, by analogy to that stated by the eleventh century philosopher William of Occam. Occam claimed that in the search for scientific explanation, *entia non sunt multiplicanda praeter necessitatem* (entities should not be multiplied beyond necessity).

What reason do we have to believe Occam's or Einstein's more complete formulation? First, let us note that simplicity has had a good track record in the history of science. For example, as it turns out, Galileo had more problems with the church than with competing secular theorists, but one can easily imagine a colleague laughing in his face: "Hah, you say that everything falls at a constant acceleration. Have you never seen a feather fall? Why just the other day, a gust of wind came along and my hat actually flew upward and danced in the air like a butterfly. That is in direct contradiction to your so-called theory." Of course, the proper specification of gravity must discount wind resistance, or equivalently, be stated as taking place in a vacuum, thereby preserving its simplicity. Einstein himself was responsible for a further simplification in the notion of gravity when he realized that the equivalence of gravitational and inertial mass follows from the fact that matter bends space. In general, we can say that successful theories invariably are usually superior not only in their predictive powers, they are also more elegant than their predecessors.

Next, the simpler the theory, the greater its ability to generalize to new phenomena. Complex theories arise because they tend to have many special cases, each encompassing a disparate set of results. Each aspect of the theory covers one and only one such set. However, if a commonality can be found that ranges over these results, it will often be the case that it also explains as yet unseen empirical data.

Finally, there is the notion that the universe is inherently simple, especially when it comes to fundamental quantities. Theories governing nonfundamental objects in contrast may take arbitrarily complex forms. There is no reason to expect, for example, that the functioning of the spleen conforms to a compact equation, or that the economy of Bolivia can be reduced to a few differential equations. These are compositional effects, far removed from the inner core of the universe, and as such are under no obligation to be simple or even comprehensible. However, as one penetrates closer to this core, it is invariably the case that simplicity becomes a dominating characteristic, Einstein's relation between matter and energy being just one among many examples of this effect. And if consciousness is not a fundamental quantity, it is unclear what should be considered as such. In fact, consciousness may be considered *the* fundamental quantity of the universe *par excellence*, in the sense that our only knowledge of the world comes about through consciousness; everything else is an inference, with varying degrees of likelihood.

Thus, Einstein's razor, always a valuable guide, assumes paramount importance when evaluating theories of consciousness. Table 4.1 shows the six theoretical candidates to be considered in this chapter and their corresponding evaluations for the two components of the razor, plausibility and simplicity. We now consider each in turn, arguing that the highlighted theory, and reduced functionalism, best meets both demands of this principle.

TABLE 4.1. The relative plausibilities and complexities of three physical theories and three process-based theories of consciousness. All of the physical theories have relatively low plausibility and have relatively few adherents. Of the process-based theories, only two are tenable, and one is considerably simpler than the other; hence Einstein's razor favors the shaded theory.

	Physical Theories	
Theory	**Plausibility**	**Simplicity**
Eliminative Materialism	very weak	not applicable
Noneliminative Materialism	weak	high
Quantum theories	empirically weak theoretically questionable	indeterminate
	Process-based Theories	
Theory	**Plausibility**	**Simplicity**
Functionalism	problematic but tenable	very low
Reduced Functionalism	**problematic but tenable**	**moderately high**
Statism	untenable	high

Eliminative materialism

We begin our survey of theories of consciousness with a suggestion that is not so much a theory per se as an attempt to reduce the need for one. As we shall discuss, it is not widely held in the scientific community, but comes up sufficiently often, especially in philosophical circles, that it is worth addressing. Boldly stated, the idea is that the thing we are trying to explain, mental states, have no objective existence. If this were the case, then of course there is not an attendant scientific problem of showing how they arise.

How could this be so? The standard argument runs as follows. It is natural, as thinking organisms, that we apply our analytical powers not just to physical events but also to other humans. The collection of theories that develop in this manner may be termed folk psychology, to distinguish it from the putatively more accurate theories developed by those in the psychological sciences. This is not to say that folk psychological theories carry no explanatory weight. Consider the following example: John is in love with Mary; John also believes that Mary likes Fred. We explain John's antipathy to Fred by invoking the concept of jealousy, yielding some but not full predictive power. We cannot, for example, without the possible invocation of other folk or other psychological principles predict whether John will ignore Fred, whether he will send him a nasty email, or whether he will slash his tires. But if any of these things do happen, we have a framework for understanding them, and we can state that it is more probable that these will happen than John buying Fred a new car.

The central claim of the eliminativist, then, is that concepts such as jealousy, and the attendant concepts needed to make the preceeding explanation work, such as that of belief and that of love, are approximate notions that do not fully capture the underlying dynamics and that will be eliminated once a full psychology is in place. One way this could occur is if sounder neural principles replace their weaker folk psychological counterparts; this is the thesis of an influential 1986 book by Patricia Churchland (1986). To bolster the intuition behind this argument, eliminativists often provide analogies to other areas of science where conceptual frameworks have been drastically altered or simply replaced as new evidence and ways of thinking evolved. For example, at one time it was considered natural to talk of a life force, an *élan vital*, present in living things and absent in others. We now consider this as a manner of speaking rather than an ontological claim, that is, one that presupposes a special substance present in living things. Moreover, questions about the life force are scientifically vacuous. For example, no one considers it productive to try to discover if viruses or prions *truly* possess a life force. They share aspects of both living and nonliving things, but it is of little interest in which category we place them, because this categorization is not indicative of an underlying substantial difference.

The eliminativist idea, then, is that claims about mentality are like claims regarding the life force, a weakly explanatory concept that will be eliminated as science progresses. This, on the face of it, seems too extreme, but to systematically counter this assertion, we must distinguish between two types of mentality:

a) Intentional states, or states that make reference to other things, such as beliefs, desires, and the like
b) Qualia, or as defined previously, the raw feels of psychic life

The history of the notion of intentionality in the philosophy of mind is a long one, and we cannot begin to summarize it here. Suffice it to say, it is still not clear if intentional states can be eliminated, nor is it entirely clear what it would mean to eliminate them. Certainly, intentional states appear to be only crudely scientific; if this were not the case, we would have a better handle on why people do what they do. We can all agree that one of the things that makes life both interesting and exasperating at the same time is the sheer unpredictability of our friends and acquaintances. On the other hand, philosophers and others are still divided over whether a sub-folk-psychological analysis is either necessary or desirable. It could be that a sufficiently strengthened folk psychology could explain at least the broad details of the relation between our thoughts and actions.

The situation with respect to qualia is arguably not as ambiguous. Take the case where we explain John's hopping on alternate feet as he walks on a sandy beach on a very hot summer day. We would naturally say that John is attempting to avoid the painful sensation on the soles of his feet. The eliminativist would counter that our ascription of pain to John is also a kind of folk psychological notion in two senses. First, it is only weakly explanatory of John's actions, and that a neurally grounded explanation could do better. More importantly, that John's pain is something going on in our heads, that is, it is a folk theory that we have developed to explain these sorts of behaviors, but it is not necessarily something going on in his (unless he is also trying to explain his actions). What is *"really"* happening is that the heat of the sand particles makes contact with pain fibers in his feet, cause them to expand, which then transmits a message to pain centers in his brain, which in turn then determines how to minimize contact with the sand, etc.

It is here that the eliminativist runs into trouble. It may very well be that a more fine-grained, neurally based account does better than the folk notion of pain in explaining John's odd behavior, but it is not hard to see how this matters little in the current context. If one is to be intellectually honest, one must come to grips with the fact that John is feeling something when his feet touch the hot sand (and we are not, unless we are also on the beach). He is feeling something different when they touch cold sand, and they feel something else again when his toes reach the water's edge. *These feels, which we have termed qualia, are something for John as much, if not more so, than something for us*. What they are for John may be different than what they are for us, as observers of John's behavior, and in fact our protoscientific notions may not fully or correctly capture them, but they cannot be eliminated by this sort of reasoning.

In fact, as evidenced by the presentation in the last chapter, the scientific community is now broadly committed to the notion of qualia as ineliminable aspects of a fuller psychological science, and it is likely that the neuroengineering community will follow suit in recognizing the importance of qualia to their enterprise. To reprise our discussion in the introductory chapter, no one would wish to be downloaded into a vessel not capable of generating qualia—this would be no different than dying. This philosophical view will also not be able to tell us just how or whether a brain-machine interface is affecting the contents of consciousness, or carrying out its effects without doing so, a topic to which we will return later. In summary, elimativism thus has little scientific appeal to those who take consciousness seriously, and is also unlikely to capture important aspects of the effects of neuroengineering in the near and distant future.

Noneliminative materialism

By noneliminative materialism, we mean that qualia are real, and are the direct causal product of some purely physical aspect of the brain (or the brain plus the body). How can matter, in its purely physical capacity, produce consciousness? There are two possibilities: either a) qualia "emanate" from some special substance in the brain, or b) some dynamic but still physics-based property of the brain yield qualia (purely causal theories based on the brain's actions but not tied to physics are considered separately in the following). Part b can itself be divided into two possibilities, that of "ordinary" physics, such as electromagnetic fields and the like, and that of quantum mechanics. The latter is sufficiently interesting that it is treated in a separate section following this one.

Substance materialism states that there is some special substance in the brain that produces consciousness. This could be something contained in neurons, or, as is less likely, something in the glia or the brain's supporting vascular network. Immediately, we run into a problem with this conception, however. We know that neural activity is intimately tied to conscious content. Not all activity yields content, as was discussed in the context of binocular rivalry, but without activity, there will be no content. Therefore, this substance must arise when neurons fire, and be absent otherwise. But, the firing of a neuron involves the transfer of ions, not the creation of new molecules (see chapter 2 for a review). In any case, it seems improbable that a special substance is created during brain activity, goes away in the absence of such activity, and is also responsible for consciousness. Furthermore, what if we were to extract this substance and inject it into a baloney sandwich. Would that sandwich then become conscious?

Most would concur that a piece of baloney is condemned to being a zombie (in philosophical parlance, a being completely lacking consciousness), regardless of its constituent parts. If the physics of the brain is capable of supporting consciousness, then, it does so because of a physical *process* rather than a physical substance. If this were possible, it would be an attractive proposition, because then there would be a home for consciousness in the natural world.

Here also, we run into immediate difficulties, though. The most compelling substrate for consciousness among physical processes is one of the four fundamental forces, and among these, electromagnetism (we can ignore gravity as being too weak, and the short and weak forces as being too short range). A key feature of electromagnetic fields is their dependence on distance and geometry. However, there is a disconnect between these fields and what may be termed term the perceptual field, or the field of qualia in a given sensory modality. Simply put, there is no natural correspondence between the geometry of the neural representation and the geometry of the perception.

Figure 4.5 presents a schematic of this argument. A regular lattice of neurons (and by implication, a resultant regular electromagnetic field) is shown in 4.5.A. However, although the visual cortex is retinotopic, and preserves relative position to a large extent, the actual situation is more akin to that of 4.5.B. If the perceptual field were a function of the electromagnetic field, then it would be distorted. Perhaps one could argue that the perceptual field somehow rights itself, and that these distortions are smoothed out. The thought experiment represented by 4.5.C shows that this unlikely possibility will still not work. Here, we have taken a neuron and moved it some distance apart, while preserving its connectivity and the transmission time to its neighbors. The field theory would predict that the perceptual field would be accordingly affected. Yet by construction, the perceiver would not be able to tell us anything different, because the dynamic properties of the network are identical. This disconnect between behavior and perception could be made arbitrarily worse by scrambling all of the

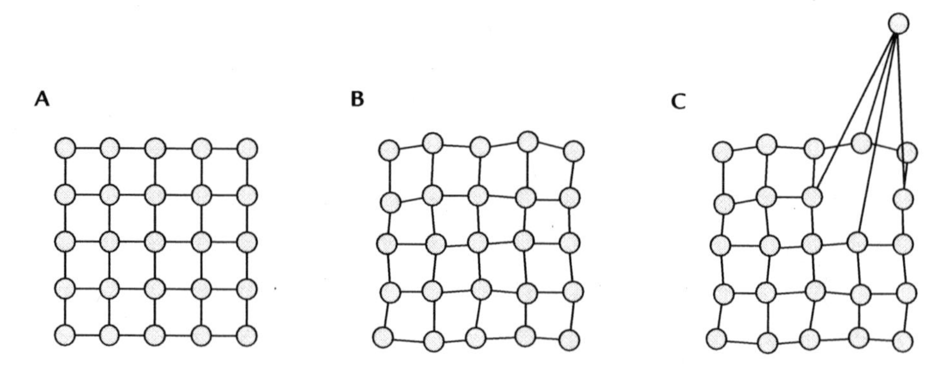

FIGURE 4.5 **A: Neurons in the visual cortex arranged in a hypothetical regular lattice. B: In reality, spacing will be irregular. C: A thought experiment in which a neuron is moved and its connectivity and other properties preserved.**

relevant neurons in space while maintaining the same connectivity pattern. The perceiver would also act as if the presented image was regular, but "inside" he would see a jumble of pixels. While not impossible, this is a disturbing proposition, and should only be accepted as a last resort.

A related problem is just as unsettling. If electromagnetic fields are all that matter, what happens if we literally knock two heads (or brains) together? Do the overlapping electromagnetic fields form a single consciousness? Two lovers kissing (or rubbing noses) certainly feel close, but no one supposes this is a literal melding of minds. This is a special instance of what may be termed the *demarcation problem*. At a minimum, a scientific theory of consciousness must tell us when and why multiple consciousnesses arise. We presume that each person possesses a unique consciousness, and any proposed theory must be consistent with this.

As we shall see, both the demarcation problem and the related perceptual field problem are strong indicators that causal interaction determines the contents of consciousness, rather than the attendant physical processes that carry this communication.

In summary, it is difficult to produce a coherent theory showing how a substance within the neuron suddenly "releases" qualia in the presence of neural activity. When energy fields are considered, the issue does not improve considerably. Conscious content does not seem to be related in a one-to-one fashion to field distribution, there is no way of demarcating separate consciousnesses if energy is the carrier, and finally, as if this were not enough, the entire universe would be permeated by mind if it were a function of energy.[3] For these reasons, materialism as a scientific program has very few adherents, although it still maintains advocates in philosophical circles. There is one exception to this rule, discussed in the next section.

Quantum theories of consciousness

A number of thinkers both within the formal academic environment and outside of the ivory tower have suggested that the anomalous qualities of consciousness somehow dovetail with counterintuitive aspects of quantum mechanics. There can be no doubt that both sets of phenomena have their oddities. That three-and-a-half pounds of organic matter could generate the taste of cotton candy or the bittersweet emotion of falling in love is one of the great mysteries of the universe. Likewise, the fact that the electron is both a wave and a particle, depending on how it is measured, is still somewhat problematic, almost

[3] This problem, known as panpsychism, is a problem that besets almost every theory of consciousness. However, the identification of consciousness with energy is particularly acute, as energy is everywhere, even in deep space, in the form of background radiation.

one century after quantum mechanics was introduced. Furthermore, as argued in the previous section, whatever theory of consciousness in the end turns out to be right, it will be a dynamic one, not one that is identified with a particular static substance, and quantum mechanics is preeminently dynamic, in that its main governing rule, the Schrödinger equation, describes how a particle evolves over time.

However, the mere fact that two phenomena appear to be mysterious is a poor basis for establishing a link between them. For example, having a drawer full of socks that mysteriously don't match, shouldn't bear on your consciousness in any way. Of course, your sock problem is not one of the fundamental mysteries of the universe, and quantum mechanics is, and therefore has a greater chance of being of use to that other great mystery, consciousness. But there are other problems of equal or greater magnitude. String theory continues to be both problematic and controversial, and it has not been linked with consciousness, at least not yet. In short, arguments based on the congruence of human ignorance in two distinct fields alone tell us little about why these fields should be intertwined.

In this section, we will very briefly examine two more direct means of connecting quantum mechanics to consciousness. Either one, if treated in all its complexity, could easily fill a volume on its own, and thus only the most cursory treatment is possible here. Nevertheless, it is hoped that sufficient material will be presented to show that the probability of such a connection is small; sections following this one will argue that is also unnecessary.

The first possible link stems from a popular interpretation of quantum mechanics, namely, that consciousness causes a collapse in the superposition of states. At the heart of quantum mechanics, unlike its classical counterpart, is an indeterminacy or randomness that affects all variables at the microscopic or quantum level. Until a given variable is measured, the value of this variable can be in numerous states at once. Only after measurement will this variable resolve into one value or another. Einstein's famous dictum, that God does not play dice, was issued to throw doubt on this counterintuitive state of affairs, but the fact remains that most physicists accept that this is not merely an epistemic claim, that is, a claim about our inability to find the "true" state of nature, but a fact about how nature is fundamentally organized.

A long-standing problem in the interpretation of quantum mechanics involves just what is meant by a measurement. It is unclear how to define a measuring device, and a theory of the universe that said, for example, that the device has to be so big or so accurate would be even more disturbing than the unusual effects that it is trying to explain. This has led some to conclude that it is not the measurement per se that is important but the fact that a conscious entity observes the result of the measurement. However, this produces a new set of problems, which we will illustrate by invoking Shrödinger's famous cat.

In Figure 4.6, the radioactive isotope in the box on the right has a 50% chance of decaying and emitting a quantum of energy. According to standard quantum mechanics, both the state in which the energy is emitted and the state in which it is not are both present, or superposed until a measurement device (such as a Geiger counter) is placed in the room. Although this is counterintuitive, as Einstein's words suggest, this is not a real problem yet, because this is a microscopic effect. What if, however, the result is amplified by having a vial of poison broken if the element decays, and not broken if no decay occurs? In addition, let us stick a cat in the room. Under the theory that consciousness causes a collapse to one state or another, the cat will both be alive and dead until a conscious observer peers into the box (we assume that the cat's consciousness plays no role in these events). But this is absurd: the universe, at the microscopic level, may hold unresolved alternative states, but how is it to house macroscopic effects such as multiple cats?[4] It is not just the two cats, alive and dead that it must contain, but also possible splitting points further down the line. Suppose, for example, that the cat is pregnant, and that each of the offspring of this unfortunate animal are themselves placed in the box. In this fashion, we can multiply the possible contents of the box *ad infinitum*.

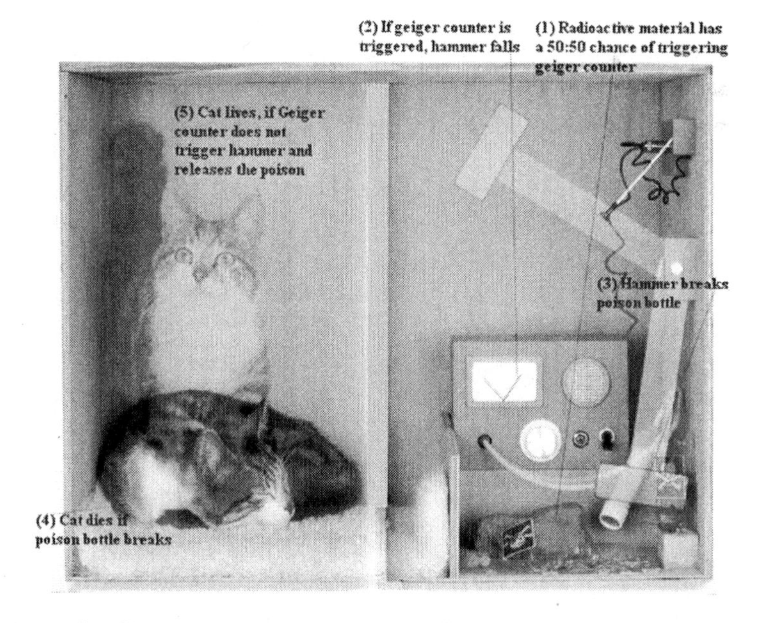

FIGURE 4.6 **Schrödinger's unfortunate cat. Like the radioisotope, this scenario places the cat in two superposed states at once, alive and dead.**
SOURCE: **http://commons.wikimedia.org/wiki/Image:Hobuaoeuj3424n.png.**

Despite its renown in the popular press, invoking consciousness to resolve the measurement problem in quantum mechanics is probably the least favorite option among physicists. Furthermore, at best such an interpretation, if valid, would merely establish a link between quantum mechanics and consciousness. It says nothing in itself about how consciousness arises in the brain, which is the central question being considered here. In summary, it is probably too much to hope for that this interpretation of quantum mechanics will form the basis for a theory of consciousness as a whole.

A more recent and more direct connection between quantum mechanics has been offered by the physicist cum mathematician Roger Penrose in conjunction with the anesthesiologist Stuart Hameroff. Penrose has given two book-length treatments of his ideas (Penrose and Gardner, 1990; and Penrose, 1996), and we will summarize only the highlights here as they pertain to the generation of consciousness. Penrose's first claim is that neither algorithmic processes, such as those implemented by computers, nor algorithmic processes plus randomness can account for the variety and creativity of human thought. His arguments are complex, but boil down to the fact that humans, unlike machines, are assumed to be able to evade Gödel incompleteness, or limitations on reasoning systems that are sound (that draw valid conclusions from valid facts). People are therefore, at least partially, non-algorithmic thinkers, and cannot be modeled by algorithmic methods alone. His second claim is that an as yet undeveloped theory of quantum gravity could generate these sorts of non-algorithmic processes, and this could serve as a substrate for human thought and consciousness. Finally, Penrose claims that microtubules, proteins in the cytoskeletons of neurons (and other cells) can act as a host for quantum gravity and therefore consciousness. He supports this claim by referring to Hameroff's work suggesting that microtubules are made inactive by anesthetic drugs that render the recipient unconscious, and also on Hameroff's work that shows how microtubules may have the right properties to support the right quantum effects (sufficient isolation from the environment of the neuron, sufficiently small size, and other factors).

In examining the likelihood of the Penrose-Hameroff hypothesis, we will draw heavily on the criticism by Grush and Churchland 1995. In their felicitous turn of phrase, this hypothesis does not quite make the transition from "campfire to scientific possibility." Lest the reader judge this as too harsh, perhaps because of exposure to less-critical accounts in the popular media, the vast

[4] An alternative interpretation claims that the universe cleaves into two for each of these possibilities However, this is as extreme a violation of theoretical parsimony as one could conceive, as a unique universe must be generated at every junction of possible outcomes, and should be taken seriously only as a last resort.

majority of working scientists, whether in AI or the cognitive sciences, in physics, or in neuroscience, concur with this assessment. Grush and Churchland first point out that there is no evidence that human beings engage in nonalgorithmic computations. There is no doubt that humans make judgments that are not logically sound, primarily because there are very few situations that arise in the actual world where sound, demonstrably true reasoning would be useful. It is almost always better to make decisions that are approximately true than to attempt to prove something is the case, which is both a lengthy and overly restrictive practice. This sort of reasoning is known as default or common sense logic, and creativity and other similar processes can be viewed as more relaxed versions of the same. Grush and Churchland then make what ought to be an obvious claim, that Penrose is piling mystery upon mystery to build a theory of consciousness on an as yet unsupplied theory of quantum gravity. Finally, they cast doubt on both Hameroff's empirical as well as theoretical claims regarding microtubules. For example, they argue that anesthesia is best explained by large-scale effects at the level of neurotransmitters, in accord with the Neuron Doctrine (see Chapter 2, and the following discussion on unconscious states), rather than lower-level quantum effects.

None of this rules out an interaction between quantum mechanics and consciousness, and indeed many feel that a real understanding of the foundation of quantum mechanics will aid us in understanding the nature of causation, which in turn is critical for understanding consciousness. At the very least, however, we need to exhaust "conventional" sources of explanations before building theories on shifting foundations. For example, one of the auxiliary claims of Penrose and Hameroff is that quantum coherence between microtubules may yield the apparent unity of consciousness by a nonlocal process known as superradiance. But what about ordinary channels of communication between neurons, could these not serve as the means of binding together different processing modules? This idea is explored in the following two sections, in which the brain is viewed as a computational device, and then viewed as a propagator of causal effects, respectively.

Functionalism

By concentrating on process rather than substrate, functionalism attempts to avoid many of the aforementioned problems of materialism. In an oft-repeated phrase, in this view the mind is just the software of the brain. Just as two different machines or two types of machines (Mac vs. PC, for example) can run the same program, different physical systems can realize the same series of events by running the same algorithm. Functionalism claims all aspects of the mental, including qualia, are a function of this algorithm and that the underlying physical implementation is not important. This view accords well with contemporary intuitions that hardware, with the exceptions of efficiency issues and the like, is in many ways irrelevant to the net behavior associated with a particular series of computations, and it is no accident that the rise of functionalism to its current position of dominance has paralleled the rise of information technology.

More specifically, functionalism claims that to be in a mental state is to be in a state that transforms inputs to the outputs appropriate for those inputs. These states serve a functional role; hence the name of the theory. As an example, let us consider stubbing one's toe. Evolution has kindly provided us with what often seems like an inordinate amount of pain relative to the force of impact in these situations, no doubt because of the importance of the toes in walking and running properly. Thus, the pain is playing a functional role, connecting the triggering of the pain fiber in the toe to a variety of other states. Some of these may be behavioral, such as saying "ouch," and others may be purely internal, such as saying "you idiot!" to yourself or, for example, triggering an unconscious adaptive processes which will act so as to reduce the probability of your stubbing your toe again. Note that being in pain is not equated with being merely in a given state, but also includes the causal relations that this state has with these other internal and behavioral states.

To make this account somewhat more precise, it is necessary to invoke a more formal model of the transformation from inputs to outputs. Putnam (1960) and other early proponents of functionalism typically used a Turing Machine, a general-purpose model for computational processes, for this purpose. However, Turing Machines describe computational processes in terms of elementary operations, and thus are awkward when attempting to convey the actions of a high-level operation. A more flexible approach is provided by the ordinary flowchart, typically used for abstracting a program's algorithm from the particular language in which it is expressed. Our flowcharts will consist of inputs, states, decisions points based on either input states or the results of subroutines, and outputs.

For example, Figure 4.7 shows an admittedly vastly simplified functional account of pain. Here, a number of possible inputs in the environment can trigger the pain state. This state has behavioral consequences, including an unconditional flinching and a conditional verbal reaction that depends on another state of the system computed by a subroutine that determines whether the pain recipient is alone or not. In addition, the pain initiates an adaptive response corresponding to its punishment value. Among other things, in the "real" pain flowchart we may wish to distinguish between different types of pains and their consequences, we could include the effect of pain on cognitive processes (if strong enough, it would suspend them, for example), and we could also allow for the possibility that under the appropriate circumstances, there may be no behavioral response at all.

Although highly simplified, this diagram captures many of the advantages of functionalism over material theories. First, unlike in at least some versions of materialism, there is a clear reason for the demarcation between different

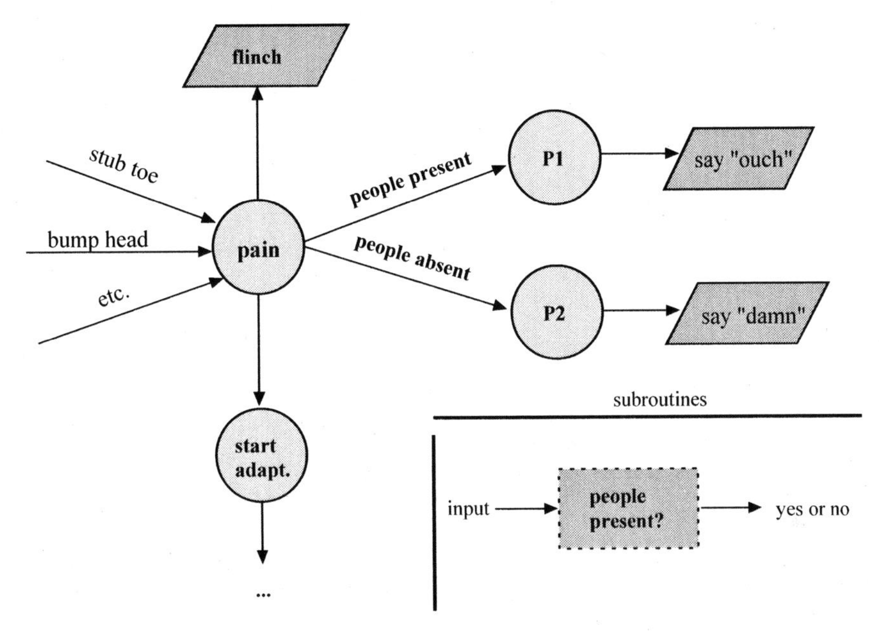

FIGURE 4.7 **A flowchart for the "computation of pain." A number of inputs in the environment can trigger a painful state, the response to which will also be a function of environmental factors (from Katz, 2008).**
SOURCE: **Katz, Fixing Functionalism.**

conscious entities. When you stub your toe the reason that you are in pain and not someone else is because the causal chain of events that the pain triggers is local to your brain and not someone else's. The identification of causality as a critical mental variable can rightly be said to be one of the triumphs of functionalism. In contrast, although physics *presupposes* causality, it does not treat it as an independent variable, and therefore runs into the demarcation problem as well as others, as argued previously.

The second reason for favoring functionalism is that it potentially gives us a reasonable answer to the chauvinism/liberalism question. This issue concerns the distribution of consciousness in the universe. To say that humans, and humans only, experience qualia is probably too chauvinistic. At the very least, most would include not only mammals but also other animals, perhaps with a more limited type of awareness. Furthermore, we have no reason to believe that consciousness must be coupled with a system with a carbon-based chemistry. This opens the door to machines of the appropriate complexity and configuration, and alien life forms, if such exist. On the other hand, we do not wish to be too liberal and populate inanimate matter with thoughts. A glass of still water, most would agree, should not have conscious thoughts bubbling up inside it as if it was carbonated with qualia. Functionalism prevents this from happening because, although the states of the water molecules are causally interactive, the kinds of large-scale coordination of causal events characteristic of brains would rarely arise.

The third reason that functionalism has gained so many adherents, especially in the cognitive and AI communities, is the close correspondence between the conscious contents of the mind and the processing therein. This is not a coincidence; functionalism by construction attempts to model the mental via the functional role that it plays. In Figure 4.7, for example, the causal role that pain plays leads to both the appropriate behavior (flinching, verbal utterances, to internal adjustments necessary to punish the behavior leading to the pain, etc.), and also to the psychic discomfort associated with pain. This leaves open the problem as to why at least some states of the brain with causal or even behavioral consequences do not produce qualia. Nevertheless, at least functionalism tells us why it is not a magical coincidence that mental qualities correspond to functional roles. Pain feels bad precisely because it is something that we should avoid; it doesn't feel purple or salty because these are affectively neutral sensations and we have no good reason to shun them.

The prospects of functionalism thus seem good, but as in any theory of consciousness, the negative side of the ledger is not insubstantial. There are three primary conceptual problems with functionalism:

a) Inverted qualia
Let us imagine that seeing red is a functional quality that transforms light of the appropriate wavelength falling on the retina to behavior that indicates that red has been seen. Would it not be possible to achieve the same result but with the conscious agent seeing green instead? Of course, such an agent, when seeing red, would say "red," because they would have learned early in childhood that is what their green is called, but their experience would be unlike ours. The problem is that functionalism by itself doesn't do enough to pin down experience; it is compatible with a vast set of experiential frameworks as long as they result in the appropriate input-output relations (Lycan, 1973).

b) Absent qualia and the hard problem
For that matter, why should a functionally identical organism to another experience any qualia at all? As previously stated, Chalmers (1986) has argued that even an atom-by-atom copy of a human need not necessarily be conscious. Of course, if there is some law of the universe, either physical or functionalist, that dictates that humans and similar organism are sentient, then the copy will be also. However, the mere fact that we can imagine a so-called nonsentient but properly behaving zombie means that we don't know why there is a need for qualia. Chalmers

calls this the hard problem of consciousness, to distinguish it from other problems that may be technically extremely difficult but are, in principle, soluble, such as creating artificial intelligence.

c) The knowledge argument
Jackson (1986) imagines a brilliant neuroscientist Mary who knows everything there is to know about the functional aspects of color vision but who is fitted with special glasses at birth that renders everything in black, and, white. One day she removes the glasses and she perceives her previously drab world in its full Technicolor glory. Should we not say that Mary has gained knowledge about the world that she did not previously possess? But this implies that she did not know everything in the first place. It would appear that direct, first-person experience of color or any set of qualia is something that cannot be fully rendered into a third-person account, functional or otherwise.[5]
It is not the purpose of this treatment to discuss these conceptual difficulties at length; the interested reader can consult the indicated references and the associated minor philosophical industries that have grown up around these critiques of functionalism. As previously stated, our concern here is of a more scientific nature, and in particular, we wish to examine the concordance of functionalism as a theory of the mental with Einstein's razor. It is not difficult to show that functionalism is the very opposite of scientific elegance. To see this, let us imagine, as Kripke has said, that God took an extra eighth day to build the laws describing how the mental arises from the physical; we will designate this as the psychophysical correspondence relation (PCR). Functionalism then amounts to the claim that the algorithm the brain runs is transformed by the PCR to one or more qualia.

This is illustrated in Figure 4.8 for the previously described case of pain. Note that there are two distinct algorithms present here: the first is the one being run in the brain, contained in the dotted box, and the second is the one that takes *this* algorithm and returns a quale or set of qualia, in accord with the PCR, as illustrated by the entire figure. The reason that the PCR takes the whole pain algorithm as an argument, and not just the pain state, is that functionalism specifies that it is the entire set of causal relations transforming input to output states that determines conscious content (see the discussion under statism to see why

[5] Compare this with a theory of physics, say the behavior of electrons. Suppose Mary knows everything there is to know about these elementary particles. Then there is no scenario in which she could gain extra knowledge about them, unlike in this example.

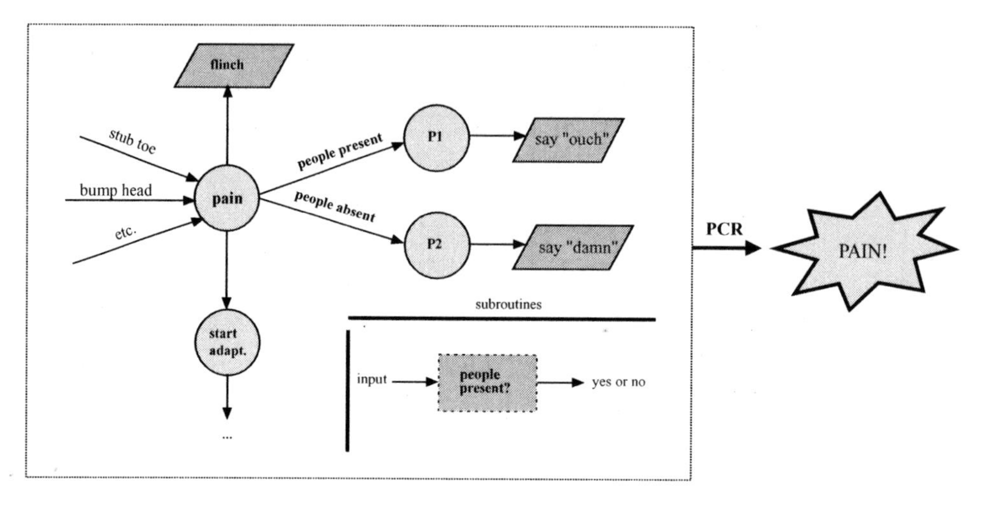

FIGURE 4.8 **The flowchart in Figure 4.7 has been embedded in a further algorithm, the psychophysical correspondence relation, that transforms it into the appropriate quale, in this case, pain (from Katz, 2008).**
SOURCE: **Katz, Fixing Functionalism.**

this must be case). Note also that the pain inside the dotted box is a very different quantity than the one to the right. The first is a brain state, the second is a hypothetical quale returned by PCR; in other words, the one on the right is the one that hurts, as the jagged edges indicate.

By any standard, the PCR will be far too complex to effect the relevant transformation in an elegant fashion. There are three reasons for believing this to be the case:

a) The complexity of the input (to the PCR)
As shown in Figure 4.8, the input to the PCR is itself an algorithm. As any programmer knows, it is possible to have complex programs corresponding to simple inputs, but complex inputs almost guarantee complex programs. As a matter of fact, an algorithm is the most complex input that can be conceivably input to a program; all other inputs are subsets of an algorithm. For example, a database can be thought of as a program that returns a value given a set of indices: given a name and purchases over $500 made in the last 6 months it returns "plasma TV."

What makes an algorithm, as opposed to a database, so complex, is that it is typically peppered by conditionals, such as the "people present" one in the pain example. Instead of a static set or ordered series of inputs, we have semantically charged notions that evaluate to true or false. But programs are notoriously poor at understanding meaning; this is why, for example, that search engines work by association, not by extracting meaning. You can type "theories of consciousness" into a search engine and get a reasonable set of results, but you will not get a reasonable set of results if you type "theories of consciousness developed in the last twenty years that eschew materialism." Semantics remains the Achilles heal of Artificial Intelligence for good reason: meaning cannot be circumscribed by a few or even dozens of rules, and by its nature it consists of a large set of special cases.

Therefore, if complexity of processing is correlated with the complexity of the input, then functionalism is in trouble, not for conceptual reasons, but for methodological ones. There are simply too many possibilities for the PCR to contend with. Moreover, each of these possibilities will not necessarily correspond to a compact algorithmic description. For example, the flowchart in Figure 4.8 is vastly simplified, really a caricature of a pain processing program. The full program must include notions of degree of pain; the fact that distraction may reduce pain but that pain itself is also a strong distracter; that pain in the service of a cause, such as that experienced in the heat of war or in childbirth becomes more bearable; etc. Somehow our assessment algorithm must take into account all of these variables if it is to produce the right output. In short, there seems little chance of coming up with a functionalist account that is both compact and does the job at hand for any given quale, and certainly not for the full range of experienced qualia. But the problems do not end there.

b) The role of inputs and outputs

Functionalism is beset by many technical difficulties, not the least of which is the role of the inputs and outputs (behavior) in the theory. For example, one can easily imagine a so-called super-Spartan that feels pain but never expresses it, who has itches but never scratches, and has the same expression on his face when watching the evening news as when he experiences an orgasm. If functionalism describes the transformations of inputs to outputs, it is not clear in this case what the transformation is for: there are no observable outputs. Suppose, however, we are being conceptually generous and grant that in these cases, there is still the tendency to behave in certain ways, even when no output exists. A serious problem still remains, however: there are no limits to the types of inputs and outputs to the system, and to limit these artificially would be parochial. We would not want to say, for example, that a system that saw in the ultraviolet range, that detected weak magnetic fields (as do many animals), or that communicated by changing patterns on its body (as do squid) are not conscious because they do not have the same input-output devices as we do. Within the pain domain alone there are stubbed toes, bent thumbs, poked eyes, and nausea, pains of loss, separation, anxiety, and humiliation to name just a few, each with its unique input-output characteristics. Once again, the PCR faces an unbounded range of possibilities, preventing a simple or even a moderately complex theory from arising.

c) Counterfactuals
All algorithms of any complexity include conditionals such as that represented by the subroutine in Figure 4.8. If the assessment algorithm is to produce the right qualia, it must see both paths leading out from the conditional, both the path taken and the "counterfactual" path not taken. Only by taking into account both these paths is it able to judge the full algorithm the brain is running. But by construction, only one path is taken at any given time. This means that it must remember previous instances where the counterfactual path was pursued and the consequent states and behavior; otherwise the full algorithm is not represented. This leads to the singularly odd conclusion that what we are feeling at any given instant is partially a function of what we felt in the past. Once again, the argument is not that it is impossible—it is certainly conceivable that the assessment algorithm stores all previous branches of the choice points and the conditions under which they arise—rather this implies that functionalism as a theory would be too complex to work in practice.

In summary, if the PCR could indeed be described, a dubious proposition in itself, it would rank as one of the most complex theories ever described, and certainly many orders of magnitude more complex than that which is typical for the physical sciences. It is thus no accident that there are no full-fledged working functionalist theories of consciousness, despite the fact that it has been around in one form or another since the 1960s. We do not, for example, have a flowchart that corresponds to the perception of red, nor has anyone been able to produce the algorithm for ecstasy, seething, or pins and needles. There do exist a few functional-like theories with broader brush strokes, but none that map particular algorithms to particular qualia.

Given these difficulties, one wonders as to why functionalism has held its ground for so long. The answer is that there really is no other game in town. As previously mentioned, the advantages of functionalism include providing a clear reason for demarcating consciousnesses, a reasonable answer to the chauvinism/liberalism debate, and above all, a good explanation of the correspondence of cognition to consciousness. The next section addresses the following question: Is it possible to have one's theoretical cake and eat it too?

CAN THIS THEORY BE SAVED?

Like many a troubled marriage, functionalism totters forward, partly because at is core, there remains a remnant of its original glory, and partly because there doesn't seem to be any real choice. None of the obvious alternatives is remotely plausible. Eliminative materialism solves the problem by discarding it, and this is just implausible on the face of it. Consider the implications, discussed in much greater detail in Chapter 8, of adding a chip to the brain. One of the first things we would want to know is which of the following three possibilities hold: a) the recipient will be completely unaware of the effects of the addition, despite the fact that there may be detectable behavioral responses, b) the recipient is aware of the results of the computations of the chip only, or c) the recipient is also aware of at least some of the intermediate steps of those computations. One of three must be correct, but if we eliminate consciousness as a scientific variable, there is no way the problem could be posed, let alone solved. Alternatively, assuming that consciousness is a respectable scientific variable, but is aligned solely with a substance in the brain, as noneliminative materialism claims, seems entirely too chauvinistic. We have no reason to believe that possibility c cannot hold; that is, we have no a priori reason not to believe that consciousness can only be embedded within systems with noncarbon chemistry. Finally, a quantum mechanical explanation of consciousness remains, for the previously discussed reasons, the scientific equivalent of a shot in the dark.

Can we, then, build upon the core strengths of functionalism to construct a theory that retains its positive effects without the encumbrance of undue complexity? This section will provide a positive answer to this by leveraging the one characteristic of functionalism that cleanly separates it from its competitors. This is the notion of causality. A fact that is so obvious that it is easy to overlook is that a nonfiring brain (dead or merely in deep sleep) is an unconscious brain. This could be because such a brain is not carrying out any computations, as functionalism would state. But the more direct claim is simpler: it is that there are no causal interactions in a nonfiring brain. Furthermore, the simple notion of causality can also tell us why your consciousness isn't mixed up with someone else's; this is another "preempirical" fact that still, however, demands an explanation. The reason is that each consciousness is a causally self-contained system. This idea will be developed in further detail in the context of the reciprocal corollary, introduced in the following text.

First, however, we introduce a bare-bones functionalism that stresses the importance of causality, but jettisons a number of complexities associated with the traditional formulation of this theory. More specifically, the critical element in the new proposal is what may be termed causal currents. Causal currents may be defined informally as the degree to which one processing module influences another at any moment in time. It may also be defined at a lower level as the network of neural interactions at any given time.

We will call this theory that is based on causal currents reduced functionalism (Katz, 2008), or rF for short, to contrast it with the full-fledged functionalism that was described earlier. We now revisit the three sources of complexity that arose in this prior theory, and show how they can be largely alleviated in the new formalism.

a) Simplifying the input to the PCR

As previously argued, one of the primary advantages of functionalism is that it shifts the explanatory burden from substance to causal relations. One way of doing so is to work with the causal transformation effected by the algorithm the brain (or a machine) is running. However, an algorithm is too unconstrained a representation on which to build a compact theory. The natural alternative then is to retain the essential component of causality and jettison the superfluous aspects of the algorithmic process that serve to complicate the life of the PCR.

This can be done by looking only at causal relations without reference to the notion of the kinds of processing the brain is performing. As previously discussed, we will speak of the causal interactions between neurons, but leave open the possibility that variants of the current theory could more profitably look at current flows over larger collections of neurons. A neuron can be characterized by a firing rate and the synaptic efficacy between it and the neurons it influences (see Chapter 2 for a review). The degree of causal influence will be proportional to both these quantities (if a neuron is not firing, it can have no influence on its neighbor; if it is firing but only weakly connected to its neighbor its influence will also be minimal). As a means of making the notion of causal interaction more concrete, let us therefore designate the causal current between two neurons as the product of this firing rate and the synaptic efficacy, relative to all the other influences the target neuron receives.

Figure 4.9 illustrates this process for a simple case. Let us say we wish to know the current from unit 2 to unit 4, and let us assume that the relative firing rate of unit 1 is 0.5, that of unit 2 is 0.75, and that of unit 3 is 0.25. Then the relative contribution to unit 4's state by unit 2 is just $0.75 \times 1.0 / (0.25 \times 0.25 + 0.75 \times 1.0 + 0.5 \times 0.5) = .71$.[6] The causal network with all normalized influence indicated is shown on the right of the figure.

[6] For the purposes of simplicity, the contribution of the inhibitory weight is assumed to be proportional to its magnitude. Hence, the absolute value of this weight is used.

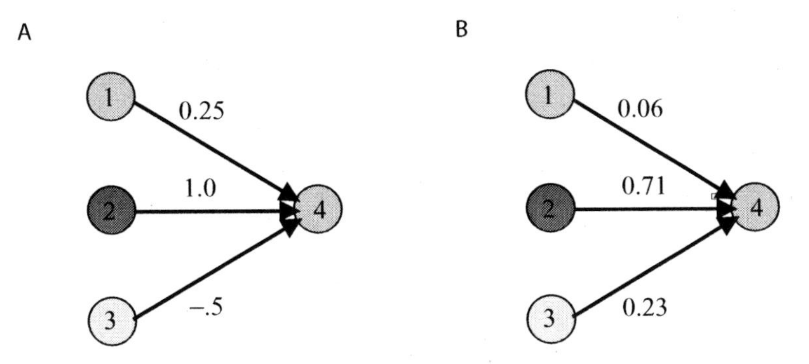

FIGURE 4.9 **A: A neural network, and B: The corresponding causal network.**

The net result of performing these calculations is a graph, or network, with the nodes of the network representing individual neurons and the edges of the network representing causal influence (not weights as in a typical neural network). This network, while still potentially large in scale, is a considerable simplification over an algorithmic representation. There are no subroutines, and there are no semantics, explicit or implicit, associated with each node.

b) Ignoring inputs and outputs
Theory rF makes an assumption that vastly simplifies processing by the assessment algorithm. This is that inputs and outputs to the processing system, in and of themselves, don't matter. This may seem like a radical step; after all, functionalism is founded on the notion that the mind results from the ways in which inputs are transformed into outputs. However, as we have seen, this makes the theory completely unmanageable. What is an input and what is an output? There are infinite varieties of each, and each would have its own unique semantics that a functionalist account would have to consider. In contrast, rF represents inputs and outputs implicitly, in the initial and final layers of causal processing respectively. By abstracting out the semantic content of the inputs and outputs, and reducing these to strict causal relations, rF removes a major stumbling block in the eventual production of a theory of consciousness.

c) Restricting the time window
The final problem that we saw with traditional functionalism was the nature of counterfactuals. A considerable apparatus must be included in the theory to keep track of the paths *not* taken at any given point. These are parts of the algorithm and therefore, in principle, influence

mental content. Once again, rF makes a simplifying assumption: only causal currents at any given moment in time matter. Therefore there are no counterfactuals, because only a single effective path from the initial to the final processing layers is being considered. There is one slight technical constraint that needs to be considered to make this work in practice. The degree of influence that a given neuron has over a target neuron must be computed over a small time window, rather than at an instantaneous point in time. The reason for this that the degree of causal influence will be a function of the spiking rate of the influencing neuron. Therefore, a time window must be used to assess this rate. In practice, determining the size of this window is something that we can leave to empirical research for now. The key assumption, however, is that this window be small enough that the assessment algorithm need not remember past events in order to determine currently experienced qualia.

The theory resulting from these simplifications compares favorably with standard scientific theories, such as those of physics, in that its independent variable, the set of causal currents is simple and well-defined, it eschews arbitrary semantics, and it is time-delimited. The remaining question is whether rF is sufficiently rich to capture the scientific data of consciousness. Before broaching this question fully, however, let us examine a direct consequence of this approach.

The reciprocal corollary

It will prove fruitful to further clarify the underlying assumption of theory rF, namely, that causal currents form the foundation of a theory of consciousness. It is an elementary exercise to show that if this is indeed the case, then we need only consider reciprocal as opposed to unidirectional causal flow. The key to understanding this argument is that the brain does not have special status as an organ of consciousness once one abandons the idea (as we have for the reasons stated earlier) that there is some special substance in the brain that generates qualia, and that instead causality determines conscious content. Let us repeat this, because it is both important and somewhat counterintuitive. In any causal-based theory, one needs to at least entertain the possibility that it is not just the action of the brain that generates consciousness, but also the environment. For example, in functionalism proper, any algorithm implemented by the environment on a stimulus prior to processing by the brain potentially influences conscious content. Why? Because the claim is that consciousness is a function of such processes, not something special in the brain itself. Likewise, in reduced functionalism, we need to consider not only causal currents in this organ, but also those leading into and stemming from its actions. This could prove disastrous, in that it would potentially make one's consciousness a function of the entire universe.

Fortunately, not only can the set of potential influences be delimited to the brain, it can also be shown that it is confined to a subset of brain processes. The reasoning proceeds by the three arguments represented by the diagrams in Figure 4.11. Consider a light ray that hits the eye at time 3, but before doing so bounces off surfaces at time 2, and time 1, as in 4.11.A. Whatever happens to the ray before it hits the eye, that is, before time 3, will not affect visual perception. Clearly, the only thing that matters is its wavelength and intensity at time 3, right before striking the retina. This argument can trivially be extended to any input in any modality; the causal path of all events prior to being registered by a transducer is irrelevant from the point of view of conscious content. But, and this is where the assumption regarding the nonprivileged status of the brain comes into play, if causality alone is what counts, then this reasoning must apply to events inside the brain also. No afferent-only processes outside of the brain, *or inside the brain*, can have any direct effect on conscious content. That is, events in which causal flow is incoming may indirectly influence the generation of qualia because of their downstream effects, but the PCR need not look at these in performing its calculations.

Figure 4.12.B shows an alternative argument that buttresses this idea. Here, we have a single light source hitting two eyes from two different brains. If consciousness is a function of causal currents, and unidirectional currents play some role in this function, then the contents of consciousness of the two brains 1 and 2 would necessarily be related. But in fact, they are only *contingently* related. That is, if it the two eyes work in a similar fashion and feed inputs to similar brains, then the same color may be seen. However, the second person may be color blind and see a different color, they may be synaesthetic and "hear" colors, or they be in a binocular rivalry experiment and see nothing at all in the given eye. There is no reason to suppose that there is any *necessary* connection between the visual experiences; if they are similar, this is coincidental.[7]

A similar argument to that in Figure 4.11.A with respect to causal currents exiting the brain can be made. Behavior, which can be thought of as the causal consequences of brain activity, has no direct influence on conscious contents. It can only influence these contents if this behavior is fed back to the brain via transducers that sense the results of such behavior, as in the case of proprioception.

[7] Of course, this is a coincidence that occurs often because human brains tend to be similarly structured. If brain 2 was a Martian brain, however, we would have no idea what it was feeling based solely on the nature of the input. If causal flow in this direction helped determined consciousness, we would know at least a little about Martian experience; in fact, we know nothing whatsoever without further knowledge of Martian brains.

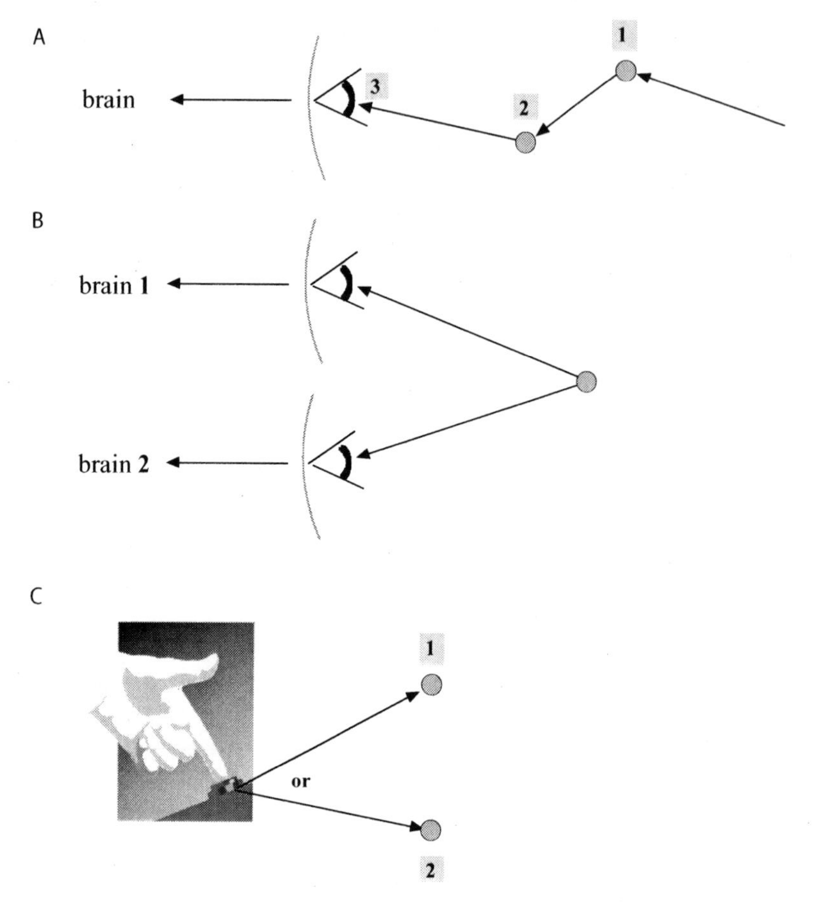

FIGURE 4.10 **A: Events in the path of a light ray striking the retina do not affect conscious content, B: Two brains looking at the same item do not necessarily share conscious content, C: The consequences of any motor action by the brain do not influence conscious content (from Katz, 2008).**
SOURCE: **Katz, Fixing Functionalism.**

Consider pressing a button as in Figure 4.10.C, and then one of two actions occurs according to a flip of a coin. Either a light goes on in the next room, or a jack-in-the-box halfway around the world pops up and says, "My name is Yani the Magnificent, and I alone have solved the riddle of consciousness." The conscious contents of the button presser are in no way influenced by which event happens to takes place (and it would be a very strange world if it did).

The conclusion from these arguments is that unidirectional causal currents, either entering or exiting the brain, make no contribution to contents of consciousness. Furthermore, given the assumption that it does not matter where these currents occur, inside or outside the skull, we can generalize this result to the brain itself. Causal influences on the brain from afferent neurons without top-down influences can play no role in the generation of conscious content; likewise, efferent neurons that do not feed back into the system also play no role. Taken together with the previous assumptions yields the following four constraints in addition to those traditionally present in the functionalist thesis:

i) The representational constraint:
Only the network formed by looking at causal currents matters;
ii) The I/O constraint:
The inputs to the network and the outputs from this network can be ignored;
iii) The temporal constraint:
Only currents with a narrow time window centering on the present are considered; and
iv) The reciprocal constraint:
Vertices in the network that are not reciprocally connected to other vertices in the network (i.e., in graphical terms, are not part of a cycle) are pruned from consideration.

In the following sections, we evaluate theory rF with respect to previous scientific conceptions of consciousness and the data developed in the previous chapter.

Theoretical comparisons

The most direct method of determining the validity of a theory is to see how well it comports with the empirical evidence. However, it is often desirable to first show that the theory is broadly consistent with previous thinking on the matter. Scientists have not been sitting quietly on the sidelines watching the philosophers getting all the theoretical action. A number of attempts have been made to naturalize consciousness, that is, to bring it into standard scientific discourse. This section introduces three related prominent lines of thought that are largely empirically driven—reentry, synchrony, and accessibility—and shows how rF, although derived from first principles, turns out to be consistent with these notions.

i) Reentry
Gerald Edelman, along with his colleagues, has been at the forefront in stressing the importance of reentrant connectivity in the generation of consciousness (Stephen Grossberg (2001), a pioneer in the field of neural networks, has developed a similar line of thought). Edelman's early Nobel-winning work concentrated on the adaptive qualities of the immune system; however, like many scientists in recent years, he has felt the considerable tug of the problem of consciousness, and has been of late concentrating on this conundrum. His main claim is that sensory

processing modules are involved in a reentrant loop with categorical and other high-level processes located in the frontal, temporal and parietal areas. Consciousness arises not via the original sensory processing, but from the interactivity that arises within 500 ms between these low- and high-level modules (Edelman, 2003).

There are at least three associated claims attached to this view. First, this interaction is proposed as a means of providing superior cognitive abilities. We have already seen in chapter 3 how interactive top-down and bottom-up activity can provide better decisions than either operation alone. Edelman also shows, for example, that color constancy can be achieved with such a reentrant model (Wray and Edelman, 1996), and how schizophrenia may be explained by postulating the existence of multiple reentrant cores (Tononi and Edelman, 1998). These sorts of results are not about qualia per se but about processing, so we will ignore them for the purposes of the current discussion.

The remaining two claims bear more directly on the subject at hand. The first is that reentry leads to a synchrony of firing between large groups of neurons. Synchrony, in turn, is hypothesized to result in the binding together of otherwise distinct elements in the perceptual field. Reentrance as a cause of increased synchrony and the purported effects of such on conscious content are taken up more fully in the following text. Reentrance is also hypothesized to result in the contents of consciousness being made widely available to other computational modules. This notion of the accessibility of conscious contents is also developed more fully in the following.

It is easily seen that reentry and reciprocity are closely related concepts, the key difference being that the former involves interactive feedback between modules rather individual neurons. However, recall that rF is neutral with respect to granularity; causal flow may be described between neurons or sets of interacting neurons. Furthermore, Edelman's concept of a dynamic core, or the subset of neural activity involved in reentrance, and alone responsible for the generation of consciousness, is closely related to the notion of the conscious kernel, which will be introduced in the discussion of the implications of the reciprocal corollary for neuroengineering. In summary, although different justificatory routes have been taken in their production (the current work concentrates more on first principles), both rF and Edelman's account of consciousness is not the result per se of computations in the brain, but rather the set of immediate and interactive causal currents.

ii) Synchrony

An associated claim of Edelman's view is that reentry facilitates temporal binding via mutual excitatory feedback. In standard neural models of pattern recognition, the disparate elements of a stimulus are assumed to become bound together through convergence, such that in each successive level of processing there are recognition units for groupings of features from the layer below. However, this leads to the well-known problem of combinatorial explosion, because it implies that there would need to be higher-level units for each combination of lower-level features. An alternative solution is temporal binding (Engel et al., 1999) whereby synchronized firing between neurons subserving the elements of the stimulus is hypothesized to form the basis for the grouping of features.

Recall that one of the constraints introduced both to eliminate counterfactuals and to give rF the flavor of a standard scientific account was to make consciousness a function of contemporaneous causal forces. rF thus provides a strong hint as to why firing synchrony implies phenomenal binding; it is simply because the mental supervenes on the physical *plus* the temporal. In other words, from the point of view of the PCR, two neurons sequentially influencing each other in successive time windows is equivalent to there being no reciprocal influence at all; only reciprocal currents in the same time window count. In summary, if rF is correct, one would expect that firing synchrony will be a paramount factor in determining which stimulus elements are joined together phenomenally, and which are entertained in separate phases of awareness.

iii) Accessibility

A final correspondence may be drawn between rF and theories suggesting that, using Dennett's (1992) term, the contents of consciousness enjoy a degree of cerebral celebrity. That is, once at item enters in the conscious arena, regardless of its origin, it becomes accessible to all other processing modules. A closely related notion is Baars' (1988) cognitive theory of consciousness, in which working memory plays the role of a central store, and the contents of consciousness are identified with its contents. Once in working memory, an item may then be modified by other modules operating beneath the level of consciousness, and then optionally placed back in the central store. Both models gain impetus from long-standing cognitive models, such as production systems (Klahr, Langley, and Neches, 1987), as well as the informal introspective observation that once an item is conscious, it seems that we can act on it at will—it can form the basis of a behavior, such as a speech act, or it be drawn upon to make further inferences.

Theory rF and, in particular, the reciprocal corollary also predict that accessibility will be a property of consciousness. Consider that if a set of neurons are subserving some quale, then by the reciprocal corollary there is either a direct or indirect route from this set back into the system as whole; otherwise they would be efferent only. Thus, the information generated by these elements is available, either directly or indirectly (modulated by intermediate elements in this case) to the system. Furthermore, if one assumes a model, such as that of Baars, in which processing modules revolve around a central store, this route will be direct in the sense that each module will place its contents in this store, which will then be one step away from every other module.

An outstanding question regardless of implementation is whether all conscious contents are "broadcast" to the rest of the brain, and if so, this is a contingent fact or a necessary property in order for the item to become conscious in the first place. Here, too, rF may be able to provide some insights. One counterexample to complete "celebrity" would be when two or more distinct kernels arise in the same brain. Presumably, only one would have access to the linguistic center of the brain, and only one would be reportable, but there would still be two noncommunicating consciousnesses. Tononi and Edelman (1998) make a similar point with respect to their model of schizophrenia, and others (e.g., Sperry, 2001) have made similar claims with respect to split-brain patients. Severing the corpus collosum immediately removes the possibility of any causal contact between the hemispheres, and rF would therefore predict the development of separate consciousnesses.

In summary, rF makes many of the same predictions as those thinkers who believe that consciousness is a respectable scientific variable, but are not themselves under the sway of the computational paradigm. These include the notions that reentry is essential to the generation of qualia, that conscious content is a function of the here and now and does not supervene on past events, and that the connectivity constraints imposed on modules that subserve conscious content make it likely that these modules are able to broadcast their results to the rest of the aware brain.

Consistency with results

We now examine how rF could explain many of the results presented previously, including those relating to binocular rivalry, blindsight, masking effects, contextual effects, and finally, perhaps the most fundamental of all results, those pertaining to the difference between the sleeping and waking states.

i) Blindsight

As previously discussed, blindsight provides extremely important clues as to the neural basis of consciousness. As in any detective story, no clue is definitive, but some are more pertinent than others. In this pathology, the stimulus, to at least some extent, is processed, but it is not consciously perceived. Summarizing the results from our prior discussion, we have:

a) Blindsight may follow from lesions to all or part of V1, the primary visual cortex (although not all patients with V1 lesions have blindsight).
b) The most notable structure that may serve as a relay point from the retina to secondary visual cortex in the absence of V1 functioning is the superior colliculus (SC), although other structures may also serve or aid in this role, including the dorsal and lateral LGN and the pretectum, among others.
c) Blindsight appears to be qualitatively different than degraded normal vision. That is, it is better to describe it as pattern detection without awareness rather than pattern detection in the presence of a noisy stimulus.

Given a and b, one may be tempted to conclude that V1 is the locus of visual awareness, but this is directly contradicted by the fact that activity correlated with the unattended eye *is* present in V1 during binocular rivalry, as well as the masking phenomena results to be discussed next. What these facts do appear to tell us is that V1 activity is a necessary condition for visual awareness, at least under ordinary circumstances. But just what is it about this area that makes it so? One possibility suggested by rF is that V1 is involved in the reciprocal trade of activity about a given pattern with higher visual areas, and that other structures are engaged in merely the afferent processing of visual stimuli. By the reciprocal corollary, such activity should not reach awareness.

To test this hypothesis, let us examine the operation of the SC. The SC is a midbrain structure that sits right below the thalamus and is implicated in the generation of *saccades*, fast, largely involuntary movements that the eye makes to reposition the fovea on the salient parts of a visual scene. The retina projects directly to the SC, which in turn makes contact with secondary visual processing areas. The SC does receive additional information from the cortex, including the parietal cortex, which guides the saccades to places of interest, as well as from the frontal eye fields, which may serve an inhibitory role in saccade generation. However, it does not appear to engage in the sort of resonance between bottom-up and top-down processing of a stimulus that typifies V1 and downstream visual structures. Hence, from the point of view of a given pattern, that pattern is processed as a purely afferent stimulus, and therefore, by the reciprocal corollary, the pattern is not brought into awareness.

Similar arguments need to be made for other possible pathways into the cortex of visual information in order to flesh out this argument. We will not do so here, except to state that given current knowledge, the rich representation of a visual stimulus and its interplay with more abstract processing structures appears to be uniquely concentrated in the interaction between the primary and secondary visual areas. Under this view, it no longer appears so strange that primary cortical lesions would suspend the ability to perceive a stimulus, but allow other detection facilities to survive through alternative routes to the cortex.

ii) Binocular rivalry

As discussed previously, the phenomenon of binocular rivalry is an extremely useful litmus test for any proposed theory of consciousness. In binocular rivalry, a stimulus clearly registers on the retina, and even activates corresponding neural structures in the primary visual cortex, yet somehow does not make it into consciousness. A candidate for a theory of visual consciousness must therefore address why this early activation is not sufficient for awareness. Summarizing the three main findings with respect to this effect, we have:

a) All levels of visual processing appear to subserve the maintenance of the dominant, consciously perceived image.
b) The competition is between eyes rather than between patterns, that is, it is not a categorical rivalry but a stimulus-based one, except under special circumstances.
c) There is some evidence that there is increased synchronization between primary and secondary cortical processing areas for the perceived image relative to the degree of synchronization for the unperceived image.

Taken together, these facts implicate both lower- and higher-order visual areas in the conscious perception of visual stimuli. Result b bolsters this idea. If the competition were between patterns, then it would be subserved by the pattern recognition areas in the secondary visual cortex. However, it appears to be the result of ocular rivalry, and separate ocular representations predominate in the primary visual cortex. Alternatively, assume that the primary visual cortex alone is responsible for visual awareness. How then to account for the almost complete absence of firing in the secondary visual cortex for the unseen stimulus, and for the result in c?

In summary, although not the only possible explanation, a good guess as to what is happening is that the seen image is the result of resonance between the primary and secondary cortical representations. But this is just what theory rF, and especially the reciprocal corollary, claims. Consciousness cannot occur in the present of feedforward activation only, it must involve a feedback system of interacting processing elements. Furthermore, because rF holds that consciousness supervenes on the temporal also, that is, is a function of what the brain is doing right now, one would also expect a result such as that in c. Further enquiry is clearly needed in this critical area, but for now, the neurophysiological basis of binocular rivalry is at the very least consistent with rF, and to a large extent, is predicted by this view.

iii) Masking phenomena

Recall that a briefly presented stimulus followed by another stimulus in the same location of the visual field will often mask the earlier stimulus; that is, only the second stimulus will be perceived. This is one of the most venerable and widely tested phenomena in psychology, but a fully satisfactory account has yet to emerge. According to the traditional explanation, visual patterns are first processed by the primary visual cortex, and then by higher visual areas, such as the inferotemporal cortex (IT). In this view, in order for a stimulus to become conscious it must reverberate in these higher areas for a sufficient length of time (Turvey, 1973).

However, it is difficult to account for all of the masking effects in Figure 4.3 by a feedforward model alone. Suppose instead, as rF would suggest, that a reverberation between lower- and higher-order processing is necessary in order for an item to become conscious. Only when there is a match between these two modules would the neural assemblies subserving the stimuli in these modules be involved in reciprocal activity, and thus only in this case will a pattern be perceived. Figure 4.11 illustrates how such an account can explain all of the masking effects illustrated in Figure 4.3. Shown are the hypothetical contents of V1, or the primary visual cortex, and area IT, responsible for pattern recognition among other things, at some short time after 200 ms.

In the following explanation, we will assume that a pattern will reverberate in a given brain area unless replaced by another pattern, and that patterns take approximately 100 ms to migrate from V1 to IT. In the topmost instance, the target has migrated to IT but does not match the mask that has entered V1 (and vanquished the target), and thus the target is not perceived (the mask eventually migrates to IT, and this matches with the reverberating mask signal in V1; thus the mask *is* perceived). In the second case, both target and mask migrate to IT in tandem, and these match with the identical reverberating signal in V1 and hence both are seen. In case three, both mask and target enter IT together, as in case two, but by the time they get there, only the mask remains in V1 (the presence of the mask in the stimulus after the target is gone eliminates the target); and hence there is a mismatch. Finally, in case four, the mask enters IT first. But when the target joins it in V1, both migrate to IT together, and this then matches the reverberating signal in V1. Hence, both are perceived.

There are certainly other ways of jointly explaining these four effects, but reciprocity between low- and high-level processing provides a particularly clean account. Furthermore, this theory is consistent with the known timing of visual processing (Lamme, 2001). It takes approximately 100 ms for an image to reach awareness, and this is close to the time it takes for neural activity to spread from the primary to secondary visual areas and back again.

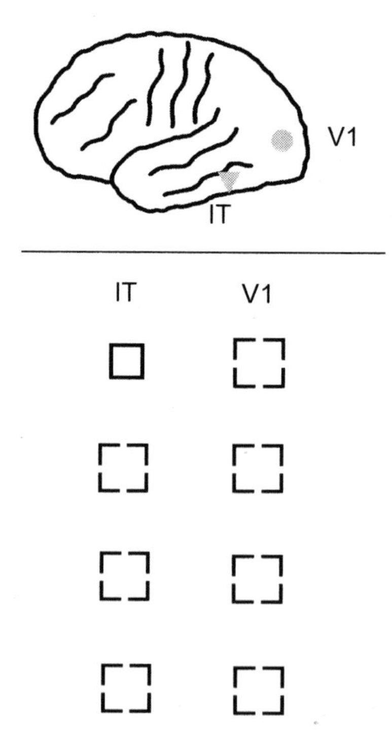

FIGURE 4.11 **Top: Patterns enter V1 (circle) and migrate to IT (triangle). Bottom: The contents of these areas after processing; only when there is a match between the two areas will the target be perceived.**

iv) Context-driven perception
There are three possible explanations for the effect whereby the contents of consciousness are modified by their contexts. First, consciousness does not have access to primary sensory modules, but only the processed information in secondary sensory ones. Second, consciousness is a function of the steady-state response of the brain in which the context has been allowed to act on the original stimulus. Third, and in agreement with rF, consciousness is the result of resonant reciprocal activity between higher- and lower-order modules. We consider each in succession.

If sensory qualities only enter into awareness from secondary processing areas, then it follows directly that all perception is context-dominated, because by the time information gets to these modules, it has been already been heavily influenced by top-down contextual information. To take the exemplary case of color constancy, color as registered immediately in the retina by cone cells is not adjusted to account for the illuminant (further processing in the retina may perform some constancy adjustment). However, by the time it reaches V4, and certainly IT, it has been transformed into the "correct" perceptual color.

There is one fatal flaw the view that we are only aware of the aftereffects of high-level processing, however, and it is a flaw that is shared by all theories that push awareness into these areas alone. This is that secondary (or later) processing areas do not contain the resolution of primary areas. Each successive layer of processing involves a convergence of activity from multiple units below to single or fewer units above. Therefore, if consciousness is "grabbing" its data from higher-order areas alone, it could not provide the necessary fine-grained perception. For example, IT cells are designed to be sensitive to the presence of specific shapes or configurations of features, such as faces, and will respond to a given face over a wide range of spatial positions, and in a variety of configurations (with a hat, without a mustache, etc.). If spatial resolution is so low in this area, how is it possible that we are also aware of fine-grained features, such as wrinkles, in addition to the fact that this face belongs to our neighbor Mrs. Smith?

One way of attempting to rescue this situation is to stipulate that we are only aware of relatively long-term stable states of the system (on the order of 100 ms or longer, for example). This explains contextual effects because, in effect, we would only have access to the final results of processing, not the initial or intermediate states where the raw stimulus still dominates. In addition, a number of other lines of evidence suggest that in order for a stimulus to reach consciousness, it must result in a persistent neural state. However, by a similar argument to that given for the reciprocal hypothesis, it is easy to show that steady state activity in the absence of reciprocity is not sufficient to generate conscious content.

Suppose, as in the left of Figure 4.12, that there is a system, *outside the brain*, that is in a steady state, the outputs of which are the inputs to the brain. The mere fact that it is in this state (mathematically represented by the fact that the next state is the same as the previous state) will not inject this process into the consciousness associated with this brain; as before, only the signals striking the transducers matter, not what happens beforehand. Likewise, the fact that some output of the brain results in a steady state further downstream is irrelevant; we are not aware of any purely efferent effects on our brains.

Thus, in order for an item to enter consciousness, as argued before, it must be present in the conscious kernel, or a set of neurons in a mutually influential relationship. This does not rule out the possibility that persistent or suprathreshold activity within the kernel is necessary, but it does mean that the key factor is the presence of reentry. The most compelling picture, therefore, of the fact that all perception is context-driven is as follows: The relatively raw stimulus enters into the primary processing module, after some initial processing at the transducer level. This is then forwarded to secondary processing, which adjusts elements of the stimulus pattern on the basis of the surrounding context. This context may include other features present in the stimulus, as in vision, or past (and possibly future) events, as in auditory processing. In either case, top-down knowledge in the form of memories or implicit computations are brought to bear on this contextual data. This processing then interacts with the iconic memory of the original stimulus, still resounding in the primary module, until the two modules, bottom-up and top-down come to a mutual agreement about how to interpret the original incoming stimulus. At this point, the stimulus enters consciousness, and only the interpreted stimulus is perceived, not the original, nor the means by which it was transformed into its contextually modulated form.

v) Unconsciousness in sleep and anesthesia

One might think that the degree of consciousness is simply a function of overall cortical activity, and to some extent this is true. In the

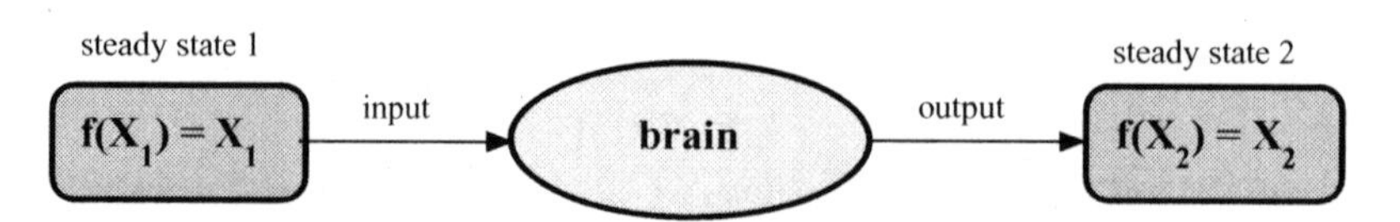

FIGURE 4.12 **Two steady states are in causal relation to the brain, one providing input, and one postoutput. But neither could possibly influence the contents of consciousness.**

extreme case of full cortical shutdown, the mind is also completely shut off. However, as the evidence reviewed earlier suggests, anesthesia-induced sleep only weakly correlates with decreases in brain metabolism. Something else must be going on, and clues to that extra process are provided by that fact that i) the awake state is characterized by low voltage and desynchronized EEG, and the unconscious state the reverse, and ii) as first suggested by John (2002), coherence between brain areas appears to be a better indicator than metabolism or activity alone of the state of awareness.

At first glance these two sets of effects appear to be contradictory. The first states that the unconscious state is correlated with high-voltage synchronized activity. That is, the neurons generating such activity appear to be firing in tandem, possibly due to the regulating effect of bursts of thalamic activity, and the awake state is desynchronized relative to this state. The second states that the awake state is characterized by relatively coherent or synchronized activity. Which, then, is correct? The answer is, as it must be, that both are valid. High-voltage synchrony in the EEG signal is the result of large numbers of coordinated firings, *within a given brain region*. The John (2003) result additionally suggests that it is *interregion* coherence that underlies awareness; this large-scale spatial coordination is notably absent in the unconscious state.

The significance of this is as follows: When we are awake, information is being shuttled forth between sensory, associative, and higher-order brain modules, and between the left and right hemispheres. This allows these regions to derive a quasi-unitary view of both sensory events and endogenously generated thoughts. Another way of saying this is that the representations in these modules become bound together, possibly via traveling signals at approximately 40 Hz, in the gamma range. In unconsciousness, processing still takes place within modules assuming activity is not too far reduced, but each region acts as an independent fiefdom, ignoring the results generated in other modules. It goes without saying that this is precisely what theory rF would predict. Consciousness in this view is only achieved when causal currents are reciprocal, that is, when there is resonant activity between the modules processing a stimulus.

Much more work is required to confirm this hypothesis. For example, the sole evidence that neural modules are isolated from each other during unconsciousness comes from EEG coherence studies; ideally, one would like a more direct means of showing the same result. Nevertheless, in conjunction with the prior evidence regarding blindsight, binocular rivalry, masking, and context-driven perception, a robust case has been made for theory rF. Recall also that this theory was derived independent of these results from three key ideas. First, it was argued that functionalism, by linking consciousness to causality, had the best chance of accounting for consciousness. Next it was shown, however, that Einstein's razor does not allow functionalism to survive in its most popular form, that which connects computations in the brain with conscious states. Instead, a bare-bones causality was introduced, with the added constraints that the inputs and outputs to the system do not matter, and that the temporal window in which causal effects are examined is highly restricted. Finally, it was argued that one-way causality cannot result in awareness, and this addition to the theory turned out to be crucial in explaining many of the empirical findings.

What is missing from rF is what is missing from most theories of consciousness, namely, a precise formula for converting from states of the brain to psychological states. One would like, for example, to say that such and such a brain condition will result in the perception of blue, or more ambitiously, that another brain state correlates with the bittersweet emotion of being in love. rF could be extended to these more specific cases, but we will not attempt to do so here—this would entail a large-scale diversion from the purpose of this book. Instead, it will be argued, theoretically in this chapter, and in practice in later chapters, that rF as currently formulated can provide a basis for understanding the way in which neurotechnology can enhance not just the output of the brain, but also the psychic content of such. However, we need to briefly consider one more hypothesis before turning to these concerns.

STATISM?

If simpler is better, why not take the next step up from reduced functionalism and identify states of consciousness solely with activity states? In fact, neurophysiologists, and even scientists investigating perception and consciousness, loosely talk about identifications of this sort. For example, a scientist investigating face recognition may informally state that "lighting up" of the fusiform area of IT in the temporal cortex is responsible for face perception. Presumably, if we had imaging devices with sufficiently high resolution, we may even see that George Bush would activate a given set of cells in this area, and Hillary Clinton another set. It may be natural to conclude, then, that the varieties of conscious states can be mapped onto the activation vector that corresponds to the firing state of the brain at a given moment in time. This is a much simpler account than that provided by functionalism, and indeed, than reduced functionalism, as a vector is a less-complex object than a network (reduced functionalism states that consciousness is a function of the network of causal relations instantiated in the brain).

Recall, however, that Einstein's razor does not favor the simplest theory in absolute terms, but the simplest one that is viable. It is here that statism, as the just-described theory may be termed, fails to meet the constraints of this dictum. To begin with, there is a serious empirical problem that must be addressed, namely, that mere activation of a structure does not appear to be sufficient for consciousness. We have seen this, for example, in the case of binocular rivalry. Recall that the image presented to the unattended eye is not seen, but that at least some neurons in the primary visual cortex fire in response to this image. The reciprocal corollary provides many more examples along these lines—purely afferent and efferent neurons should have no direct effect on conscious contents, no matter how many there are, no matter what their firing pattern.

But we do need to look at the data to reach the same conclusion. Statism, for purely conceptual reasons, cannot serve as the basis for a theory of consciousness. To see this, let us consider a hypothetical simplified situation. Let us say that a pepperoni pizza is represented by the following five neurons with the corresponding binary firing vector: [0 0 1 1 0], and sausage pizza by the vector [0 1 1 0 0]. Here "pizzaness" is represented by the middle neuron, and the toppings by the neurons bordering this cell. So far, so good, as the representation seems to distinguish between the two cases. However, there is an implicit fallacy lurking in this representational scheme. Without a means of ordering the neurons, the two representations reduce to the identical situation, that of two firing cells amidst three nonfiring cells. One way of providing the necessary distinctions is to designate the fourth neuron as the pepperoni neuron, the center one the pizza neuron, etc. *But neurons do not come labeled in the brain*—this is a convention that the neuroscientist, as an external entity, applies to the brain to make sense of it for his purposes.[8] Therefore, there must be some information intrinsic to the brain that creates this ordering. This information, however, is not present in the activation vector, which just tells us what is firing and what is not. What is needed is a means of ordering not based on arbitrary designations, but neural dynamics. Including the causal links between the activations, as in rF, is one way of providing this information.

In summary, conventional functionalism is too complex to be a working theory of consciousness, and statism does not provide enough theoretical leverage to act as such. By simplifying the former, or by replacing what is missing in the latter, a new theory emerges that has been termed reduced functionalism. This theory states that consciousness supervenes on reciprocal quasi-instantaneous causal currents in the brain, and nothing else. In the next section, a framework

[8] Usually, the scientist uses labels that correspond to physical locations in the brain. But in this case the physical location is simply acting as a reference mechanism; no one is claiming that position in space, by itself, determines the phenomenal import of a group of firing neurons.

for incorporating consciousness into neuroengineering will be given, and in later chapters, the implications of this framework for brain machine interfaces and the crucial importance of these ideas for the uploading of the self to a machine will be discussed.

REDUCED FUNCTIONALISM AND NEUROENGINEERING

In the following discussion, we will assume an rF, plus the reciprocal corollary, although many of the comments and overall framework could be adapted to conventional functionalism, albeit with greater difficulty. The key premise, as you may recall, of reduced functionalism is that only causal currents matter in determining conscious content, and the key premise of the reciprocal corollary is that only reciprocal currents matter. This produces the overall framework shown in Figure 4.13.

The framework comprises three regions corresponding to the three types of causal influences on the brain. Type I causal influences are feedforward only, and

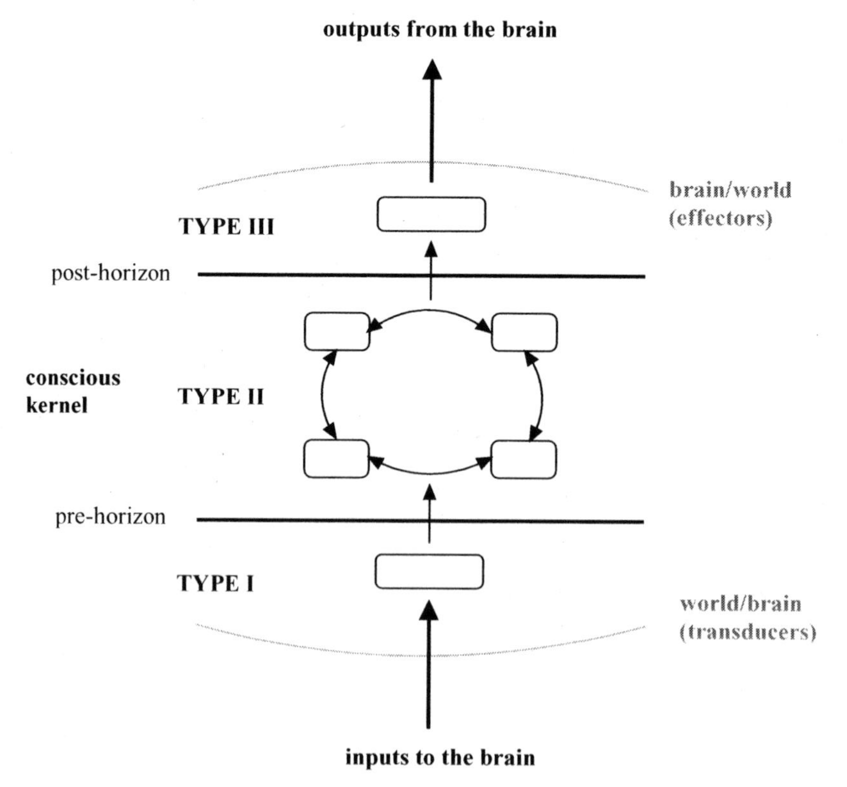

FIGURE 4.13 **Three types of causal influence on the brain.**

operate at the prehorizon of what may be termed the conscious kernel. That is, they can influence the contents of consciousness by affecting the firing of neurons in the conscious kernel, but are not themselves generators of conscious content. This is consistent with reciprocal corollary, which states that causality must be bidirectional for qualia to result. Type II neural units do have this property, and therefore are potentially within the kernel of consciousness. The associated properties, as described previously, of these units include reentry, synchrony, and accessibility. Type II neurons, in turn, may influence Type III neurons outside the posthorizon of consciousness. Like Type I neurons, these do not have reentry into the kernel, and thus cannot be part of consciousness, except indirectly via feedback from the environment. For example, suppose you desire to move your arm. You are conscious of the thought (a Type II process) but not the motor program to carry this out (a Type III process). You may become conscious of the movement itself, though, via proprioception, (a Type I process influencing a Type II process).

Note that the pre- and posthorizons of consciousness need not be identical with the brain/world barrier, and in fact are generally not so. For example, Figure 4.11 shows that Type I processes may be after the world/brain barrier but before the conscious kernel, and likewise Type III processes may be prior to the brain/world barrier but lie outside of the conscious kernel. An example of the former is the visual processing associated with blindsight, and the latter motor programs and other unconscious actions driven by thought but not feeding back into it.

What, then, does this have to do with neuroengineering? Recall that according to reduced functionalism, and any theory that involves process rather than substance, the brain does not play a special role in the generation of consciousness. Therefore, the previous comments apply not only to causation within the brain itself, but also to devices that may augment this organ. As we shall see in the coming chapters, the current state-of-art concentrates almost entirely on Type I and Type III processes. Type I processes include artificial transducers, such as cochlear inputs and other deep-brain stimulation devices (remember, it is not where the stimulation occurs, but the causal direction that matters). Type III devices include brain-machine interfaces enabling "thought control" of the external world. What Figure 4.13 suggests, however, is that Type II devices are possible, at least in principle, if the causality is reciprocal.

In future chapters, we will simplify this scheme and label devices that are directly integrated in the conscious kernel $\mathbf{C}^+$ devices, and devices that feed into the kernel or are fed by it $\mathbf{C}^-$ devices. In Part II of this book we will turn our attention to devices that enhance the power of the brain's power without becoming truly intertwined with its conscious functionality ($\mathbf{C}^-$ devices); the description of true integration of machine with mind affected by $\mathbf{C}^+$ devices awaits Part III of this book. However, in preparation for this venture, we first need to describe not just what could make a brain possess consciousness, but also what gives it the feeling that it is different from other consciousnesses; this is done in the following chapter.

REFERENCES

Baars, B. (1988). *A Cognitive Theory of Consciousness.* Cambridge: Cambridge University Press.

Blake, R. (2001). A primer on binocular rivalry, including current controversies. *Brain and Mind,* 2, 5–38.

Breitmeyer, B., and Ogmen, H. (2000). Recent models and findings in visual backward masking: a comparison, review, and update. *Perceptual Psychophysiology*, 1572–1595.

Chalmers, D. (1997). *The Conscious Mind: In Search of a Fundamental Theory.* Oxford: Oxford University Press.

Churchland, P. (1986). *Neurophilosophy: Toward a Unified Science of Mind-Brain.* Cambridge: MIT Press.

Cowey, A., and Stoerig, P. (1995). Blindsight in monkeys. *Nature, 373,* 247–249.

Dennett, D. (1992). The Self as a Center of Narrative Gravity. In F. Kessel, P. Cole, & D. Johnson (Eds.), *Self and Consciousness: Multiple Perspectives.* Hillsdale, NJ: Erlbaum.

Edelman, G. (2001). Naturalizing consciousness: A theoretical framework. *Proceedings of the National Academy of Sciences USA, 100*, 5520–5524.

Engel, A., Fries, P., Konig, P., Brecht, M., and Singer, M. (1999). Temporal binding, binocular rivalry, and consciousness. *Consciousness and Cognition*, 128–151.

Fries, P., Roelfsema, P., Engel, A., Konig, P., & Singer, W. (1997). Synchronizatin of oscillatory responses in visual cortex correlates with perception in interocular rivalry. *Proceedings of the National Academy of Sciences USA, 94,* 12699–12704.

Gazzaniga, M., Fendrich, R., and Wesainger, C. (1994). Blindsight reconsidered. *Current Directions in Psychological Science, 3,* 93–95.

Grossberg, S. (2001). Brain learning, attention, and consciousness. In B. Baars, W. Banks, & J. Newman (Eds.), *Essential Sources in the Scientific Study of Consciousness.* Cambridge: MIT Press.

Grush, R., & Churchland, P. (1995). Gaps in Penrose's Toilings. *Journal of Consciousness Studies,* 2, 10–29.

Jackson, F. (1986). What Mary didn't know. *Journal of Philosophy, 83,* 291–295.

John, E. (2002). The neurophysics of consciousness. *Brain Research Reviews, 39,* 1–28.

Katz, B. (2008). Fixing Functionalism. *Journal of Consciousness Studies, 15*, 87–118.

Klahr, D., Langley, P., & Neches, R. (1987). *Production System Models of Learning and Development.* Cambridge: MIT Press.

Kolb, F., & Braun, J. (1995). Blindsight in normal observers. *Nature, 3777*, 336–338.

Kripke, S. (1972). *Naming & Necessity.* Cambridge: Havard University Press.

Lamme, V. (2001). Blindsight: The role of feedforward and feedback corticocorticalconnections. *Acta Psychologica, 95*, 209–228.

Logothetis, D., & Schall, J. (1989). Neuronal correlates of sujective visual perception. *Science, 245*, 761–763.

Lycan, W. (1973). Inverted Spectrum. *Ratio, 15*, 315–319.

Marcel, A. (1998). Blindsight & shape perception: deficit of visual consciousness or of visual function? *Brain, 8*, 1565–1588.

Penrose, R. (1996). *Shadows of the Mind: A Search for the Missing Science of Consciousness.* Oxford : Oxford University Press.

Penrose, R., & Gardner, M. (1990). *Emporer's New Mind Concerning Computers, Minds, and the Laws of Physics.* Oxford: Oxford University Press.

Persaud, N., & Cowey, A. (2007). Blindsight is unlike normal conscious vision: Evidence from an exclusion task. *Consciousness and Cognition* .

Polonsky, A., Blake, R., Braun, J., & Heeger, D. (2000). Neuronal activity in primary visual cortex correlates with perception during binocular rivalry. *Nature Neuroscience, 3*, 1153–1159.

Putnam, H. (1960). Minds and machines. In S. Hook (Ed.), *Dimensions of Mind: A Symposium.* New York: New York University Press.

Ro, T., Shelton, D., Lee, O., & Chang, E. (2004). Extrageniculate mediation of unconscious vision in transcranial magnetic stimulation-induced blindsight. *Proceedings of the National Academy of Sciences USA, 101*, 9933–9935.

Sperry, R. (2001). Hemisphere deconnection and unity in conscious awareness. In B. Baars, W. Banks, and J. Newman (Eds.), *Essential Sources in the Scientific Study of Consciousness.* Cambridge: MIT Press.

Tononi, G., & Edelman, G. (1998). Consciousness and the integration of information in the brain. *Advances in Neurology, 77*, 245–279.

Turvey, M. (1973). On peripheral and central processes in vision: Inferences from an information-processing analysis of masking with patterned stimuli. *Psychological Review, 81*, 1–52.

Weiskrantz, L. (1986). *Blindsight: A Case Study and Implications.* Oxford: Oxford University Press.

Weiskrantz, L. (1996). Blindsight revisited. Current Opinions in Neurobiology, 6, 215–220.

Wray, J., & Edelman, G. (1996). A model of color vision based on cortical reentry. *Cerebral Cortex, 6*, 1996.

Chapter 5 Personal Identity

Questions regarding personal identity, like those regarding consciousness, are as old as human intellectual inquiry. The literature on identity is considerably sparser than that for consciousness, however, perhaps because identity is seen as a difficult problem but one that could conceivably admit a solution once enough is known about the mind. That mere matter could produce thought is, on the face of it, astounding—as stressed in the previous chapter, we would have no reason to believe that it could unless we were conscious ourselves. That a being already in possession of consciousness, especially one imbued by evolution with the will to survive and reproduce, would divide the world into self and other is merely perplexing.

As we will see, however, there are a number of important conceptual and technological difficulties that sit at the junction of neuroengineering and identity. These arise immediately when we attempt to address the following two questions:

1. A sentient "occupant" of vessel V1 (such as their original body) gets uploaded into vessel V2, possibly of nonorganic construction. Under what circumstances, if any, would this transfer of identity be considered a success?
2. A sentient being has an operation to repair, enhance, or otherwise modify their neural processes. Under what conditions, if any, will this operation be deemed unsatisfactory, because in the process of providing otherwise salutary benefits the operation has fundamentally altered the identity of the being in question?

Question 2 is of almost immediate concern. Neural technologies are on the verge of not merely correcting brain-based pathologies, such as the tremors of Parkinson's disease, but of altering the functioning of the brain in more fundamental ways. Nevertheless, we will concentrate primarily on the first question, because any answer to this more general query will likely strongly constrain the sorts of responses to the second question. As we will have frequent occasion to refer to this important problem, let us label it the upload identity (UI) question.

It is easily seen that an absolute minimal condition for a positive answer to the UI question is that the receiving vessel is capable of sustaining consciousness; this is one of the reasons that we spent so much time on this topic in the previous two chapters. If this were not the case, then it would be nothing at all to be ensconced in this new body. In the terminology of the previous chapter, this vessel would be a zombie, no matter how much of a success the operation appeared to others. An identity transfer operation is thus an anomaly among medical procedures. It is not enough that the doctors are completely happy with the operation—it is the patient who is the ultimate judge. Certainly, Leonard Moscowitz, our hypothetical patient from Chapter 1, would not be content if he knew that he would not regain consciousness in his new body. He would claim, with excellent justification, that from his first-person viewpoint, this no different than dying, despite what others might think.

But, and this is the crux of the matter, the receiving vessel must do more than sustain a consciousness, it must sustain the same consciousness. It is not enough that Moscowitz regains consciousness; he must also retain the same consciousness as before. In short, from his point of view, the identity transfer operation must be no different than if he had his appendix removed. In both cases, he first succumbs to the anesthesia, and then some hours later he awakes, feeling somewhat groggy but still like himself. Naturally, changing one's body could lead to a dramatic set of altered sensations, especially if the new vessel was a mechanical one, and one may argue that these new perceptual inputs in themselves constitute at least a mild change of identity. For the purpose of conceptual clarity, however, we can stipulate that Moscowitz gets transferred into a younger clone of his old body (well before the point that he contracted ALS to minimize this effect, and concentrate on the essentials of the identity transfer operation).

Let us, for the sake of argument, grant these generous assumptions, namely that the new vessel can sustain consciousness, it can sustain the old identity, and that the operation of the vessel itself does not radically alter this identity, all of which are necessary for the UI question to be answered in the affirmative. Unfortunately, these assumptions taken to their logical conclusion generate a host of new difficulties, the resolution of which will be necessary before definitively concluding that it is indeed possible to exchange one body for another and yet maintain the same identity.

Consider the scenarios depicted in Figure 5.1. The first, at the top of this figure, illustrates the default case. Here, the dying Lenny is transferred from his

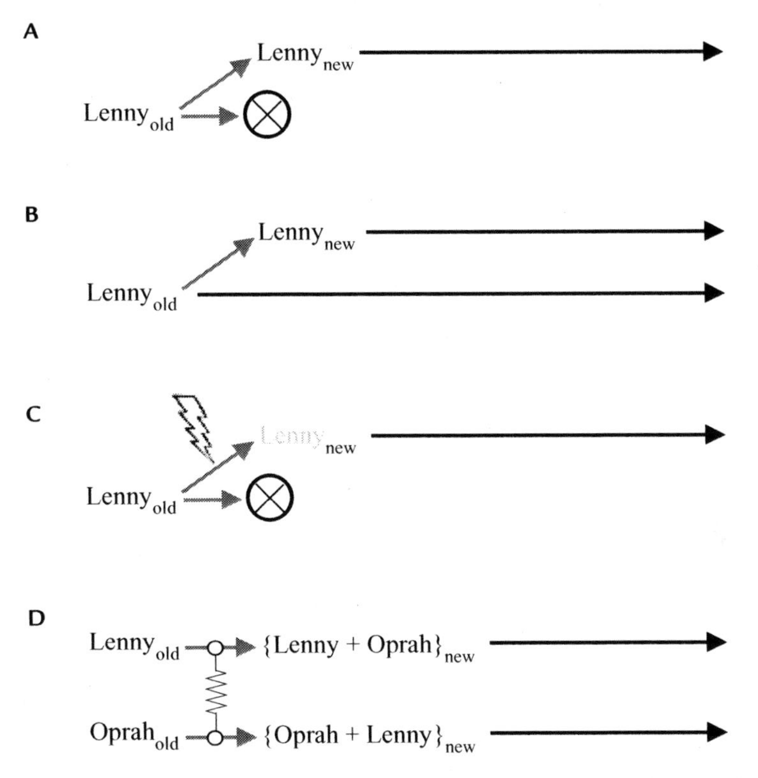

FIGURE 5.1 **A: Scenario 1. Lenny is transferred to a new body, and the old body dies shortly after the transfer takes place. B: Scenario 2. Lenny is transferred to a new body, and the old body is saved—preserving the original Lenny. C: Scenario 3. A malfunction causes Lenny to be only partially transferred to his new body. D: Scenario 4. Lenny and Oprah are undergoing a body upgrade at the same time. A short circuit somewhere in the middle of the procedure causes Lenny's identity and Oprah's to be mixed.**

old body to a new, healthier vessel. Before the original Lenny in the sick body has a chance to wake up, this body is terminated. Assuming that identity transfer is truly a possibility, this should not matter. Lenny will wake up in his new body and feel, by hypothesis, just like the old Lenny (and in fact better, because he is no longer suffering from ALS). What if, however, during the time of the transfer, a miracle cure is discovered for this disease? We can make this somewhat more realistic by assuming that the transfer takes a full fortnight because of the quantity of information involved. In the middle, after one week, the cure is found. The doctors are naturally obliged to treat the original Lenny in the original body for ethical reasons. But this means that at the end of the procedure, we have two people, both of whom feel like the real Lenny, both of whom with the same memories of the same life, both of whom love Mrs. Moscowitz and

Lenny's daughters, and in fact, both of whom would be deeply hurt should they be deprived of the ability to interact with their loved ones.

Who, then, is the "real" Lenny? The original? Neither? Both? If one concludes that only the original is the true Lenny, this violates the assumption that identity transfer is possible. If fact this may be the case—nothing that we have said so far has proven that identity transfer is possible. A full justification that such a transfer is indeed possible awaits further treatment, but for now let us simply state that this scenario is unsatisfactory. We don't want to identify Lenny with his body, but if Lenny is some collection of psychological attributes, what is to stop us from instantiating these attributes in another machine? Furthermore, let us imagine an atom-by-atom copy of Lenny. If consciousness supervenes on the physical, and identity also does, we have no reason to believe that this copy will not have the same identity as the original Lenny. But identity must be some function of the physical, unless one supposes identity resides in a nonmaterial soul. As with consciousness, this is something that we are reluctant to admit, and in fact as discussed in the following text, consideration of pathological states and other neurophysiological evidence provides a good deal of confidence that identity is determined by the brain and is altered as the brain is altered.

This leaves the possibility that neither is the real Lenny, or both are. The first case is absurd. We wouldn't want to say that the original Lenny somehow loses his Lennyness if it so happens that his original body is preserved. Assuming that the operation is nondestructive, as in say, the atom-by-atom copy, then nothing changes from the first Lenny, and he must therefore have the same identity. The only remaining possibility, then, is that two Lenny's coexist. But this contradicts one seeming incontrovertible principle of identity, namely its uniqueness. If you are not unique, then what are you? Remember, we are not talking about two people who merely share the same memories, beliefs, desires, and so forth. By construction we mean two people who believe and feel like the original Lenny. Each one, not merely the original, has genuine claims to be Leonard Moscowitz.

To get a better feel for the counterintuitiveness of this situation, consider it from Lenny's point of view. Let's say that he knows in advance that a copy will be made. If both are truly himself, or should we say his self,[1] maybe he should divide his savings in half and bequeath one half to the copy. But after the transfer takes place, it also seems that the original would say that "there is another guy in

[1]The inadequacy of English to describe this situation is a good indirect indication that our intuitive notions of the self are not adequate to the task at hand. In fact, as discussed in the section on deflationary accounts of the self, ordinary language may be positively misleading when it comes to identity.

the room that certainly acts like me, but I see no reason to give him my money as opposed to any other person." Who then is right, the original before the copy, or the one after?

Or consider this. Suppose that a somewhat malicious benefactor comes along, before the copying takes place, and gives the original Lenny the following option. He states that if the operation goes ahead, he will agree to: a) cure the original Lenny of ALS and give him 10 million dollars, and b) force the copy to work as a middle school janitor eighty hours a week for the rest of his life. Lenny will naturally be tempted to go through with the operation, reasoning that he, as the original, will benefit. But is this correct? If the operation truly works, then the situation ought to be symmetrical. That is, it will be as if the original were forced to spend all his waking hours cleaning up after preadolescents, and the copy were given a life of luxury. How, then, can Lenny choose between this and other similar scenarios?

Finally, consider the following situation. The transfer operation is a success. However, Lenny's company discovers that in order to pay for the operation, he embezzled the firm out of two hundred thousand dollars. Who, among the two Lenny's, should be punished for this crime? The copy's lawyer will of course try to implicate the original Lenny (and the jury may very well fall for this argument), but this is unsatisfactory. Criminal justice implicitly depends on the continuity of personal identity through time—we punish the future person for the crimes of the same person in the past. If the copy has the same identity as the original, he stands in relation to the original as a future self stands in relation to a past self, and therefore he bears the same blame for the crime. Do we then punish both? What if the crime was murder, and the sentence was the death penalty? Do we execute both copies?

The scenario in Figure 5.1.C provides a host of different but equally troubling questions. In this hypothetical case, a lightning strike or other malfunction causes Lenny to be only partly transferred to his new body; as in scenario A, the original body dies before there is a chance to extract out what is missing. Just how much of the transfer needs to take place before we can say that it was successful? Clearly, we cannot hold that personal identity is equivalent to complete identity, and that any tiny malfunction renders the operation invalid. Otherwise, we would not be identical to ourselves from month to month or even minute to minute, as old memories are lost, new ones are gained, and the brain is molded and reformed to meet new circumstances.

Let's assume, for example, that the transfer was almost complete, but because of some quirk in the copying process, Lenny lost all his knowledge of the reruns of *Hogan's Heroes* that he watched as boy. It is unlikely that the excising of this aesthetically challenged drama would pose a catastrophic loss to Lenny, given their minimal impact on his personal development. We can also agree that if he lost the memory of his first kiss with Janice Newhart under the bleachers in his middle school gym, and all that this entailed for his identity, that this would be potentially more disruptive. As we will see, certain life events are more likely to be recalled than others, and these usually have strong emotional overtones. However, it would have to be some kiss for its absence to create a brand new Lenny; he would still have *a* first kiss, it would just be with someone else (possibly his wife).

We have a big problem, though, once we start down this path. We can ratchet up the deterioration to any desirable level, from small memories to the kind of devastation seen in Alzheimer's. Starting with the loss of memory of his first girlfriend, we could move to the loss of his first wife, his current wife, his children, house, the fact that he is an accountant, etc. At some point, the being that we have created can no longer be said to be Lenny Moscowitz. Certainly if only 1% of the old Lenny is preserved, then the operation will be a failure. The difficulty is where to draw the line between creating a copy with defects but identity intact to something that is little more than a sentient shell, aware of sensory input but without its prior sense of a definitive self. Any percentage, or for that matter, any purely qualitative account (losing first base adolescent activity but not second base activity, for example), appears arbitrary and capricious, but at the same time we know that if the upload is seriously degraded, then identity will not survive. Without a plausible answer to this question, we cannot answer the upload identity problem, and without answering this, we cannot begin to propose a technological solution to the preservation of identity in digital form.

In the final scenario in Figure 5.1.D, aspects of identity are both subtracted and added by the transfer process. Oprah Winfrey, who is not ill but merely wants to undergo a body renewal for business purposes, is having the same operation in the room next door. A hardware error cause a short circuit between the wires leading from the old to new Lenny, and from the old to the new Oprah, such that Lenny receives some of Oprah's identity and she some of his. As in the prior case, the proportion of the "false" identity received is crucial. If Lenny simply was unaltered but became a slightly better listener, it would be reasonable to conclude that the transfer was a success. But what if Lenny now remembers having a difficult childhood in Wisconsin and Tennessee, and in the process of becoming a black beauty queen overcoming many of these difficulties? The problem is not simply that Lenny may now have the conflicting memories "I was a beautiful woman from the South" and "I was an awkward teenager with glasses from Long Island"; it is that Oprah's rise from early travail to pageant fame to television star is part of her essence, and to the extent that Lenny now has that essence, he *is* Oprah, or at the very least, a fusion of Lenny and Oprah. He may, for example, have the sudden desire to start a magazine called "L" that concentrates on articles about the love lives of suburban middle-aged men. Conversely, Oprah may find herself without the empathy chops that she now has in abundance, and in that case could she really and truly be the same Oprah?

In summary, well before one confronts the technical difficulties of uploading a personal identity to an alternative vessel, a number of conceptual problems arise that must be addressed. These include the fission of one identity into two, the partial copying of an identity, and the fusion between identities. The purpose of this chapter is to review the current literature on personal identity, both philosophical and scientific, with the aim of creating a synthesis of these various accounts. The insights gained in the process of forming such a synthesis will prove invaluable in providing a solution to the corresponding engineering task, as will be discussed more fully in Chapter 11.

PHILOSOPHICAL FOUNDATIONS

Historically, strategies for defining personal identity have centered on three approaches: the simple view, in which the self is considered to be an irreducible and unfragmentable object; the view based on psychological continuity, which states that the self is a thread or collection of threads comprising psychological variables; and organic continuity, which connects identity to the brain and/or the body. We will also examine two other strategies, the first of which includes deflationary accounts, which either deny the existence of the self or at least try to minimize the ontological status of the self. We will then take a slight detour from the Western philosophical canon and look at the concept of self in Buddhism, which shares certain features with the deflationary view, but is sufficiently unique and well-developed as to provide its own set of helpful insights.

The simple view

The simple view states that personal identity is not something that can be defined in terms of any "ordinary" scientific variable, such as the state of the body, or the workings of the mind. Identity, if it is reducible to anything, is reducible to a thing outside our ken, a nondestructible essence not following natural law. It is this invariance that permits identity to persist over time, as changes occur in the body, in the brain, and even in the mind of the owner of the identity. Thus, the adjective "simple" in the phrase "simple view" refers to the nature of the identity—it being without parts, and incapable of being fragmented—and not the complexity of the view itself, which as it happens, is anything but simple.

The idea of an invariant and persistent self is consistent with the notion of the soul as found in many religions, and also with popular views of the self.

In the movie *All of Me*, for example, a bowl containing the soul of the Lily Tomlin character falls from an office building and strikes the Steve Martin character on the face. He gradually comes to realize that he no longer has full control over his body, and that Tomlin's character—her soul if you like—is now competing for his muscles and joints. This is confirmed when he looks in the mirror and he sees her face instead of his. Leaving aside this far-fetched cinematic device, the scenario as a whole is consistent with the simple view. The self, in this conception, is an object that exists apart from the body, and following unknown laws, attaches itself to a given body some time after conception. In principle, it could also be liberated from its native host, and enter into the body of another, as in the conceit of this film.

Thus the simple view appears, superficially, to be congenial to the goal of this chapter, namely to argue that identity transfer is a very real possibility. However, there are serious problems with the simple view, both from the point of view of its philosophical foundations, and also as the basis for technological manipulation of the self.

In particular, most versions of the simple view inherit many of the difficulties of the kind of substance dualism discussed in the last chapter with regard to the mind/body problem. If there is a soul that exists apart from physical and mental acts, then how and why does it attach itself to the body? Does it float into the body sometime after conception, and then float away after death, possibly to be resurrected in an afterlife? Does each chicken and each rabbit and each beetle get a soul also, or perhaps more aptly for current purposes, at what point would a machine, or a cyborg-like extension of the brain get a separate soul to itself? Finally, how could we possibly approach the identity transfer problems posed at the start of this chapter? The simple view claims that a transfer occurs if, and only if, the soul makes the leap, but in that case we may as well conclude that we need to find a guru to affect soul transfer as in *All of Me*; we have no scientific basis for proposing any other mechanism.

The simple view precludes any conceivable supervenience relation on either the body, brain, or as functionalism would have it, the workings of the brain. It is not just that this impedes scientific and technological progress; it is that this view is inconsistent with the facts on the ground. Consider one common result of a stroke in which the stroke victim considers the left half of his body to no longer belong to him. It is fair to summarize this as organic damage resulting in an alteration of identity. In addition, a stroke of sufficient magnitude and placement can and does alter personality, body image, memory, preferences and beliefs, and all the elements that we ordinarily take to be constitutive of identity. Furthermore, if the resulting lesions are sufficiently severe, we are forced to conclude that the unfortunate victim is no longer the same person as they were prior to their affliction.

Thus, the nature and quality of identity does in fact supervene on the brain and its processes. If for metaphysical, religious, or ethical reasons[2] one still wants to maintain the notion of a soul, that is a matter beyond the scope of the current discussion. But it should be clear from the preceeding text that such a view is not doing any work for us, and in particular, says little or nothing about the kinds of technological issues regarding identity with which we will shortly have to wrestle if the predictions of the coming chapters hold.

Psychological continuity

If the simple view is mistaken, as it most likely is, then the most natural fall-back is a view based on psychological continuity: we are who we are because of some sort of continuity in our psychological makeup. This view is historically associated with John Locke (2007), although it has many modern proponents (Noonan, 1989), and like functionalism in the field of consciousness studies, it is the default for those in the cognitive sciences. All elements of human psychology are fair game for the maintenance of identity, including hopes, desires, beliefs, and personality characteristics, although the continuity of memory is most often cited as the means by which an identity coheres over time.

Why is this the case? Consider the following motivating scenario, which plays out every day for every human, and possibly some animals. Jane Dozer wakes up from sleep and is initially groggy, possibly at that time not knowing where she is, or even who she is. But shortly thereafter her senses recover, and she begins the day by brushing her teeth, making the coffee, etc. Two related questions that we can pose as identity theorists, not as difficult as the thought experiments given at the start of the chapter, but by no means trivial, are as follows: What makes Jane the same person as when she went to sleep? and why does she not begin to question the fact she is the same person (unless Jane happens to be an identity theorist herself)?

We cannot pin our hopes on spatial continuity. Let's say Jane is a student at MIT, and her friends pull a prank on her. The night before, they slip her some sleeping pills so that she doesn't wake up, and during the night they move here bed and her entire belongings to the main quad. Certainly she will be startled on waking—that is the purpose of the prank—but not because she thinks she is a

[2] One of the reasons that many religions advocate a simple view of identity is so that the persistent immaterial identity can be rewarded or punished in the afterlife. Regardless of one's driving philosophy, however, notions of punishment and reward are intimately bound up with identity. For example, we will someday need to decide if someone should be punished for a crime they committed when they inhabited a different body; this turns on what it means to be the same person over time.

different person. Likewise, because of sleep, coma, and other unconscious states, we cannot claim that continuity of consciousness alone is necessary for identity. A break in such continuity does not constitute a break of identity. Finally, we probably do not want to claim that her body need remain largely identical between the night before and this morning (but see the next section). There will always be small changes in bodily identity over time, and moreover, if Jane were unlucky enough to lose a finger during the night she would still be the same Jane in the morning, although naturally distraught.

What does remain largely invariant, regardless of these external circumstances, is Jane's psychological makeup. We can distinguish two aspects of this makeup that remain more or less constant. The first is the external, or third-person view of Jane. She appears to have the same personality on waking as the night before, and indeed, today as the same day last year. The other is Jane's view of herself, or her first-person point of view of herself. Memory plays a large role in this conception. When Jane wakes up one of the reasons she believes she is Jane is because she remembers going to bed as Jane the night before. It is possible, of course, that her memory has been erased by alcohol or other means, but in these cases she still remembers a not too distant time in the past when she was Jane, and not Jean or Jake. Memories can be continuous where consciousness is not, because memories are not bound by real-time constraints, although they are typically influenced by the flow of time (we would not expect Jane to have the same feeling of continuity after waking from a seven-year coma as she normally does in the morning). Memories, as brain-based events, are also independent of the body, and can thus serve as a bridge between aspects of the self as the body changes, or if it changes rapidly.

Locke, for example, imagines a prince waking up in a cobbler's body. Locke's intuition, probably sound, is that the prince is still a prince if he remembers being a prince the night before, and has enough other memories of his former princely life, even if the villagers refuse to bow down to him as he walks through the muddy streets in the morning trying to make his way back to the castle. Given these considerations, the psychological view, and especially the notion of continuity of memory, have a ring of plausibility to them. Despite the concordance with intuition, this view still faces major challenges. These include:

a) Transitivity

   Jane remembers first taking an interest in philosophy in tenth grade, and as a tenth-grader she remembers her third-grade piano teacher, but as an adult, in her current state, she cannot remember her piano teacher. This implies, given that memory is the strand that connects identities over time, that current Jane is identical to tenth-grade Jane, and that tenth-grade Jane is identical to third-grade Jane, but that current Jane is not identical to third-grade Jane. But this appears to lead to a contradiction, given that identity is transitive: if a is identical to b, and b to c, then a must be identical to c.

b) Circularity
Jane has many memories of the tenth-grade, including playing on the lacrosse team. She also remembers her best friend and teammate Jean scoring the winning goal in the league championship game. That memory may be part of Jane's identity, but only so far as she remembers what it felt like to see Jean score the goal, and obviously not because she remembers scoring the goal herself. Therefore, memories must be separated into two camps, autobiographical, and memories about other people or objects, if they are to be informative about identity. But if this is the case, we are appealing to a special property of memory to solve the identity problem, and we have created an apparent circularity: identity depends on memory, but memories must be classified on the basis of identity to be useful to the construction of identity. To put it another way, what stops your hard drive from having an identity? It clearly does not, but we could certainly trace the continuity of contents from moment to moment. What it is lacking then is not continuity, but the feeling of continuity, but this is something different than a memory-based property.

c) The prior *Gedanken* experiments
The previous two difficulties, although not fully resolved in the literature, have the aura of semantic confusion, in the sense that if we could merely describe identity in the right fashion, they might fade away. This is not the case for the thought experiments that we introduced at the start of this chapter: there is a deterministic mater of fact about whether one will survive a given identity transfer or not, and whether a transfer could ever, in principle, achieve such a result. The problems posed by these hypotheticals are particularly acute for a psychological approach, precisely because this approach strongly suggests that transfers are possible. If identity is a mater of psychological variables x, y, and z, transferring x, y, and z to a different vessel should make that vessel have the same identity. But then we run against our old friends: how much psychological stuff needs to be transferred in order to achieve a successful transfer, and what are we to make of multiple copies of the same identity.

The simple view, however implausible, at least does not suffer from this latter difficulty. In this account, the soul presumably attaches to a single body, one at a time, and if it is dislodged from that body by any means, it will carry itself whole to its new body, leaving its former body without a soul (unless one rushes in from somewhere else to replace it). Another approach that is also immune from this difficulty is considered next.

Organic continuity

It is all very well to consider the sorts of experiments that may or may not be one day possible, but what of the normal default case asks the organic identity theorist (see, e.g., Snowdon, 1991, and Olson, 1997). Is it not usual for people to be identified with their bodies[3]? While it is true that the body constantly changes, there is a substantial amount of overlap from one moment to the next, and in the same way that psychological continuity need not rely on exact identity over time, bodily continuity could do the same in order to achieve its theoretical aims. One problem that this common sense theory is meant to address is the fact that we do not, as a matter of course, think that we are distinct from the person that we were when we were an infant, or a fetus. But it is clear that we do not have memories extending back this far, and therefore psychological theories cannot explain this identity relation. Likewise, if someone slips into a coma, we do not think of them as a different person, anymore than we do if they are sound asleep. But in both of these states, there is no psychological activity, and therefore there cannot be psychological continuity.

Despite the undeniable simplicity of the organic view, it has one fatal flaw, and this is what drove us into the arms of psychological view in the first place, whatever its defects. If we transfer Jane's brain into Jean's body, the organic view is committed to claiming that the new person will be Jean, not Jane. But how could this be if the new person acts like Jane, talks like Jane, and most importantly, believes herself to be Jane.

We can also imagine a bionic Jane, in which her body parts are slowly replaced. Does she attain a bionic identity when enough of these replacements take place?

The intuition is that Jane will remain Jane in this case, just as giving someone a prosthetic arm does not change his or her identity. The same intuition would likely claim that the organic view smacks of philosophical desperation: in order to avoid what are admittedly unruly difficulties, it seeks to eliminate or minimize the importance of the variables that comprise identity. In this sense, this view

---

[3] The organic view must be distinguished from another view that is superficially similar. This is the notion that at least some of the psychological variables relevant to identity are stored in the body rather than the brain. For example, heart transplant patients sometimes claim to have acquired the personality traits of their owners (e.g., one hears things like, "After my transplant I suddenly started preferring sweet to dill pickles, and lo and behold, I found out a year later that my donor also liked sweet pickles"). This is really a variant of the psychological view, only with psychological elements supervening on aspects of the body other than the brain. Identity is still a function of variables, such as the preferred pickle variety, regardless of what body part is generating this preference.

is the theoretical counterpart of eliminativism with respect to consciousness. Interestingly, what may be thought of as the more direct counterpart of eliminativism, the deflationary account that is presented next, is more difficult to dismiss out of hand, and in the synthesis of the views presented, it will retain a prominent role.

Deflationary accounts

By a deflationary account of identity we mean one that either thinks of the self as a kind of conceptual illusion, or treats the self as real, but as a mere collection of elements. To be sure, the members of such a collection are conceptually related, and may even be causally intertwined, but this does not imply that the collection itself is a real object, and certainly not one that has an ontological life apart from its individual elements. As an analogy, consider a waterfall. As a matter of conceptual convenience we may think of the waterfall as a distinct object, and when it is sufficiently important or large we may even give it a name, such as Niagara Falls. But a waterfall has no real existence beyond the collection of moving water molecules moving more or less in unison that are its constituents; the term "waterfall" is a wrapper that enables us to carry and manipulate an otherwise unwieldy set of elemental parts. Likewise, as the identity minimizer will argue, the self is a related collection of memories, beliefs, fears, hopes, etc., but nothing whatsoever beyond this, except as a conceptual wrapper. Crucial to the minimizer's argument is the notion that this wrapper is something that is constructed not just by external observers, but also by the person supposedly in possession of the identity himself. At no point and for no one, then, is the self anything more than a convenient fiction.

The chief motivation behind a deflationary account of identity, like its counterpart in the study of consciousness, eliminative materialism, is to reduce conceptual problems not by addressing them head-on, but by minimizing the significance and import of the concept itself, in this case, that of a notion of identity above and beyond its constituent parts. If identity is not quite real, then problems arising about its nature cannot be quite real either. Consider, for example, the difficulties arising in the thought experiment in Figure 5.1.D, in which the identity of the transferee became mixed with that of Oprah Winfrey. The deflationary advocate might claim something like, "Not a problem, and we can throw in a chunk of Regis Philbin for good measure". The resulting identity may be unusual, but it is no different *in kind* than any other personal identity, which is always an admixture of genetic determinants of one's personality plus the life events that further modify or constrain these determinants. Thus, any answers regarding the success of identity transfer cannot be given because the question is not well-formed in the first place. Or, to return to our earlier analogy, it makes little sense to try to determine when one waterfall becomes another sort of waterfall via erosion of the underlying rocks. It is always a collection of independent objects united only by virtue of the common movement of those objects.

The words of the great Scottish philosopher and historian David Hume give some idea as to why the skeptical account of identity has so much more force that its deflationary counterpart in the mind-body problem, eliminative materialism, and why it continues to have resonance in the philosophy of mind to this day:

> "For my part, when I enter most intimately into what I call *myself*, I always stumble on some particular perception or other, of heat or cold, light or shade, love or hatred, pain or pleasure. I never can catch myself at any time without a perception. When my perceptions are removed for any time, as by sound sleep; so long am I insensible of myself, and may truly be said not to exist" (Hume, 1978).

In other words, Hume claims, introspection does not reveal anything other than a succession of perceptions. Try as we might, there is no direct perception of the self, although naturally in any given person thoughts about one's self will be plentiful, or even predominate, depending on the level of egoism present. But these thoughts are, in essence, just like any thoughts, although they are centered on a putative object mistakenly believed to have an independent existence over and above this stream of consciousness. As Hume also stated, then, what we call the self is just a "bundle or collection of different perceptions."[4]

A natural corollary to Hume's thesis is that the simple view, such as that advocated by his predecessor Descartes, is based on a misconception. Hume has this to say in the paragraph following that quoted earlier:

> "The mind is a kind of theatre, where several perceptions successively make their appearance; pass, re-pass, glide away, and mingle in an infinite variety of postures and situations. There is properly no *simplicity* in it at one time, nor *identity* in different; whatever natural propension we may have to imagine that simplicity and identity."

If the simplicity of mind is an illusion, why then is it such a powerful one, and why is it one to which common sense invariably succumbs? This is properly a psychological question, and will be treated in more depth later, but the famous

---

[4] Although it is certainly a psychological and therefore contingent truth that introspection does not reveal the existence of a self, we can also pose the question as to whether it could be any other way. That is, is there any possible world in which one could think a thought such that that thought reveals anything about the ontological status of the thinker other than minimal conditional fact that if such a thinker did exist, he has just thought a thought?

analogy of the mind to Oxford or Cambridge offered by the twentieth century philosopher of mind Gilbert Ryle offers a high-level explanation:

> "A foreigner visiting Oxford or Cambridge for the first time is shown a number of colleges, libraries, playing fields, museums, scientific departments, and administrative offices. He then asks 'But where is the university?'... It has to be explained to him that the university is not another collateral institution, some ulterior counterpart to the colleges, laboratories, and offices which he has seen. The university is just the way in which all that he has already seen is organized. When they are seen and when their coordination is understood, the university has been seen. His mistake lay in his innocent assumption that it was correct to speak of Christ Church, the Bodleian Library, the Ashmolean Museum, *and* the university... The mistake is made by someone who does not know how to wield the concept *university*... The puzzle arises from an inability to use a certain term in the English vocabulary" (Ryle, 2000).

In the same way, Ryle argues that the mind as a reified entity is a conceptual mistake. Like any university, and especially like one of the Oxbridgian variety, comprising a number of more or less independent colleges, the mind is a collection of entities to which an appellation may be applied, but only for convenience. Note in particular that this sort of view is antithetical to the provision of a genuine answer to the question posed at the start of the chapter regarding the success or failure of identity transfer. In order to say that person X is still person X after transfer to a new vessel it must be more than an issue of convenience to give the new entity the label "X"; it must have something to do with how the new entity feels about the operation. That is, it is more of a first-person issue than a third-person one; this is an issue that will be reprised in the synthesis section that follows.

Daniel Dennett, a former student of Ryle, has assumed the present-day mantle of the identity skeptic. Dennett's (1992) stand on identity cannot be cleanly separated by his deflationary approach to consciousness in general, although we will, for the sake of argument, do so here. His main claim, echoing Ryle and Hume before him, is that the self is a useful abstraction, but an abstraction nonetheless. Dennett's illuminating analogy is that of the notion of the center of gravity of an object. It is a mistake to think of the center of gravity of a chair, say, as an object in the universe apart from the constituent particles of the chair and the laws which drive them. Nevertheless, it is an extremely convenient abstraction: if we were asked how difficult it would be to knock the chair over, this would depend on how high the center of gravity was off the ground, and also on how off-center this location was relative to the direction in which we strike the chair. In principle, the same calculation could be done by simulating the interaction between the strike and the elementary particles in the chair, but only by an extremely fast computer taking a very long time, and in the end, the result would be the same.

Likewise, we do not reason about people by simulating a neural network emulating their brains. Even if we had access to this information, for all but the most fine-grained predictions we are content to create what might be called a narrative center of gravity, with the occupant of this center the person in question. Crucial to Dennett's claim is that we perform this process not only for others, but also for our own selves. Or, to put it in Humean terms, there is no thinker, just thoughts, although the content of at least some of those thoughts are about a hypothetical person who exists apart from such thoughts, who persists over time, who may survive bodily death, or, as in the central problem posed by the UI question, could be transferred to an alternative embodiment.

Insights from Buddhism and Eastern religions

If Western religions are concerned with the elevation of the self or soul to a higher spiritual plane, in either this life or the next, it is left to Eastern religions, and especially Buddhism, to deny the very reality of the self. This is similar to the deflationary view, except that the deflation here is considerably more severe—the self is relegated to the mere wisp of the wind, a transitory concept ultimately without import or significance. In addition, and this is as valid for Hinduism as well as Buddhism, the illusion of self is something to be overcome, not simply recognize. It is not sufficient, as was presumably the case for Hume, to realize fundamental inconsistencies in the concept of self, and then continue on with one's personal ambitions. If Hume were a true Buddhist, for example, he would not attempt to achieve philosophical fame and ego gratification via his skepticism; why reinforce something that is not real? One must work in this lifetime to remove the veil of Maya, or illusion, that hangs over quotidian perception, and realize the transcendent self (Brahman in Hinduism) or nothingness (Buddhism and Zen). In both traditions, this leads to stopping the cycle of reincarnation, or Samsara, and to the start of nirvana or enlightenment. Compare this also with the statement of the Taoist sage, Chuang-tzu:

> "The perfect man has no self; the spiritual man has no achievement; the true sage has no name." (Ho, 1995).

Be that as it may, at the very least there is the illusion of self, and here also Buddhism provides useful insights. In the Pali Canon, the collection of Buddhist scripture from the fourth or fifth century BC, the distinction is made between discourse regarding ultimate reality and discourse that is merely conventional (Giles, 1993). The latter form of description is in play whenever we speak of ourselves or another. A name or the pronoun "I" is simply a convenient term that places a number of distinct elements in the same conceptual packet, but has no ultimate significance beyond this utilitarian reference.

The story is often told in this context of the confrontation between the Indo-Greek king Menander and the Buddhist monk Nagasena. The king begins by asking Nagasena about his name, and is astonished to hear Nagasena claim that the sound "Nagasena" does not stand for anything, and what the world calls Nagasena is a mere collection of parts. The king asks how this can be, given that Nagasena is standing there, responding to the king's questions. Nagasena then turns the table on the king by asking him about what constitutes his chariot, whether it is the axle, the wheels, the seat, etc. When the king replies that it is none of these components, but that the term "chariot" is a practical designation for this collection, Nagasena praises the king for his insight. The king then understands that the world is not composed of chariots or people who ride in them, and it is the mind that imposes this order on things for its own convenience.

Thus, the insights of Buddhism and its theological cousins regarding the self predate those of Hume, Ryle, and Dennett by many centuries, and they provide a sketch of why the concept of self persists, despite its illusory nature. Still, one is left with an uneasy feeling about these views, and not merely because they denigrate the one thing most in one's evolutionary interest to protect, the ego. Taken at face value, these views are exceedingly sketchy, and as such, cannot provide answers to the identity transfer problems discussed previously. They cannot, for example, tell us what admixture of Oprah and Regis results in the one or the other, other than to say that both are unreal. But surely it is like something to be one rather than the other, and when Oprah wakes in the morning she believes herself to be Oprah and Regis, Regis. However ineffectual and unnecessary the concept of self from the point of a fully enlightened being is, to the rest of us it is exceedingly real. *Amour propre* drives our emotions, and thereby our thoughts and deeds, and the landscape of both social and individual life would be radically different without it. To make both scientific and technological progress, we must somehow capture what *is* real about the self, although this discussion must await a brief diversion into the psychology of identity.

PSYCHOLOGY AND IDENTITY

The self has traditionally been a poorly researched topic in comparison to other areas of academic psychology, perhaps because it has been considered too difficult a subject to manipulate in the laboratory. The same cannot be said for clinical psychology, and of course, psychologists from Freud on have given the ego a central role in their theories. However, in most of these cases, the self is taken as a given, albeit one that is subject to numerous pathologies and the consequences thereof. It is only recently that the self has been subject to the critical rigor granted other areas of psychology; here we look at a number of topics of emerging importance, including pathologies of identity, the plasticity of the self, autobiographical memory, and finally, some of the latest finding on the neurobiology of self.

Pathologies of identity

The colorful story of the railroad worker Phineas Gage has been retold so many times that it is now a permanent fixture in neuropathological lore. In September of 1848, at the age 25, Gage suffered a horrific accident when a 1 meter tamping iron with a diameter of 1.25 inches passed directly through the front part of his skull. Remarkably, he survived this accident, and after an initial bout with infection, went on to live another 12 years. However, as noted by his attending physician, the effect on Gage's personality was severe:

> *Gage was fitful, irreverent, indulging at times in the grossest profanity (which was not previously his custom), manifesting but little deference for his fellows, impatient of restraint or advice when it conflicts with his desires, at times pertinaciously obstinate, yet capricious and vacillating, devising many plans of future operations, which are no sooner arranged than they are abandoned in turn for others appearing more feasible. A child in his intellectual capacity and manifestations, he has the animal passions of a strong man. Previous to his injury, although untrained in the schools, he possessed a well-balanced mind, and was looked upon by those who knew him as a shrewd, smart businessman, very energetic and persistent in executing all his plans of operation. In this regard his mind was radically changed, so decidedly that his friends and acquaintances said he was 'no longer Gage'* (Harlow, 1868).

There is some controversy regarding both the locale of the injury and the extent of psychological damage; the referenced account, for example, was written a full 8 years after Gage's death. There does seem to be a large degree of truth to the story, however, if parts of it may have been exaggerated. For example, the alteration of his personality was severe enough that his former employer refused to take the formerly dependable foreman back, and his friends also agreed with the main claim of his doctor's account that after the accident Gage "was no longer Gage." Thus, we may safely conclude that a significant change in his personality followed the accident, and it is noteworthy that the damage was primarily frontal in nature.

It would be rash to conclude from this alone that identity can be localized in the frontal cortex. Unlike speech or vision, there is no known loss of identity associated with specific cerebral lesions, although significant and widespread damage, such as that in late-stage Alzheimer's, may largely eradicate the notion of self. There are a number of pathologies of significance in this context, though. The first of these is hemineglect and an associated syndrome known as asomatognosia. *Hemineglect* is typically the result of right parietal damage and in this condition patients will ignore most stimuli from the contralateral left side, including those of the left visual field. *Asomatognosia* is the additional denial that the paralyzed left arm and sometimes the left leg belong to the patient (Feinberg, 2001). When asked whose limb it is, the patient will either plead ignorance or else produce confabulations, such as claiming that it is the doctor's arm, or more startling still, that it belongs to a dead husband (women with asomatognosia tend to think that the useless arm belongs to their husband for some reason; men rarely make the same claim regarding their wives). Left parietal damage does not usually lead to these conditions, as the intact right hemisphere has greater powers to direct attention to both sides of the body.

Why, in the cases of asomatognosia, does the patient simply not say the hand is theirs, although it is paralyzed, rather than denying ownership? We can only speculate as to the reason, and in that we must also be cautious because the brain damage may be accompanied by cognitive as well as perceptual difficulties. Nevertheless, it appears that the active inclusion into the self involves an additional mechanism apart from noticing the spatial-temporal contiguity of the arm with the body. For example, Ramachandran (1998) invokes a similar principle to explain Capgras syndrome, a rather bizarre pathology in which one believes that one's family members or acquaintances are not themselves but look-alike copies. The syndrome may even extend to the patient in question, and he or she may deny that they are the person they see in the mirror. Ramachandran argues that this is the result of a blockage of interaction that ordinarily takes place between face recognition in the IT area and the gateway to the limbic system, the amygdala. The patient is able to recognize the similarity between the "imposters" and the actual people, but because this information does not flow to the limbic system this is not accompanied by the emotional quality of familiarity. One piece of evidence for this hypothesis is that Galvanic skin response, modulated by limbic activity and indicative of emotional arousal, is flat for these patients in the presence of these body doubles.

Other more extreme disassociations between brain and body, though rare, may also manifest themselves depending on the extent and location of brain damage. At the start of the twentieth century, the German neurologist Kurt Goldstein described a female patient who claimed that her hand had a will of its own. The hand would rip off the covers of her bed while she was trying to sleep, and even put its hand around her neck and try to choke her. Feinberg (2002) also described many other cases of so-called alien hand syndrome. He describes one patient, Stevie, who referred to the hand that he could control, the left, as the good boy, and the one that was alien, the right, as the bad boy. The bad boy would often perform inappropriate actions, such as sexual acts, on its own, while the left would try to control it. These hands would then engage in fights, with each attempting to achieve dominance.

Later in this chapter, a theory of self will be proposed in which the self is first and foremost a kind of emotional sensation, as oppose to a purely cognitive one. One does not so much perceive the self as feel that it is present, whether this presence consists of bodily image, memory-laden concepts, or some combination of the two. These syndromes are consistent with this notion, in that they show that if this sense of belonging is removed, then the item in question is perceived as other or foreign, regardless of the strength of the objective evidence indicating that it belongs to body.

The plasticity of self

Not only may the self may be contracted, it also may expand to arbitrary boundaries in either physical or conceptual space. Let us first consider this topic from the point of view of common sense. Two representative examples will suffice. The phenomenon of road rage is one that appears to be increasing, although whether this is a result of the general decline in civility or is an artifact of increased reporting is unknown. In any case, it has probably always been with us in one form or another since people began driving. It is natural that one wants to protect one's possessions, including expensive ones like a car. Nevertheless, the sheer viscerality associated with road rage, and its more common variant, road irritation, strongly suggests that these reactions are more than simply those related to ownership. It is almost as if, when driving, we are not sitting behind the wheel and steering our possession, but rather it is our egos that are cruising down the highways, in constant danger of getting bruised by collision with other egos on wheels. This is why other drivers are not merely careless or rude, they are in idiots, bums, or worse, if the shouting is taken at face value.

Another extension of the ego is even more striking. As is readily evident, the sports fan attaches his ego to his favored team, with their losses becoming his losses, and their victories his triumphs, also. What is remarkable in this case is the sheer abstractness of this attachment, especially when one considers professional sports. The only apparent link between the fan and the players he roots for is the fact that their so-called home field happens to be located in his town (or, more remotely still, in the town where he was born). These players themselves are unlikely to be from this town, may be traded at a moment's notice, and it is unlikely that the fan has had any personal contact with any of them. Yet the passion that a true sports fan feels may exceed anything else in his life.

One possible explanation for this type of empathy comes from a recent result showing that so-called mirror neurons fire when either when one performs a particular task, or merely when one watches someone else perform a similar task (Rizzolatti et al., 1996). This suggests that it is not necessary to participate in a sport directly to obtain a similar emotional reaction, and in fact, the higher level of performance of the trained athlete may yield greater satisfaction than the fan's own poorer attempts in the same circumstances. But this cannot alone explain the fan's attachment to his home team, because the mirror neurons should fire regardless of whom he is watching.[5] The simplest explanation for the level of emotion involved is that the boundary of the fan's ego partially extends to encompass his team; this is why he is glum when they lose and why he marches through the streets when they are crowned champions.

A more systematic experiment on ego extension has been carried out by Ramachandran and his colleagues (Ramachandran and Blakeslee, 1999). They had subjects reach out and touch the nose of a collaborator who was sitting in front of the subject, according to a random time schedule, and in synchrony with this action the subject's own nose was brushed by the experimenters. After a relatively short period of time, in up to two-thirds of the subjects who undergo this procedure, the subject experiences the illusion that their nose has grown several feet, Pinocchio-like, to the distance of the collaborator's nose in front of them. This extraordinary effect can also be achieved with dummy hands substituting for the real hands, and even objects that bear no resemblance to the original body part, such as tables and chairs.

The best explanation for cases such as these, as Ramachandran stresses, is not that the illusion is the result of an extraordinary process, but rather that the body is always a phantom that is under constant construction by the brain. Under normal circumstances, one's body image will be coincident with one's actual body. However, the brain can almost as easily be fooled into forming an altered body image when tactile feedback cooperates with visual information to suggest a lengthened or otherwise altered source of contact. As far as the current discussion goes, these sorts of results are significant in that they give credence to the notion that the self is an inherently malleable concept, neither fixed in space, nor, as the sports examples indicate, tied to a single body.

Autobiographical Memory

If the self is, at its core, a psychological notion, the question then becomes what particular psychological aspects are necessary for the construction of this concept. As will be argued in the following, this process critically depends on the mind's ability to bind together a number of related elements into a unified whole. A significant, if not dominant, component of these elements

---

[5] It would be interesting to measure the strength of mirror neuron response as a function of the degree to which the fan is rooting for his team; as far as the author knows this has not been done.

is what may be termed autobiographical memory. This form of memory is characterized as the reexperiencing of an episode that occurred in one's past in conjunction with the belief that this episode was personally experienced (Brewer, 1986).

A crucial distinction is that between episodic and semantic memories. The former are memories of life-events; the latter general knowledge. For example, if you know that snow is white and that if falls in the winter, this is an example of semantic memory; in contrast, your memory of the joy of taking out your Flexible Flyer during the first snowfall of the season as a child is a clear example of autobiographical memory. Furthermore, the collection of such memories, especially when woven into the kind of narrative that Dennett, among others, identifies with the self, are key distinguishers of self from others (cf., Rosebud in Citizen Kane).

One may also distinguish between episodic memories that are autobiographical in character and those episodes that one observes from afar, and do not include the self. This is a somewhat problematic distinction, however, because observed events can be potentially incorporated into the self. For example, the music of one's adolescence often has a special place in the development of identity, especially given that it is associated with other life-altering events. In fact, when asked to produce autobiographical memories, the typical subject produces what is known as a reminiscence bump that centers on this period (See Figure 5.2).

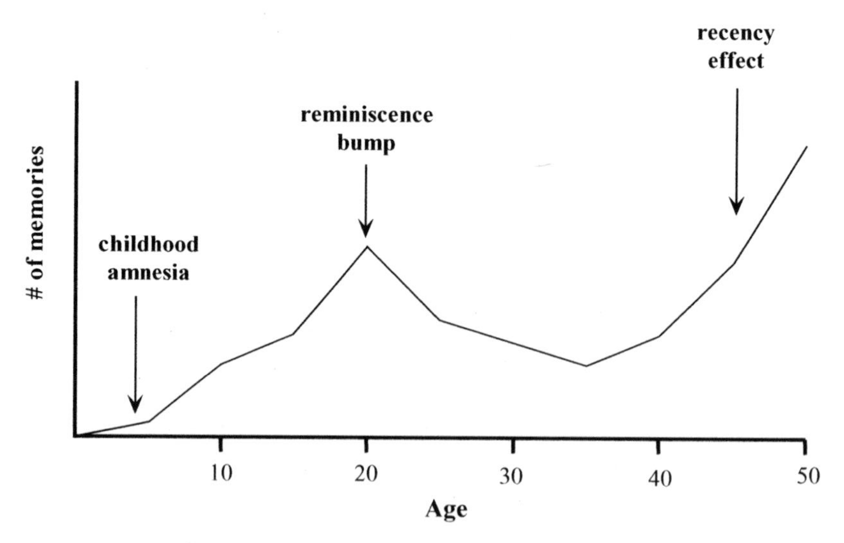

FIGURE 5.2 **In general, more recent events, and events centering on late adolescence, are remembered best. There is very little autobiographical memory for events before the age of five (adapted from Conway and Pleydell-Pearce, 2000).**

SOURCE: **Psychol Rev. 2000 Apr; 107(2):261–88. Conway MA, Pleydell-Pearce CW.**

This graph shows the hypothetical volume of memories for a fifty-year-old. There is a period of relative amnesia until the age of five or so, then an increase at about the age of twenty, and then a recency effect that causes an increase in memories close to the current period of one's life. Possible reasons for the existence of the so-called adolescent reminiscence bump include the fact that events are more novel at this period (one is typically leaving home and making one's way in the world), and/or the fact that it is at this time that the adult personality emerges.

Autobiographical knowledge may be divided into three overlapping categories. The first is lifetime periods, which encompasses distinct swatches of time in which life is more or less stable and is associated with a particular place, set of people, or events. For example, one might, at the end of one's life, describe its trajectory as a sequence of periods from childhood to high school to university to grad school to one's first philosophy professorship and first marriage and concentration on solving the personal identity problem to messy divorce and second professorship and a refocusing on the problem of evil. The second type of autobiographical knowledge encompasses general events, whether they are repeating or not. These include things such as playing tennis, going to movies, working, and the like. Robinson (1976) found a strong primacy effect for such events; for example, almost everyone remembers their first kiss; the ninth and the thirteenth kiss are apparently not as sweet, and less likely to be recalled. The third type of autobiographical memory centers on specific events, including everything from picking up the dry cleaning yesterday to catching the winning touchdown in high school. At least some of these will take the form of so-called "flashbulb" memories, in which imagery from some aspects of the event are frozen in time in the mind's eye, although almost all event-specific knowledge has a strong visual component.

What are the implications of autobiographical memory for identity transfer? Clearly, someone with little or no autobiographical memory cannot be said to have a sustained identity, and if this component is left out of the transfer, it must be put down as a failure. There still remains the question, however, of just how much of this form of memory needs to make it to the vessel before the operation is considered a success. Recall that requiring 100%, or something close to this figure, would be too stringent; on the other hand, stipulating an arbitrary threshold is without motivation. A full resolution to this problem is suggested in the following; here we will simply point out some qualitative features of a successful download. First, all major lifetime periods should ideally be captured. To leave out one's first marriage, however emotionally satisfying that might be, could potentially be detrimental to the survival of one's identity. Second, the transitions between periods are likely to be significant, especially those in which the identity is being formed. This would likely include adolescence, regardless of the painfulness of those memories. Third, firsts of everything should be transferred, including first kisses and sexual experiences, first cars, first apartments, and so forth. Finally, recurring themes in one's life are likely to be important, whether these be simply repeated thoughts, or actual events. If every August you go to a particular beach, this is likely to be an important part of who you are. In general, we can state that autobiographical memories form both the mortar and bricks of the edifice of identity, and capturing these will be critical to the successful identity download.

The neurobiology of self

Current controversy in the neurobiology of self centers on two related issues: first, the extent to which the self is localized in either the right or left hemisphere, and second, the degree to which the self is localized in the frontal or other structures in the brain. Taking the latter issue first, it is significant that among the primates, it appears that chimpanzees and orangutans can recognize themselves in the mirror; gorillas and other monkeys do not (Keenan et al., 2000). Researchers test this by making a mark on the face of the animal while they are sleeping; upon waking and looking at themselves in the mirror, animals capable of self-recognition try to remove the mark. This may be due to the increased frontal capacity of these animals relative to their simian cousins. It is also significant that no nonprimate has this ability, providing further evidence for the frontal theory.

There is also a growing body of imaging evidence implicating frontal cortex in combination with related limbic structures in self-related activities. For example, Kelly et al. (2002) report an fMRI study in which subjects were asked whether trait adjectives applied to themselves or to others. When asked about the former, the medial prefrontal cortex was more active than when judging others. A similar study by Johnson et al. (2002) produced nearly identical results. Subjects were asked either self-referential questions, such as "I'm a good friend," or "I have a quick temper," or more general questions, such as "You need water to live." The self-related queries preferentially activated the medial prefrontal cortex and the posterior cingulate cortex. In another study by Gusnard et al. (2001), when subjects were asked to state how a picture made them feel, subjects experienced increased activity in the dorsal medial prefrontal cortex relative to when they made a more neutral judgment about the picture. A study by Kircher et al. (2000) used either pictures of the subject's face or their partner's face as stimuli; thus, both could be expected to be highly familiar. However, in the self face condition right limbic structures, including the hippocampal formation, insula, and anterior cingulate, as well as the left prefrontal regions, showed increased activity relative to baseline conditions; the partner face showed only increases in the right insula.

Recent evidence also suggests that the right frontal cortex rather than the frontal areas as a whole may be implicated in the processing of self. This is supported by studies of brain pathology; Keenan and Feinberg (2005) report that out of 29 known cases of patients with significant impairment in their understanding of self, 28 had lesions in the right frontal cortex, while only 14 had corresponding injuries in the left. A more direct attempt to assess hemispheric differences was carried out by Keenan et al. (2000). They had subjects watch an animation in which either a famous image morphed from a famous person to a picture of themselves or to a familiar co-worker (Figure 5.3). They were asked to press a button with either the left hand (controlled by the right hemisphere) or the right hand (controlled by the left hemisphere) when they believed that the picture had become them or their colleague. In the latter case, both left and right hands pressed the button at a point about equidistant in the transition (Figure 5.3, right). In the former case (Figure 5.3, left), a similar result was obtained

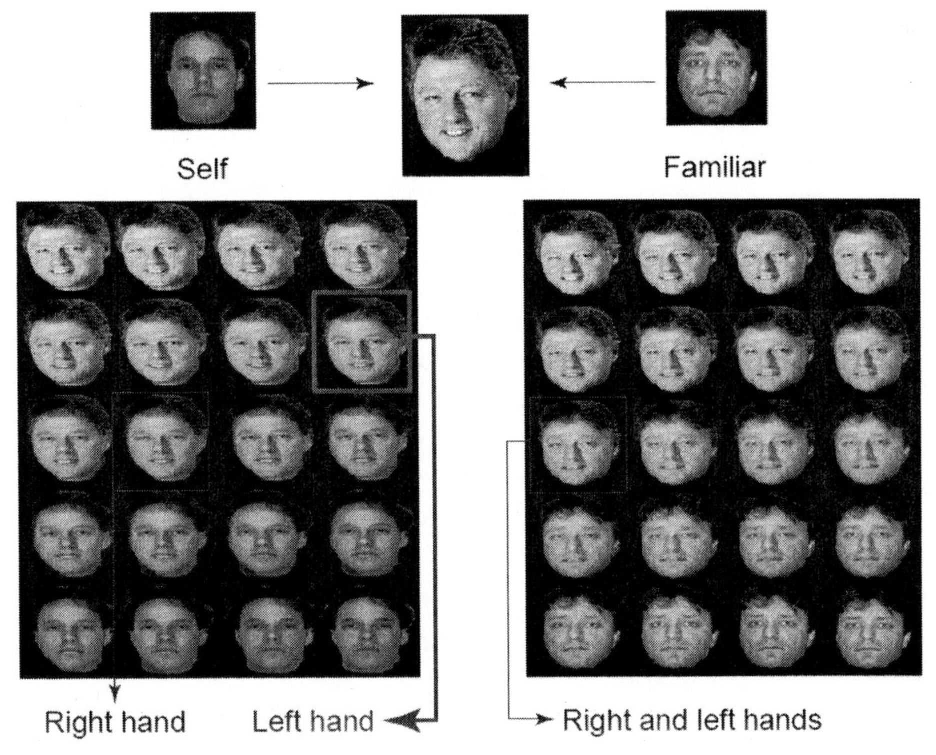

FIGURE 5.3 **When the self is blended with another face, the right hemisphere (left hand) is quicker to identify the self than the left hemisphere (right hand), as shown in the left-hand transition. In the transition on the right involving two non-self pictures (Clinton and a colleague), no such distinction is seen (from Keenan et al., 2000).**
SOURCE: **Keenan, J. P. et al. Hand response differences in a self-identification task. Reprinted from, *Neuropsychologia*, 38, 1047–1053, June, 2000 with permission from Elsevier.**

with the right hand, but the left hand (wired into the right hemisphere) stopped approximately 40% of way to the picture of the self. This would tend to indicate that the right hemisphere has a heightened sensitivity to self, and is more likely to spot the a representation of the self in the environment. Other evidence along the same lines is provided by an experiment in which epileptic patients had one half of their brain anaesthetized. When shown a picture of themselves morphed with another, upon awakening they claimed to have seen their own picture if their left brain was put to sleep, and were more likely to see the other face if their right brain was inactive.

While these data are not without significance, it is still too early to definitively conclude that the sense of self is localized to the frontal cortex as a whole and especially the right versus left frontal cortex. The latter result, in particular, could be an artifact of the experimental conditions; after all, self-recognition is probably only one component in a larger model of the self. It could also be a reflection of more general processing capabilities between the two hemispheres. Nevertheless, it is tempting to speculate that a true sense of self only emerged on the phylogenetic ladder when a species achieved sufficient cognitive power to form a mental model of the self. This modeling ability would naturally be localized to the frontal cortex, given that multiple neural modules converge here, and given that the frontal cortex is implicated in complex linguistic and pre-linguistic representations. The animal kingdom would then be divided roughly into two camps: those with an implicit model of self, and those, such humans and possibly other primates, with an explicit model of self. Organisms with implicit models preferentially perform actions that increase their chances of survival and mating, but they do not, for example, go around thinking that "I am Joe Hela monster, and my wife is Jane Hela monster" (compare with the Tale of the Monkfish in chapter 3). They cannot do this because they do not have the representational capacity to maintain a concept of this complexity, assuming they could form concepts in the first place. We, on the other hand, are generously imbued with such a capacity, enabling us to form a self, to distinguish it from others, and, perhaps uniquely among the animals, to foresee our own demise. In the following discussion, we will provide more details about how and why the model of self is formed, and how this view allows us to approach the problem of identity upload.

A SYNTHESIS OF VIEWS: THE SELF AS GENUINE ILLUSION

While current psychological research may provide hints as to how to go about constructing an apparatus for uploading the "soul" to a machine, and indeed we will use many of these results as leverage in the construction of such a hypothetical device in chapter 11, they are too weak in themselves to directly address the central issues raised at the start of this chapter. As you may recall, this included the problems of multiple uploads of a single self into different vessels, of partial downloads, of the placing of multiple selves into the same vessel, and above all, the question as to whether uploading of this sort is even possible, and, if so, how would we know.

To properly address these issues we must reexamine the philosophical treatment given earlier. Here, we hit upon a dilemma of sorts. It is unlikely that some form of the simple view holds. First, we have no direct evidence of a soul that exists apart from the brain and that can survive the destruction of the brain or its alteration to any extensive degree. Of course, we will only learn whether this is correct after our own deaths, or, less probably, if someone in the afterlife, like Houdini, is able to send us a message to that effect. Next, it violates the principle of parsimony to posit such an entity without good cause. Finally, and most importantly for the scientific argument, personal identity, like consciousness itself, is as far as we know completely supervenient on brain processes. This is indicated by pathological cases, such as the radical alteration in Phineas Gage's personality after his accident, as well as in nonpathological ones. Whatever model of the self turns out to be correct, it is likely that it is mediated heavily or exclusively by frontal circuits, possibly in consort with associated limbic circuitry and stored memories throughout the cortex. Alter any one of these modules, or the nature of their interaction, and you will produce a different personality, and if they are altered sufficiently it will be as if a different self has been produced.

However, if we abandon the simple view then we are left with what appears, in Hume's terminology, to be just a bundle of perceptions. In this view, as in the Buddhist conception, the self is but an illusion. Apart from the fact that this is unsatisfying at a personal level—we want to believe in our hopes and dreams and that striving to reach them is not more than delusional—it does little to answer the UI question nor does it square with common sense. There are better and worse illusions, for example. Some would much rather be under the illusion that they are the King of France rather than one of his pox-raddled starving minions. More germane to current concerns, there are better and worse uploads, and we need a means of distinguishing between the two.

To make this more concrete, let us say that Bill Gates discovers that he has a fatal disease in which he has only two years to live. This disease will not affect his mind until the very end, and he agrees to pay you a vast sum if he can be uploaded into a robot. He is not so much worried about the physical nature of the embodiment; it will always be possible to improve this later, assuming everything else turns out alright. His primary stipulation is as follows: you will only get paid if it is, to him, as if he went to sleep and woke up in a different body. Let us leave aside the question as to what happens if he says that the operation was a success when it wasn't, and let us say that he will only say this if this is what happens (if you didn't mind being underhanded, you could program the new vessel to say that it was Bill Gates and that you should receive your payment immediately, regardless of what really happened).

At this point it is tempting to offer the possible solution to the UI problem that is consistent with the bundle theory. One could simply say that the likelihood of transferring Bill Gates to a souped-up desktop, presumably running a proprietary version of Vista, and his assenting that this indeed occurred, is proportional to amount of the bundle that crosses over. The first problem with this answer is that there is most likely a nonlinear relation between the proportion transferred and the probability or degree of success: a ninety percent transfer of Bill Gates could very well result in a new entity that bears little resemblance to the original. Limiting ourselves to working only with memories, let us assume that we are able to transfer everything except Bill Gate's memories of computers. It seems reasonable that he would be losing on the order of one tenth of his memory content. Although his life coincided with rise of the personal computer and he was one of the chief architects of this ascension, there is his childhood before Microsoft, his family, all of the people he has known, famous and otherwise, issues of accountancy at Microsoft, memories of watching the sunset over Lake Washington (the location of his sprawling mansion), etc. Yet to leave out this crucial ten percent is most likely to diminish his identity considerably, if not entirely destroy it.

A second related issue is that not all elements of the bundle will have the same psychic resonance; some will be more important than others. Let's say that Bill Gates went on twenty-four dates with his wife Melinda before proposing on the twenty-fifth. Not all of these will be emotionally equivalent. Presumably, the first and the twenty-fifth will be of much greater importance than the others. To some extent this will already be reflected in the vividness with which they have been laid down in memory; as discussed earlier, emotionally significant events are more likely to be recalled later, and are therefore more likely to be retained. But there is no reason why some quirk of the process of the creation of autobiographical memories couldn't cause a particular date, say the fourteenth, to be remembered as well as the significant ones. But to lose this date would not alter Bill Gate's essential nature, whereas to lose the first or propositional date might just do so.

These two problems, however, pale in comparison to the final one. This problem centers on the cohesion of the bundle, and how this cohesion relates to personal identity. To elucidate this, let us return to Hume's treatment of identity. Recall that Hume turned his gaze inward and searched for a crystalline Platonic entity that contained his essence. Finding none, he concluded that the self was just a bundle of sensations.

First, let us state what was right about Hume's approach. He correctly, if implicitly, realized that identity is a first-person and not a third-person question. That is, it is a psychological fact, as the theory of psychological identity claims. It makes little difference if an imitator is able to fool others regarding his or her identity. Let's say that someone that looks a lot like Elvis starts appearing in supermarkets in the South and that people see him purchasing tubs of peanut butter and bananas. Further suppose that the resemblance is extremely good—this person is just what Elvis would like now, he has the right accent, his blue-black mane is now largely gray, and so forth. In fact, the resemblance is so good that the story manages to migrate from the tabloids to the front page of the *New York Times*. No matter how many are fooled, however, the identity of Elvis will only be preserved if indeed it is the true Elvis, and he did not die on the afternoon of August 16, 1977. It does not matter that this is the best Elvis imitator among the many thousands that have tried to be the King.

Thus identity, as Hume realized, is ultimately a meaningful entity only to the possessor of that entity. Where Hume went wrong was to assume that the only evidence of the self that could be obtained by introspection was one that delivered something beyond sensations. To see this we need to take a brief detour into how the mind binds together different elements of a stimulus or an internally generated thought. In Chapter 2 we discussed possible mechanisms for binding, including binding via convergence, and binding via synchrony. Here, we are not so much concerned with the mechanisms behind this effect, but with a high-level description of binding.

Well before the cognitive revolution that began in the 1960s, there were a group of psychologists associated with the so-called Berlin school, including Koffka, Wertheimer, and Köhler, who took the inner workings of the mind very seriously. Their methodology stood in marked contrast to the atomizing tendencies of the English-speaking counterparts of their day. For example, one of their chief concerns was how elements of a perceptual scene are bound together into a unified whole. Their main conclusion was that that the mind is governed by what they termed *Pragnanz*, or simplicity, and attempts to understand the scene by forming groupings that are as simple as possible and also consistent with the perceptual elements. Figure 5.4 illustrates a characteristic example of Pragnanz at work. The sets of dots on the left are clearly perceived as two intersecting

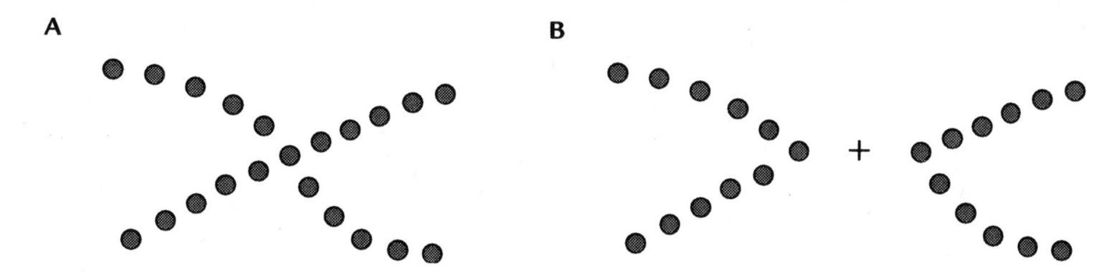

FIGURE 5.4 **A: The law of good continuation dictates that two intersecting lines are seen. B: The alternative grouping is not perceived under ordinary circumstances.**

lines, and not, for example, as the junction of "v's" turned on their sides as in the diagram on the left. Gestaltists explained this as an instance of the law of good continuation. The eye runs naturally along the contours of the two lines, as these imply minimal change from dot to dot. If the eye were to assume the organization in Figure 5.4.B, then it would need to make a more complex assumption, namely, that there is radical change in the direction of the lines in both cases, on both the right and left of the figure.

The main point about Gestalt grouping phenomena such as these is that they are genuine, albeit phenomenological (i.e., purely psychological) effects. They are consistent across people of widely varying backgrounds and personalities, and across time within a given subject. In addition, they can be manipulated to achieve certain ends. For example, one can make the effect in Figure 5.4 stronger by making all the dots in one line the same color, and the dots in the other line a different color. There are also a number of associated effects with Gestalt groupings, including the fact that items arranged with Pragnanz will be more likely to be remembered, there will be short recognition time for such items, the subjective distance between elements in such a whole will be smaller than it would be otherwise, and finally, the features in a given grouping member are more likely to falsely migrate to other members of a same group than to a member of another grouping (this is known as the illusory conjunction effect; see Prinzmetal, 1995).

It is these sorts of consistencies that generate confidence that we are dealing with a real scientific effect, and the fact that we are working in the phenomenological realm does not detract from this claim. But it is important to bear in mind that these regularities are merely the icing on the evidential cake. The most important fact is that given a picture such as that in Figure 5.4.A, we cannot help but see it as a set of two intersecting lines sharing a common midpoint. It is this fact and this fact alone that gives us the latitude to label this situation a "genuine illusion." It is an illusion because on the page, there are no groupings of dots, just a set of such, and it is genuine because it is an unavoidable psychological fact.

What does this have to do with personal identity? The claim, as you might expect, is that the continuity of the ego is also a genuine illusion similar to a Gestalt grouping. It is upon this fact that we will be able to construct a theory that will not only help us to understand identity under normal circumstances, but also to answer questions related to the UI problem and associated conundrums. Before doing so, however, one important distinction needs to be made between Gestalt illusions and identity-based ones. This is that although there are no doubt good evolutionary reasons for the origins of Gestalt effects, their strength pales in comparison to the categorical impetus forced onto identity by evolution. The reason for this is simple. Without a clear delineation between self and other, an organism will be much less likely to survive to mate. Thus, any Gestalt effects implicit in ego-construction are likely greatly magnified by this additional evolutionary mechanism. If you like, identity is a super-genuine-illusion, although for simplicity in the following discussion we will treat it as if it were an ordinary perceptual one.

Of what does the illusion of the self consist? We may distinguish a number of non-mutually exclusive loci of contiguity. First, and most obviously, bodily continuity forms an integral part of one's self-identity. From any given moment to the next, the set of proprioceptive inputs largely overlaps with those of the prior moment, and varies relatively smoothly as the body moves through space. Moreover, one's body image, barring accident, changes slowly through time. When we first look in the mirror in the morning, the image is almost identical to the image we saw when we brushed our teeth the night before.

The next element of continuity is our memory of ourselves, or what was earlier called autobiographical memory. Recall that the critical elements of these memories included memories of specific periods, memories of transitions between periods and especially emotionally significant transitions, firsts of all sorts such as first crushes, and recurring episodic events. All of these are combined by the innate agglutinating nature of the mind into a kind of narrative (cf. Dennett, 1992), which, while not being constantly recited in one's inner voice, does inform and influence a large amount of what we do. This narrative is also under constant revision, but usually in such a way as to preserve the continuity from event to event, and in such a way as to make an otherwise unrelated string of events into coherent story.[6]

Another critical factor in reinforcing the notion of self is the way in which we perceive ourselves to be different from others. This list of possible differences is almost endless, but includes to varying degrees the person's age, gender, ethnicity, intelligence, appearance, abilities and capacities, political affiliation, club affiliation, nationality, food likes and dislikes, career, official title, attitudes, gait, clothes, shoes, hairstyle, tattoos, piercings, preferred yoga positions—you get the idea. It may also include quirks and idiosyncrasies of minor importance in the grand scheme of things but of greater import in serving to distinguish the self from others, especially when there is no other basis to form distinctions.

Finally, there are the influences of more nebulous but nonetheless important elements of the self, such as hopes and fears, and dreams and ambitions, realized or otherwise. The lineage of the past melts into the projection of our hopes and fears onto an extrapolated hypothetical future, serving to further strengthen the self's narrative by giving it heft in the direction of events yet to unfold. Through this factor and the previously mentioned factors, the line of the self becomes like

[6] To the extent that talk therapies for mental illness work, they may do so because they strengthen the coherence of the self-narrative. The mere act of giving voice to the narrative may grant it a greater reality than it would otherwise have.

a set of intertwined braids, and the resulting rope is much stronger than the sum of its individual parts.

Hume's mistake, then, and one that he shares with his philosophical heirs, is to see only the constituents of the self when he could have also seen the thread that binds these together into a unified psychic entity. It is as if Hume made the following claim: When I look at a diagram, such as Figure 5.4.A, I see only pixels, neither circles nor the lines that they allegedly form. But if he said this, it would be a lie! It would be a subjective, perceptual falsehood, but false nonetheless; barring serious damage to the visual system (and in particular the ventral stream), it is not possible to *not* see the emergent groupings in visual scenes. Note also that Hume could not, with good ontological conscience, state the following: "For my part, when I enter most intimately into what *Descartes* calls himself, I always stumbles on some particular perception or other,... ." That is, in the process of denying the self, it is unlikely that Hume ever lost the illusion that he was David Hume, née Home, of Scottish descent, and that in fact he was someone else, say the Frenchman Rene Descartes. Even for identity deniers or minimizers, in practice, the self remains a powerful concept that cannot be ignored, despite its apparent lack of "objective" third-person status.

It is important to stress that the genuineness of the illusion is as important as the fact that the self is a constructed item. One case that any theory of the self needs to contend with (and especially one that relies on psychological continuity) well before considering heady issues, like identity uploads, is mistaken identity. Suppose we come across a madman who truly believes that he is Napoleon Bonaparte. He dresses like Napoleon, he tucks his right hand into his jacket, and he even will regale you with stories of Waterloo and how he could have won it had he just been able to separate Wellington's army from the advancing Prussians.

What is to stop us from concluding falsely that he is not the real Napoleon, or some sort of reincarnation thereof, if identity depends only on the illusion of continuity? The point is that our madman wasn't at Waterloo, wasn't thereafter exiled to Saint Helena, and in fact knows only what the history books (or Wikipedia) has told him about Napoleon's life. His continuity is not between his current circumstances, wandering the streets and begging for a cup of coffee, and his former Imperial glory, but between his current circumstances and some *imagined* earlier narrative. In short, he has what we could call the fake or nongenuine illusion of continuity. It is possible, of course, to say anything and truly believe it, but only the real Napoleon could have the genuine illusion of continuity with his earlier memories. Or to put it another way, the madman's line of fate extending from the past to the present is far different than that of a reincarnated Napoleon living in the present. The fact that the madman, in his delusionary state, labels his fate line with the appellation "Napoleon Bonaparte" is irrelevant; identities are determined by the constituents of one's narrative and how they cohere, and not by any labels we or others place on them, even if it is the self that is doing the labeling.

In summary, we have presented a psychological theory of identity that stresses that the self is not merely a collection of tangentially related items, but a set of dependent events that are tied together via the mind's natural tendency to impose Pragnanz, or order on the universe. This effect is boosted considerably by evolution, which has a powerful vested interest in the crafting of a notion of a self, and the means by which the self may distinguish itself from other. This account is also partially deflationary, or if one likes, Buddhistic, in that the continuity of the self is dependent on the continual construction of a binding narrative, and not on any persistent quality beyond this construction, such as that claimed by the simple view.

Personal identity would appear, then, to fall into that class of items such as love, beauty, justice, and wisdom, which are at once ill-defined and seemingly insubstantial, but whose absence is sorely felt, and therefore cannot be in any sense be thought of as inconsequential. The question remains whether such a minimal account of the self has enough meat on its bones to do the work that we originally set out to do, namely, to tell us whether identity uploads are possible, how to judge the success of such, and what to do about partial uploads and simultaneous downloads to multiple vessels.

Let us begin by considering what we might call a perfect upload. In this scenario, we transfer the dying Bill Gates to a new lightning fast machine, accompanied by robot arms and other actuators so that he can interact with the world, and we assume that from the point of view of both an external observer and, most crucially, Bill Gates himself, it is as if he simply went to sleep and awoke in his new robot body. We also assume that we have captured everything about what it means to be Bill Gates, including his memories, his hopes, dreams and fears, his preferences, his quirks and foibles, etc., and of course, that the new machine is capable of sustaining consciousness (the background assumption to the entire enterprise).

Is this scenario consistent with the genuine illusion theory of identity? That is, will the new Bill Gates have the same feeling of continuity with the old Bill Gates that he would have had had he remained in Bill Gate's original body. The answer is remarkably simple: there is absolutely no reason to suppose that he would not. By construction, we have given the new Bill Gates all the qualities of the old Bill Gates. This includes not only the elements from which to construct the lineage of self, but also the Gestalt mechanisms to do so that the original Bill Gates possessed.

If you are feeling queasy about this line of thinking, it is most probably because you are harboring some version of the simple view, in which the new Bill Gates couldn't possibly be the same as the old one because there is something missing. The question you should be asking yourself is the following: what more do you want out of personal identity other than the (true) feeling of personal continuity with a constructed self? Look at it from the new Bill Gate's point of view. As he wakes up from the operation, he feels a bit groggy, and might not even know who he is. But then as his mind clears, he starts to remember the reason he had the operation, how he kissed his wife and children goodbye just before going under the anesthetic, how he had trusted the running of Microsoft in his absence to his Chief Information Officer (CIO), and so forth. Moreover, all of these are of a piece with his current desire to see his family, his desire to get back to work and to forge ahead with his charity work, and to do all the other things that he has put aside because of his prior illness. If you ignore the fact that he now sees the world through a camera, and rolls instead of walks, there is no difference between Bill Gates in his prior incarnation and Bill Gates as robo-mogul.

The point is not merely that the simple view is wrong; it is that it does not buy you anything. Our bodies are constantly in flux anyway; in order for the simple view to hold, the "soul" of Bill Gates would have to constantly reattach itself to different pieces of matter as he moved from gangly adolescent to well, somewhat less gangly entrepreneur, to his current state. In the simple view, you need to have Bill Gate's soul jump over to his robot body in consort with the rest of the transfer, although it is hard to see why this would happen, *nor why it needs to happen*.

Perhaps the application of the theory to the problem of partial transfer will help clarify matters. It is here on this ostensibly more difficult question that the genuine illusion theory really makes itself useful. Recall that one of the problems with partial transfer is that we clearly do not require that the upload be 100%; otherwise, we would be constantly changing identities as we lost small pieces of memory or gained new knowledge. On the other hand, to stipulate a fixed

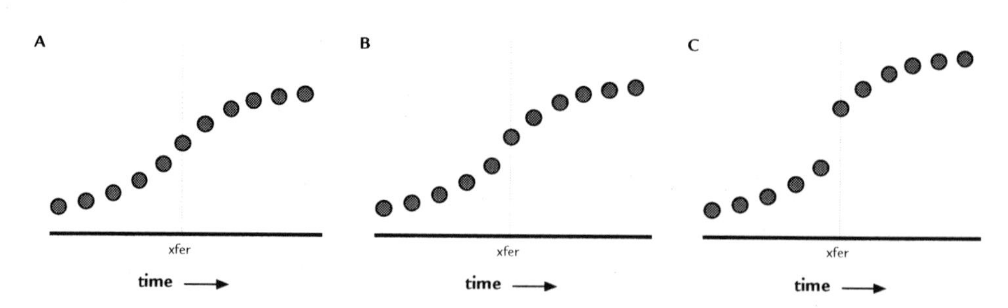

FIGURE 5.5 **A: Analog of an identity transfer in which the new self smoothly connects to the old. B: Analog where the post-transfer self is somewhat different than the old self, but the new self is still able to connect the new with the old. C: Analog where the new self cannot make the connection to the old self, and does not see itself as the same person.**

threshold, say 93%, is arbitrary and it is hard to see how any percentage figure could be motivated.

The genuine illusion theory provides a clean solution to this problem. It states simply that the transfer will be a success to the extent that the illusion of continuity of the self is maintained despite the possible loss of information. This illusion need not depend on the completely smooth perception of continuation of the past into the present, but may break down in a nonlinear fashion if this continuation is sufficiently degraded. Observe the analogous Gestalt phenomena in Figure 5.5. In 5.5.A we see a continuous line, corresponding to either a perfect transfer or near-perfect transfer. In 5.5.B, the percept is still one of a continuous line, although one with a greater amount of discontinuity at the moment of transfer. This corresponds to a case of partial transfer, in which the new identity would feel *fundamentally* like the old identity, but would notice a difference between its former and present selves. In 5.5.C there is a greater break at the moment of transfer, and it is now viewed as separate lines. Here, the new identity would have some of the same memories as the old one, but it would not feel like the original person. It would no more think so than you think you are your college roommate, simply because you shared a number of experiences together.

FIGURE 5.6 **A Dalmation in the woods. Once seen, it cannot be "unseen".**
SOURCE: **http://www.answers.com/topic/emergence-jpg-1.**

As a further comparison between the construction of self and Gestalt processes, consider another example, the Dalmatian that emerges from the background in Figure 5.6 (look to the right and center). There are three aspects of this diagram worth highlighting:

a) The dog is perceived because there is just enough "doggy" stuff for the mind to be able to differentiate between the Dalmatian and the background.
b) The dog will either be perceived or not, and the Gestalt of the dog will disappear if the dog and the background blend too much.
c) Once perceived, the dog cannot be "unperceived," although the amount of time it takes to see the dog may be greater than unusual.

Likewise, if enough of the mind gets transferred, and enough of the right sort of elements (both quantity and quality will be context-dependent), then the transfer will appear successful to the transferee; he or she will have the Gestalt of personal continuity. This will likely be a threshold process, in the sense that one will either get the feeling, or not, of being the same person. Finally, one can feel fundamentally like the same person, but at the same time notice that one's self is not quite the same as it was before, analogous to the perception of continuity in Figure 5.5.B, or to the perception of the Dalmatian.

In summary, there is no fixed, context-independent percentage below which the operation will be deemed unsuccessful, nor will a high percentage of transferred material guarantee continuity of self; it will depend on both the amount and type of elements transferred, and also on the grouping mechanisms of the person in question. In order for the transfer to be considered a success, then, there must be enough of the right sort of stuff in the new identity to cause it to believe, via Gestalt-like mechanisms, that it is in fact the old person.[7]

We will turn to the implications of this account for various mechanisms of transfer in Chapter 11. Now let us consider the application of this theory to the fusion problem. Recall that fusion occurs, by definition, when one personality is injected into another. In our original scenario, at the start of this chapter, we imagined the wires being crossed during the transfer, such that the new entity gets at least some of the psychological elements of the other person being transferred. Let us say that this happens to Bill Gates, and that person in the next room with whom he becomes mixed is Steve Jobs. This scenario is directly analogous to that of a partial transfer. Assume that the new Bill Gates gets just

[7] In practice, this is likely to be a very high percentage, but we leave open the possibility that there may be, under certain circumstances, a "core" self, such that if this core is captured, the transfer will be deemed successful by the transferee.

a little bit of the old Steve Jobs. He could, for example, feel just like the old Bill Gates, except that he awakens with an odd affinity for iPods and other Apple products. It is not hard to imagine him claiming that the transfer was successful in this case, and for all we know, it is possible that the real Bill Gates harbors secret sentiments to this effect already, although he is unlikely to make a public pronouncement to that effect. As we increase the proportion of Steve Jobs, however, at some point the new Bill Gates will lose the illusion that he is indeed Bill Gates, and a new hybrid will emerge, neither Bill Gates nor Steve Jobs but a mix of both, believing himself to be neither. Thus, fusion is qualitatively no different than partial transfer, and admits of the same solution.

Fission presents us with ostensibly a more problematic situation than fusion. If personal identity is presented as special case of general identity, then we have the following contradiction. Let us say that Bill Gates (BG_0) is split into two, BG_1, and BG_2. By hypothesis, BG_1 has the same identity as BG_0, and BG_2 has the same identity as BG_0. Therefore, by transitivity of identity, BG_1 should also equal BG_2 ($BG_1 \equiv BG_0 \equiv BG_2$, therefore, $BG_1 \equiv BG_2$). But, once the split is made, they become two different identities with two different sets of experiences, and eventually differing personalities, as these experiences mold their respective bearers. So BG_1 does not equal BG_2, but we have shown that it must.

The resolution of this contradiction, given the current theory, is not difficult. Before the upload, BG_1 and BG_2 are coincident; after, they wend their separate ways as their experiences take them in different directions. According to the genuine illusion conception of self, identity is a function of the *perceived* continuity of the present self with the old self. More formally (and simplifying), assuming that the split occurs at time t_n and the present is now t_{n+2},

$$BG_1 = G(e(t_{n+2}), e(t_{n+1}), e(t_n), e(t_{n-1}), e(t_{n-2}), \ldots; e(t_0)), \text{ and}$$
$$BG_2 = G(e'(t_{n+2}), e'(t_{n+1}), e(t_n), e(t_{n-1}), e(t_{n-2}), \ldots, e(t_0)),$$

where G is the Gestalt function generating the perception of unity, and the e's are experiences encoded as memories by the two BGs. But, as is easily seen, these experiences differ after the split, and therefore the arguments to G differ (in general, $e'(t) \neq e(t)$), and therefore BG_1 is not equivalent to BG_2. Although each BG (1 and 2) perceives himself to be the "real" BG_0, in fact they diverge as soon as the fission occurs. There is no contradiction, just an unprecedented situation: two separate people with a large body of shared memories.

We can make the identical argument graphically (and more simply) by referring to Figure 5.7. The upper line and lower line are not equivalent after the split, although they share a common segment. There is no more contradiction in a split in identity than there is in a single line branching into two. To put it another way, there never was an invariant in the universe called "Bill Gates" (as there would be with the simple view), but there was a perception on the part of

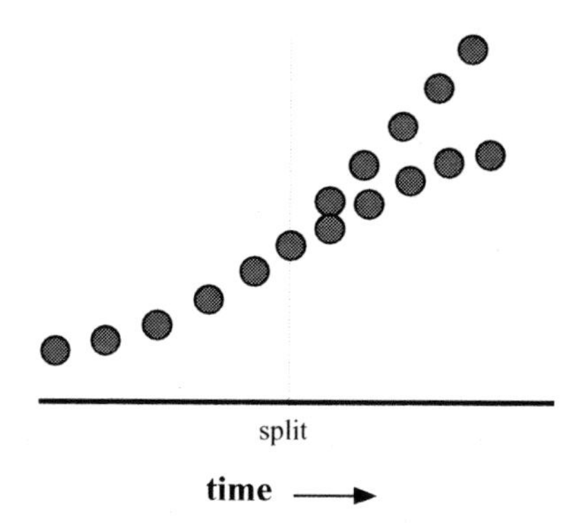

FIGURE 5.7 **Fission as represented as a split of a single identity into two, after which the identities diverge.**

a given consciousness that he was Bill Gates. Thus there is no contradiction in having two such perceptions.

In summary, we began with what appeared to be a somewhat flimsy construction to hang a theory of identity—not only are we claiming that personal identity turns on psychological variables, it is a mater of psychological or first-person perception of a special Gestalt, which we call the self, rather than an "objective" entity accessible to external or third-person observers. It is also important to realize that this is a kind of deflationary account, in that identity is not something that persists over time, although the perception of it does. To a large extent, there is inevitability to this view. Once one gives up the notion of an indestructible self that persists apart from the mind's imaging of such, one is left with a view that identity is in constant flux. Any attempt to ally such a notion with a notion of invariance is an attempt to reconcile the irreconcilable, and is bound to failure.

Despite this minimization of identity, or perhaps more aptly because of it, however, we have been able to produce a satisfactory answer to not only the UI question, but also to the crucial questions of what happens when the upload is incomplete. Furthermore, the constructed theory also tells us what happens in the case of identity fusion, and why the seemingly bizarre case of fission is a real possibility, however unsettling that may be. We postpone our full discussion about how to engineer an identity download, given this framework, to Chapter 11. Here, we will concentrate on the immediate consequences of the preceding discussion abstracted from the particular technology used to manipulate personal identity.

THE FLUIDITY OF IDENTITY

The prior theoretical treatment strongly suggests that an identity upload is indeed a metaphysical possibility, if not yet a technological one. This conclusion is not at all trivial nor is it independent of a theory of identity. For example, if either the simple view or the organic view of identity holds, then it is not clear how to proceed to realize this feat. In the case of the former, one would have to first capture the ectoplasmic self, as in "All of Me", and then attach it to its new body. If there is a science of such processes, it is unlike any science that we currently possess, and therefore technological progress along these lines is unlikely. Alternatively, a view that ties the self to the body means by construction that any other embodiment will destroy the self. Theorists along these lines, for example, are committed to the counterintuitive claim that if X's brain is transplanted into Y's body, then the new identity is that of Y rather than X (see, Olson, 1992). There is then no point to transferring an identity to a new vessel, because it would take on the identity of the vessel, and the original identity would be lost.

In contrast, the genuine illusion theory of self not only states that identity transfer is a distinct possibility, it also provides a formula for its success: it will be achieved to the extent that the new self is able to maintain the (genuine) illusion of continuity with the old self. As previously argued, this illusion will be created by Gestalt-like forces possibly strengthened by the biological imperative to distinguish self from other. For these forces to hold sway, the two selves, old and new, need not be identical, but must bear sufficient family resemblance such that new self perceives itself as a continuation of the old. Another consequence of the proposed theory is that multiple cosimultaneous copies of the same identity are possible, although these will quickly diverge as they come under the influence of new experiences. Identity then, may be conceived as a fluid that may be poured into new vessels or even multiple vessels as we see fit.

It is also fluid-like in that other liquids may be poured into it, which it may choose to dissolve or not; identity is a plastic substance and not one that is fixed over time. In particular, the ego may choose to incorporate the kinds of additional elements discussed earlier into its fold, including people, cities, countries, trains, planes, and automobiles or other objects of affinity. It will do so if: a) those elements no longer feel as if they are completely separate, and b) if the change introduced by the incorporation does not push the ego to its breaking point. Given the way we are currently constituted, and the nature of current technology, this plasticity is a semivoluntary process. To some extent, we can choose our hobbies and clubs, perhaps to a lesser extent our friends and love interests. In all cases we are bound by the range of our experiences, which are necessarily limited by who or what we come into contact with. But let us say that we make the conscious decision, given a future fusion technology, to incorporate a particular element into our egos. Would this be possible, ignoring the purely technological barriers to such?

For the sake of concreteness, let us say that we wish to remain otherwise identical, but have the ability to improvise like Thelonius Monk. Despite the largely optimistic claims made with regard to the current theory of identity, there is nothing that has been said that guarantees that this would be a possibility. The problem is that to play like Monk, to some extent you have to be like Monk. A Monk-like composition is, if you will, not merely a succession of notes, but the personality of this jazz great distilled into musical form. Thus, a skill, and particularly a skill of this character, is not something that one can necessarily divorce from the personality of the bearer of that skill. Even something like learning a language is not guaranteed to be an identity-preserving operation, and in fact, one of the reasons for doing so is to expand one's intellectual and affective horizons. And as we have seen, if this expansion is too great or too fast, then the self will not expand to incorporate its new knowledge, but pop like a balloon filled with too much air. Just how to prevent this, and the full consequences of fusion-like additions to the self is a topic that we will defer until the final chapter of this book, however.

For now, we simply note that the ordinary pace of personality development that we are accustomed to need might not be as glacial in the future—it may be possible to experience many psychological lives within a given physical life, if so desired. This is just one among many reasons that neural technology is truly revolutionary technology, as we shall see.

REFERENCES

Brewer, W. (1986). What is autobiographical memory? In D. Rubin (Ed.), *Autobiographical Memory* (pp. 25–49). New York: Cambridge University Press.

Conway, M., & Pleydell-Pearce, C. (2000). The construction of autobiographical memories in the self-memory system. *Psychological Review, 107*, 261–288.

Dennett, D. (1992). The Self as a Center of Narrative Gravity. In F. Kessel, P. Cole, and D. Johnson (Eds.), *Self and Consciousness: Multiple Perspectives.* Hillsdale, NJ: Erlbaum.

Feinberg, T. & (2002). *Altered Egos: How the Brain Creates the Self.* Oxford: Oxford University Press.

Feinberg, T., & Keenan, J. (2005). Where in the brain is the self? *Consciounsness and Cognition, 14*, 661–678.

Giles, J. (1993). The no-self theory: Hume, Buddhism, and personal identity. *Philosophy East and West, 43*, 175–200.

Gusnard, D., Akbudak, E., Shulman, G., & Raichle, M. (2001). Medial prefrontal cortex and self-referential mental activity: Relation to a default mode of brain function. *Proceedings of the National Academy of Sciences, 98*, 4259–4264.

Harlow, J. (1868). Recovery from the passage of an iron bar through the head. *Publications of the Massachusetts Medical Society*, 2, 327–347.

Ho, D. (1995). Selfhood and identity in Confucianism, Taoism, Buddhism, and Hinduism: Contrasts with the West. *Journal for the Theory of Social Behavior, 25*, 115–139.

Hume, D. (1978). *A Treatise of Human Nature.* Oxford: Clarendon Press.

Johnson, S., Baxer, L., Wilder, L., Pipe, J., Heiserman, J., & Prigatano, G. (2002). Neural correlates of self-reflection. *Brain, 125*, 1808–1814.

Keenan, J., Freund, S., Hamilton, R., Ganis, G., & Pacual-Leone, A. (2000). Hand response differences in a self-face identification task. *Neuropsychologia, 38*, 1046–1053.

Keenan, J., Wheeler, M., Gallup, G., & Pascual-Leone, A. (2000). Self-recognition and the right prefrontal cortex. *Trends in Cognitive Sciences, 4*, 338–344.

Kelley, W., Macrae, C., Wyland, C., Caglar, S., Inati, S., & Heatherton, T. (2002). Finding the self? An event-related fMRI study. *Journal of Cognitive Neuroscience, 14*, 785–794.

Kircher, T., Senior, C., Phillips, M., Benson, P., Bullmore, E., Brammer, M., et al. (2000). Towards a functional neuroanatomy of self processing: effects of faces and words. *Cognitive Brain Research, 10*, 133–144.

Locke, J. (2007). *An Essay Concerning Human Understanding.* Hebden Bridge, UK: Pomona Books.

Noonan, H. (1989). *Personal Identity.* Milton Park, UK: Routledge.

Olson, E. (1997). *The Human Animal: Personal Identity with Psychology.* Oxford: Oxford University Press.

Prinzmetal, W. (1995). Visual feature integration in a world of objects. *Current Directions in Pscyhological Science, 4*, 90–94.

Ramachandran, V. (1998). Consciousness and body image: lessons from phantom limbs, Capgras syndrome and pain asymbolia. *Philosophical Transactions of the Royal Society, 353*, 1851–1859.

Ramachandran, V., & Blakeslee, S. (1999). *Phantoms in the Brain: Probing the Mysteries of the Human Mind.* New York: Haper Perennial.

Rizzolatti, G., Fadiga, L., Gallese, V., & Fogassi, L. (1996). Premotor cortex and the recognition of motor actions. *Cognitive Brain Research, 2*, 131–141.

Robinson, J. (1976). Sampling autobiographical memory. *Cognitive Psychology, 8*, 578–595.

Ryle, G. (2000). *The Concept of Mind.* Chicago: University of Chicago Press.

Snowdon, P. (1991). Personal identity and brain transplants. In D. Cockburn (Ed.), *Human Beings.* Cambridge: Cambridge University Press.

Part 2 Neural Engineering: The State of the Art

Chapter 6 Brain Machine Interfaces: Principles, Issues, and Examples

Imagine being fully awake, and having an active and intense mental life, but being completely paralyzed, and not being able to control the world in the slightest. This unfortunate "locked-in" condition is relatively rare, but unimaginably debilitating. If somehow we could tap into the thoughts of these victims, then they would be liberated from their completely passive state, and return to a modicum of normality. Furthermore, there are a correspondingly larger number of people with partial paralysis, cerebral palsy, and other similar conditions whose control over their environments is weak or ineffective. They too could benefit from a device that read brain signals directly and that bypassed their inoperative motor control systems to effect changes in the world.

One does not want to diminish the suffering of those that cannot do what most of us take for granted. However, by far the largest set of applications that will be made possible by a "thought-reading" technology in the long-term will not be to enable disabled people to approach normality, but to allow everyone to do things that they could not do otherwise. The argument for this claim proceeds by noting that evolution designed our musculature and the accompanying neural control of such under a number of very specific conditions. Thus, we have been given the ability, among other things, to walk, run, grasp, etc. At least some of these abilities, such as grasping, are much more general than those of our animal counterparts, but all have limitations. We can only carry so many things at once, we can lift only so much, we cannot fly (except in our dreams), and perhaps most importantly, evolution clearly had no idea that we would one day be spending so much time interacting with computational devices.

To allow direct access from the brain to a device, computational or otherwise, we need to construct a Brain Machine Interface, or BMI. By a BMI we will simply mean any system comprising brain sensors and some form of actuator that affects the world external to the brain. To count as a BMI, the system must detect brain signals directly; thus, a device that reads electroencephalogram (EEG) signals and converts them into mouse movements would be included in this class, a squash racquet would not (it does indirectly respond to the brain via muscle movement, but does not interface with the brain directly). For the purposes of the following discussion, we will treat what are sometimes called Brain Computer Interfaces (BCIs), where no direct noncomputational action is performed, as a special case of BMIs. We will, however, limit our discussion in this chapter to BMIs that produce sensory feedback only as devices that provide feedback directly to the brain are in a comparatively primitive stage of development (see Chapter 8 for a fuller discussion of this topic).

Figure 6.1 shows the primary components of a BMI. First, there must exist some means of detecting signals from the brain. As described in the following, this includes three primary types of transducers: invasive, which penetrate into the brain and read neural signals directly; semi-invasive, which sit on the surface of the cortex, but beneath the skull; and noninvasive, which attempt to pick up electrical, magnetic, or light signals from the brain but are situated external to the skull. In all cases, the signals must be transformed before driving an actuator directly. Thus, a signal processing module must transform these signals into usable form. The actuator itself, pictured as robot arm schematic in Figure 6.1, can be any device in the environment of the subject, including a computer. Finally, the subject may receive sensory feedback from the environment and may use this feedback to adjust his brain signals in order to better control the actuator. The following section will introduce some general principles of BMI, and the sections that follow will consider the two fundamental aspect of the BMI environment in detail, the nature of the transducers that read from the brain, and the process of information extraction from these devices. We will conclude with a discussion of current representative BMIs.

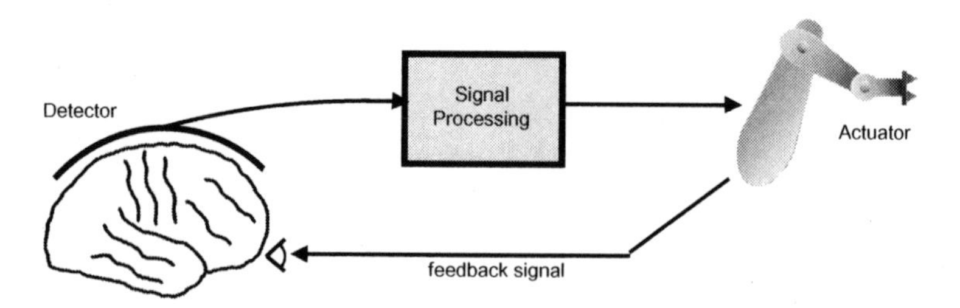

FIGURE 6.1 **The components of a BMI include some means of reading or detecting brain signals, a signal processing module that converts these signals into a usable form, an actuator that changes the environment, and sensory feedback from the environment to the brain.**

PRINCIPLES OF BMI

Although the relatively youthful field of BMI remains largely an experimental art, with researchers often resorting to ad hoc methods to attempt to get a given imaging modality to drive a given type of output, a number of important principles and guidelines have emerged in recent years. Many of these, such as the principle of transparency of action and the importance of multiunit recording, were not entirely obvious before research had begun in earnest in this field roughly twenty years ago, and others follow from careful consideration of first principles. Regardless of their origin, however, each principle described in the following text is in its own way crucial to the understanding of this emerging field, and each is worthy of careful consideration.

Bandwidth and degrees of freedom

The most fundamental requirement of a BMI is that the sensing device must contain sufficient information for the task at hand; without this, no amount of signal processing could possibly yield adequate results. More specifically, we can distinguish between two closely related concepts, that of bandwidth and degrees of freedom (DoF). The former refers to how much information per unit time is available, and the latter refers to the number of distinct channels in which this information can be produced. One might think that if bandwidth is high enough the information contained within could be partitioned into separate channels to create arbitrary DoFs, but it may not always be possible for the brain to produce these orthogonal streams. To take an analogy from piano playing, one may be able to play a fairly difficult piece with one hand and yet be incapable of playing simultaneously two very simple pieces with different hands, especially when the time signature for each hand is noncoincident.

It is also instructive to attempt to delineate just when it will be possible to produce independent streams of information and when this will not be the case. Generally speaking, if two or more tasks must be consciously attended to, this will restrict the ability to fully separate them. Alternatively, if the tasks are automatized, through frequent repetition, for example, and do not require attentional resources, then it may be possible to generate multiple simultaneous signals representing noncompeting degrees of freedom. A classic case of this is the motor control of arm movement. In order to reach to a particular position in space, multiple muscles must be activated simultaneously. This is not a problem for the noninfant, however, because this task has dropped below the level of conscious thought.

There are two complementary ways of explaining this effect. First, a fundamental empirical lesson is that tasks that are routed through consciousness are subject to severe bandwidth limitations. There are strong limits on the amount of items in short-term memory, as described by the famous 7 ± 2 limit on the number of items that can be held in that store (Miller, 1956). The limit on nonrelated tasks that are not being performed automatically may be even more severe, and in the worst case approaches unity, as argued in chapter 3. It is true that we can walk and chew gum at the same time (or true for most of us) because these are not attention-loading tasks, but we cannot concentrate on a radio broadcast and carefully read the paper at the same time; these compete for the same linguistic resources.

The other way at looking at this is through the framework developed in Chapter 4 to explain consciousness. Recall that one of the central claims of that chapter was that a condition for a neural ensemble to contribute to the qualities of the phenomenal field is that it be part of the kernel, or a set of mutually interacting neurons. It is this interaction that will constrain the independence of the ensemble. If it receives substantial input from other ensembles, it must be coordinated with them. Hence, the ability to represent multiple unique streams of information that all demand conscious attention is severely limited.

Thus, there are fundamental task-dependent limits on the simultaneous DoF the brain can achieve. A sophisticated future BMI will not, by itself, for example, allow us to become like a Hindu God, with one arm playing the flute, another making tea, a third petting the cat, and the fourth talking on the cell phone. The cognitive limitations imposed by the brain in its native state apply whether the effectors for this organ are natural or artificial. Coupled with potentially large bandwidth requirements, DoF requirements severely constrain the current set of interface technologies to perform anything but very simple tasks.

Table 6.1 gives an estimate of the bandwidth required and the DoFs for some common tasks. Mouse movement in two dimensions typically requires at least 30 readings per second with 16 bits (2 bytes) encoding each of the spatial dimensions, or about 500 bits/second. This figure is halved for movement in one dimension (i.e., only up and down or right and left on the screen), and the DoF is also halved. Please bear in mind that, as argued previously, driving a 2D mouse is not merely a matter of doubling the effective output of the sensing modality; these signals must be completely independent of each other in order to achieve arbitrary mouse position in two dimensions. Typing at 25 words per minute implies about 125 characters or bytes per minute, which is about 15 bits per second. When done with the fingers many degrees of freedom are involved but this task can also be reduced into a single stream; this matter is discussed at greater length later on in the context of actual typing BMIs. A robot arm typically has six degrees of freedom corresponding to each of the three dimensions of movement per two joints; in humans these joints correspond to the shoulder and the wrist, and with the elbow supplying another DoF. Assuming roughly the same resolution and speed of movement of a mouse, we then obtain 6×250 bits/s or approximately 1500 bits/s of bandwidth requirement. The Segway, the

TABLE 6.1 Estimates of bandwidth and degrees of freedom for various actions and the theoretical limit of the information-generating capacity of the brain.

Action	Bandwidth (bits/s)	DoF
Mouse 1D	250	1
Mouse 2D	500	2
Typing (25 wpm)	15	1
Robot arm	1500	6
Segway	240	3
Tennis	>5000	At least 19
Theoretical limit	$\sim 10^{10}$	>25

self-balancing personal transportation device invented by Dean Kamen, receives about 100 signals per second in each of three dimensions (forward/back, and left and right). This figure can probably be reduced by an order of magnitude and still provide reasonable performance; if each signal consists of a byte of information (encoding 1 of 256 possible states), then the bandwidth is 240 bits/second. Even with these generous assumptions, the amount of information needed to drive this device far exceeds that produced by any noninvasive imaging device (see the following text, and the next chapter), preventing its use in a typical BMI context. Tennis engages the entire body and entails at least 19 degrees of freedom, 7 for the arm, 3 for each of two hips, 1 for each of the two knees, and 2 for each of the two ankles. This is a conservative estimate; for example, the pressure of individual fingers also affects the way the ball is struck, and the nondominant arm is not counted.[1] The bandwidth for tennis is probably well in excess of 5000 bits per second, or at least 250 bits per second per degree of freedom.

The final row of Table 6.1 shows the maximum information production capacity of the human brain. The brain consists of approximately 10^{11} neurons, each of which fires up to 1000 times per second. However, because of noise, it is unlikely that the brain is truly encoding information at such a fine rate, nor would this rate be meaningful in driving an external device. Thus, instead of 10 bits per second per neuron ($\sim \log_2 1000$), let us say that the brain actually uses closer to 5 bits per second. This implies 5×10^{11} bits per second total. There is

[1]This arm is used for balance, for the server toss, to steady the backhand in the one-handed case, and fully in the case of the two-handed backhand and in the rare, but often effective, two-handed forehand.

another factor that also limits information capacity. Recall from Chapter 1 that each neuron is connected to approximately 1000 others. Thus, let us reduce the total capacity by an order of magnitude, to indicate that any given neuron will not be completely independent of its synaptic neighbors. The final result is something on the order of 10^{10} bits per second. Finally, as previously discussed, DoF for the brain as a whole will be highly context-dependent. There will be instances, such as tennis, where a number of highly skilled muscle actions are being coordinated in parallel, and others, such as high-level cognition, in which only one line of thinking may be maintained at any time. In the extreme case, in a topspin tennis serve, for example (an athletic feat that is fully mastered by only the most advanced tennis players), both arm movements must be included, the back is arched, the eyes are directed upward, and fine-grained toe movements may also aid in the necessary springing action, leading to a total of at least 25 degrees of freedom.

In summary, even relatively simple tasks, like controlling a mouse, require a substantial bandwidth. Also note that for tennis, the only "natural" task in the chart (that is, one in which another machine is not being controlled) there is a substantial leap in both the number of DoF and the bandwidth requirements relative to the other actions. This indicates that when BMIs do come into prominence, the first things they will be used for will be to drive other devices, rather than say, robots or exoskeletons moving freely about the world. However, to end on a slightly more optimistic note, the bottom row of the table does suggest that the information is present in the brain, and when this capacity can be tapped, a number of hitherto prohibitively difficult technologies will be made possible.

Open and closed loop systems

A key distinction in BMI technology is that between an open and closed loop system. In an open loop system, the subject either carries out a given task or in the case of humans, merely thinks about carrying out this task. The job of the BMI is to decode the signals received by the reading device under varying conditions. For example, suppose we are trying to get a primate to control a robot arm with his thoughts. In an open loop system, we would have the animal carry out various tasks in which it would reach to a number of points in three dimensions. We would then attempt, using analytic methods described more fully in the following text, to take the signal read from the brain for these different points and extract the information that allows us to move the robot arm in the same manner as the animal's arm.

In a closed loop system, in contrast, after or coincident with this BMI training phase, the subject receives feedback as to the accuracy of its ability to control the device. This feedback may be visual, auditory, or tactile, and in the case of animals, may include a reward as reinforcement. Thus, in closed loop systems two distinct types of adaptation are taking place: the BMI is trying to optimally decode the signals for the given task (as in open loop systems), and the subject is modifying its brain state on the basis of feedback in order to optimize *its* role in the process. The reason this latter process may contribute to the accuracy of the system is not entirely mysterious. The brain, while not a general-purpose learning organ, must at least be flexible enough to learn motor tasks and the like and refine them as circumstances dictate; if not, it would be locked into the capabilities it was provided with at birth (which in the case of humans, are fairly minimal). It is this native plasticity that we are tapping into with closed loop systems; the next section will consider this process in more detail.

All things being equal, closed loop systems tend to be more accurate than their open loop counterparts. The Taylor, Tillery & Schwartz (2002) result, described in more detail at the close of this chapter, is typical in this regard. These researchers had 2 monkeys implanted with recording electrodes controlling a robot arm; performance was roughly doubled in the closed loop relative to the open loop case. The scope of applicability of this methodology is still unknown. Consider, for example, a future thought-reading device that takes some brain signal and produces the inner monologue coursing through consciousness. Given the almost infinite variety that such a stream of thought could take, it is unlikely that learning on the part of the subject could do anything more than fine-tune such a device. In this case, most of the adaptive burden will be on the BMI side rather than on the subject's end. In general, the relative need for these two types of adaptivity as a function of task, and the interaction between them when both are present, is a topic that requires much more empirical and theoretical research.

Adaptation and plasticity

The brain is parceled into modules, with each having its own set of inputs and unique processing capabilities, as has been stressed in Chapter 2. It is only natural then, when attempting to interface with a particular capability that we "tap into" the brain's native resources by looking at the activity of the module corresponding to a given task. For example, if attempting to drive a robot arm with invasive signals, it will likely be most profitable to use a population of neurons in the area of the motor cortex that is already driving the organism's real arm. However, it is also the case that the brain is adaptive. This has two implications: a) first, within a neural module, it is not necessary to find the precise neuron or population already responsible for accomplishing the task, and b) in certain circumstances it may be possible to partially retrain the output of neurons within a module to perform in a nonnative mode, i.e., to accomplish a task completely different than the one that they are currently designed for.

In fact, without a large degree of plasticity, BMI would be scarcely possible. Consider trying to drive a robot arm. If adaptation were not possible, we would have to find just the right neurons within the motor cortex already responsible for arm movement, or at the least find a group of neurons encoding this information. However, unless the robot arm works in a nearly identical fashion to the real arm, then these would not be able to accomplish this task. But it is unlikely that prosthetic devices and other artificial extensions of the body can be built to precisely mimic existing capabilities[2], and in any case, we are also interested in extending such capabilities. Thus, in many cases, adaptation via feedback is not only desirable, it is absolutely essential.

Fortunately, all neurons, and especially those in the action chain, are capable of some degree of plasticity; otherwise, it would be impossible to learn new motor tasks. We can state an even more remarkable result, however:

The BMI plasticity principle

> *Given some degree of plasticity and a possibly noisy imaging device, a population of neurons will adapt such that the optimal result will be achieved, subject to the information content of this population, in a reasonably short period of time.*

The net effect of this principle is that any neurons that are capable of encoding the correct response may possibly be used for a given task; for example, contrary to intuition, sensory neurons could be used to drive a motor output, for example. Furthermore, they will work even when the imaging modality produces a noisy signal (as is invariably the case), as long as there is some information buried in the signal that reveals the state of the relevant neurons.

Let us illustrate this principle by way of a representative example. Without loss of generality, consider a binary task, where there are only two possible outputs. For example, Figure 6.2.A illustrates a BMI scenario in which the goal of the training is to move a cursor either to the right or to the left. Accordingly, the subject thinks the thought "right," this activates a population of neurons, which in turn trigger a transducing cell population, which is then detected by the imaging device. The control system (not pictured) does its best to interpret this output, but let us also assume that the imaging process is noisy, and that it misinterprets the transducer signal a certain percentage of the time. For the sake of argument, let us say it is only right 55% of the time. Thus, the subject thinks "right," and 45% of the time the imaging device moves the cursor left. If no plasticity were present, then the correct response rate would be fixed.

However, let us assume a reinforcement algorithm is in place, such that when the correct response is made, the connection between the population

---

[2] A possible exception to this rule is the use of BMI to bypass a damaged spinal column and to drive limb muscles directly. It remains to be seen whether this is a feasible undertaking.

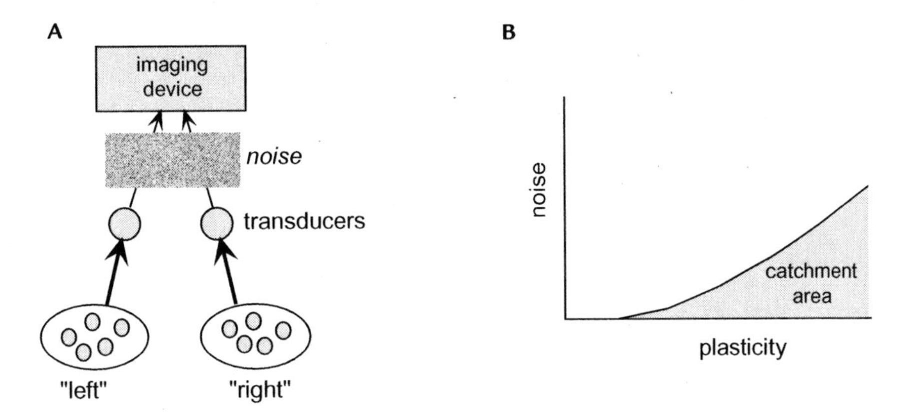

FIGURE 6.2 **A: An adaptive scenario in which the organism is learning to move the cursor left or right. B: In practice, learning will only take place in the catchment area, that is, when there is sufficient plasticity relative to the amount of noise in the signal.**

of neurons encoding the thought and the transducer neuron is strengthened whenever the correct response is made, and weakened otherwise. Even in the presence of the low initial correct response rate of the current example, learning will still move in the proper direction. For example, given 100 trials, 55 will increase the strength of the connection between the thought of "right" and the transducer that encodes this, and 45 will weaken it, leading to an overall strengthening corresponding to 55 – 45 = 10 net trials. Furthermore, and it is this fact on which the BMI plasticity principle turns, a positive feedback loop will develop, because the increased activity of the correct transducer will improve the performance of the imaging device, which will then increase the effective learning rate per trial, which will then further strengthen the appropriate connections, etc. Hence, learning will accelerate and will rapidly converge on the optimal response given the limitations of the device for the given set of neurons being measured.

It must be noted that the optimistic view of control presented in this scenario will not necessarily translate into a perfect response rate after training. First, a precondition of the plasticity principle is that the proper information must be in the signal, no matter how obscured by noise. For example, it is extremely unlikely that EEG could ever be used to produce a "thought-reading" device, that is, a machine that reveals the contents of one's inner voice, regardless of the amount of training time. The information to do so is simply not present in the EEG signal. Second, "unlearning" will take place outside the training time, in which the relevant neural populations will return to their native mode of processing. Unlearning will also take place if the same area is used for radically different tasks. For example, one could not use a

32-electrode array to control a robot arm and to drive a computer cursor; the learning of the one task will vanquish the ability to carry out the other task. Third, if too much noise or too little plasticity is present, then training will take too long to be practical, even if it is possible in principle. Finally, the fact that results will be optimal, for a given device and a transducer population, by no means implies that they will be perfect; there may simply be too much noise in the signal to reliably indicate which response is intended all of the time. This will hold even when the brain has trained itself to unambiguously trigger the correct transducers under all circumstances.

These limiting factors are captured by the graph in Figure 6.2.B, in which the catchment area indicates when adaptation is possible. Although only minimal plasticity is theoretically needed to acquire a given ability, if adaptation is to take place within a reasonable period of time, then a good deal of plasticity is required even when no noise is present. Hence, the catchment area starts well beyond the intersection of the axes. Furthermore, as noise increases, greater and greater plasticity will be required to counter the fact that many of the learning trials will produce adaptation in the wrong direction. Given that all means of obtaining neural signals, with the possible exception of invasive electrodes, involve a very noisy signal, this graph places strong limitations on what can and cannot be done with a BMI device.

In summary, some elementary theoretical considerations have produced both good news and bad news about the role of plasticity in training a BMI device to perform an intended task. The good news is that the system as a whole can harness the fact that the brain is, by virtue of its usual tasks, already a superb adaptive system. This means that, in principle, an arbitrary neural population could be used to accomplish the task, regardless of their original role in the brain, and by the plasticity principle, it will not take long to train them to do so. In practice, this rosy scenario must be tempered by the fact that: a) either insufficient information may be present in the population to carry out the task, or b) the imaging modality may be so noisy that learning is impractical, or rises to a fixed level of performance and then halts at this point.

Course coding and cell ensembles

Chapter 2 introduced the principle of coarse coding, in which a population of cells represent an intention, category, or other psychologically significant event. That chapter demonstrated the two primary advantages of this representational scheme: a) efficiency relative to more sparsely coded schemes, and b) graceful degradation and the related notion of resistance to noise. Coarse coding is also an extremely important concept for BMIs, and it is not an exaggeration to state that along with plasticity, if this principle were not capable of being leveraged, such devices would scarcely be possible. Consider the opposite extreme of population encoding, in which a single so-called "grandmother cell" was responsible for the representation of a given concept or action. Suppose first that we were attempting to read the firing pattern of this cell with deep electrodes. We would have to be extremely lucky to pick it up; out of the roughly 100 billion neurons distributed across numerous neural modules, our electrode would have to hit on this one to the exclusion of all others. It is not so much like finding a needle in a haystack as finding a contact lens in Lake Michigan. Alternatively, suppose we are trying to read the firing pattern of this cell with noninvasive technologies. As described in the following, all such technologies, to a lesser or greater extent, require the coordinated action of a large number of cells in order for a usable signal to be picked out of the noise contained in the reading. It is unlikely that we will ever be able to design an external device that could zero in on the activity of a single cell.

Fortunately, as far as we have been able to determine, all areas in the brain use at least some degree of coarse coding. The Georgopoulus et al. (1986) landmark *Science* study, for example, provides one point of confirmation for this claim. These researchers studied the firing rates of cell ensembles in this neural area while monkeys reached for targets in three dimensions. Figure 6.3 summarizes their results. The direction of optimal firing for a given cell is represented by the direction of the vector. The length of the vector illustrates the firing rate of the cell for the given 3D movement (the bold line). The sum of the vectors for the cell ensemble is shown in gray (after normalization). Note that this sum is very close to that of the target movement.

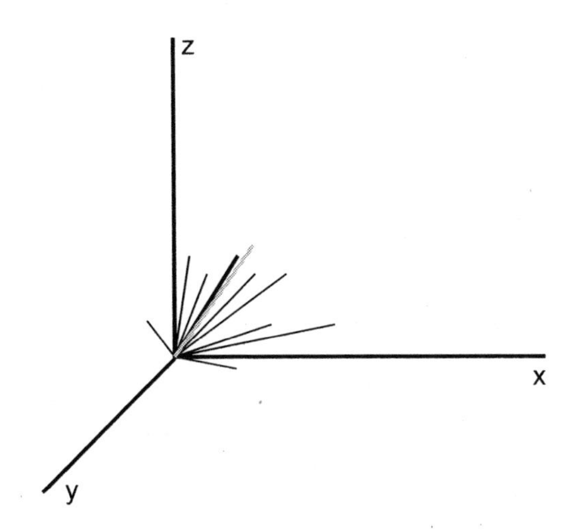

FIGURE 6.3 **None of the firing responses of the individual cells (thin lines) codes for the movement (thick line). However, the sum of the responses (thick gray line) comes very close to this movement in three dimensions.**

Thus, one can reconstruct the intended movement direction by examining the firings of a cell ensemble. Any given neuron will have a broad tuning curve and respond to a number of different movements, and therefore one cannot unambiguously determine movement direction from this cell alone. The population as a whole, however, does encode this direction to a high degree of accuracy. Moreover, subtracting a single cell or even a small group of cells from the mix will have little effect on the final result. This resistance to degradation is very useful in practice. When recording invasively with a microelectrode array, it is not unusual to lose the signal from a percentage of the cells due to shifting of the array and other factors discussed later. If this percentage is sufficiently low, then the quality of the result may not be altered significantly. It is also possible in a closed loop system that plasticity will compensate for the loss so that the degradation will be even more graceful than it would be otherwise.

In an experiment that will be described in more detail later, Carmena et al. (2003) provide additional confirmation for the importance of coarse coding in BMIs. The task here was also to control a robot arm, although in this case the monkey was expected to control both the speed of the movement and the gripping force of the arm in addition to controlling the direction of the arm. Figure 6.4 illustrates the relevant results of this experiment for both single-unit and multiunit cell recordings. There are two complementary ways of reading these graphs. Going from left to right, we can see that the greater the cell population, the more accurate the motion. Moving in the other direction, we can see that the removal of a small number of cells has relatively little effect once the predictive accuracy has reached plateau. Both effects are manifestations of the principle of coarse coding.

In retrospect, it is not a coincidence that BMIs are able to exploit coarse coding to their advantage. The brain has the same task and faces many of the same difficulties in trying to understand its own signals as we do! If the coding

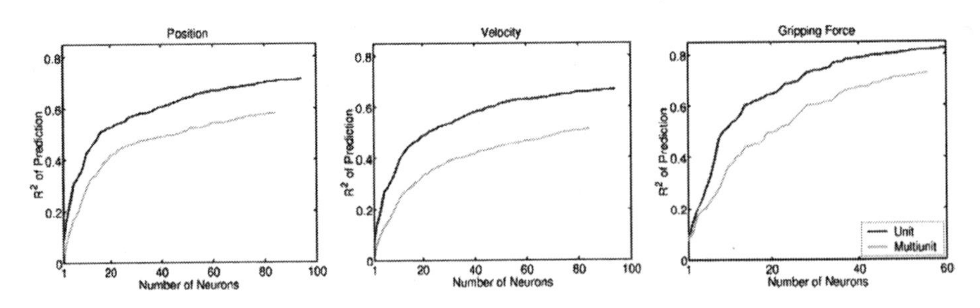

FIGURE 6.4 **Predictive strength as a function of cell population for both single-unit and multiunit recording units for arm position, velocity, and gripping force (from Carmena et al., 2003).**
SOURCE: **Carmena, JM Learning to Control a Brain–Machine Interface for Reaching and Grasping by Primates. PLOS Biology 1:2003.**

were too sparse, then downstream areas that drew on that representation would be extremely sensitive to noise in the system. We can see this once again by considering the case of the grandmother cell. A misread signal from this cell, due to random fluctuations, errors in timing, or cross-talk with other signals would result in a partial misinterpretation or complete loss of information if the noise level was sufficiently strong. Likewise, the death of this neuron, a not uncommon occurrence, would also mean a complete loss of the concept it represented. The redundancy built in to the brain by coarse coding enables it to function in a noisy and sometimes hazardous environment, and likewise, this representational scheme permits BMIs to operate in the presence of noisy or degraded signals.

Transparency of action

One of the more remarkable qualities of BMIs, frequently noted by those few who have had the first-hand opportunity of experiencing these at this early date, is what may be termed the transparency of action. Once proficiency with the system is achieved, one does not feel as if one is manipulating a set of mental variables, which then in turn manipulate the device. There is simply the thought and then the action, with no awareness of mediating neural events. For example, as discussed in the next chapter, if one is controlling a robot arm via implanted electrodes one simply perceives the desire to move the arm in a certain direction, and if one is proficient with the system, the arm will move in that direction.

A noteworthy aspect of this process is that it is virtually identical to normal motor action. One does not have the thought to send a signal to the cerebellum to compute a trajectory for one's arm to catch an incoming ball and then a second thought to send that signal to the motor cortex where it is then sent via the spinal cord to the appropriate muscles. There is simply the thought to move the arm, or, in the case of a trained athlete, just the thought of catching the ball alone. Furthermore, to spend too much conscious effort on a motor task will usually be antithetical to proficiency in this task.[3]

We can explain both the transparency of action in both the normal case and the neuroengineered case by invoking a principle discussed in Chapter 4 regarding the nature of consciousness. Recall that in that chapter the principle of reciprocity was introduced, which states that we are only aware of those neural events in the kernel, or those events that are involved in mutually causal interaction at

[3] Greg Dale, a sports psychologist puts it this way: "Some people think my job is to make athletes think more about what they're doing. My job is really to make them think less." (Waggenspack, 2007).

any given time. Motor events are outside of the kernel because these are purely efferent actions, and thus remain external to awareness. A schematic for this effect is shown in Figure 6.5.A. A conscious thought in the kernel, represented by the triangle, sends a signal to the motor cortex to initiate an action (here we simplify a complex interaction between frontal planning, parietal and cerebellar determination of bodily trajectory, and final action via motor cortical interaction and muscle triggers). The motor act itself does not feed back directly to the kernel and thus remains external to awareness.[4] The situation is no different in the case of a detector monitoring the motor cortex in order to influence an actuator (Figure 6.5.B), as would occur if a BMI was involved. There is the initiating thought in the kernel, but then no direct feedback from the motor cortex or the actuator itself, and it will appear to the user of this system that his thought directly moved the actuator. He will not be aware, for example, of any calculations done by the signal-processing unit (not shown in the diagram).

It may be that some amount of conscious thought and/or conscious effort is initially required to produce the appropriate response at the actuator. For example, you may find that when you think of moving your actual arm to the right, the BMI misinterprets this and moves the robot arm upward. Other more indirect routes from thought to action are possible, depending on the location of the detector, the signal processing algorithm, and the nature of the task. For example, you may discover by accident that whenever you recall your Uncle Frank embarrassing himself at the family Christmas party that the robot arm shoots downward. This is shown in Figure 6.5.C, in which an initial conscious thought on the left triggers another conscious thought, which in turn triggers the motor action. Thus, it is possible, at least initially, for control of a BMI device to not be completely transparent, and involve the production of other thoughts. However, eventually a direct connection will form between the intention to act and the action itself via associative learning. This automatization process, for example, is responsible for the fact that when we ride a bike, we no longer think about how to shift our weight to maintain balance and to effect a turn; we simply concentrate on the intended actions. Likewise, the situation to the left of Figure 6.5.C will eventually convert to that of the right of Figure 6.5.C, where adaptation has produced a more direct connection between thought and action, and the unneeded intermediate thought (of Uncle Frank, for example) drops out of the processing sequence.

[4] There is, of course, sensory and kinesthetic feedback accompanying any motor action. Thus, when catching a ball, you are aware of the original intent to do so, and you can feel your arm move through space, but you are not aware of calculations of the trajectories of hand and arm that made the catch possible.

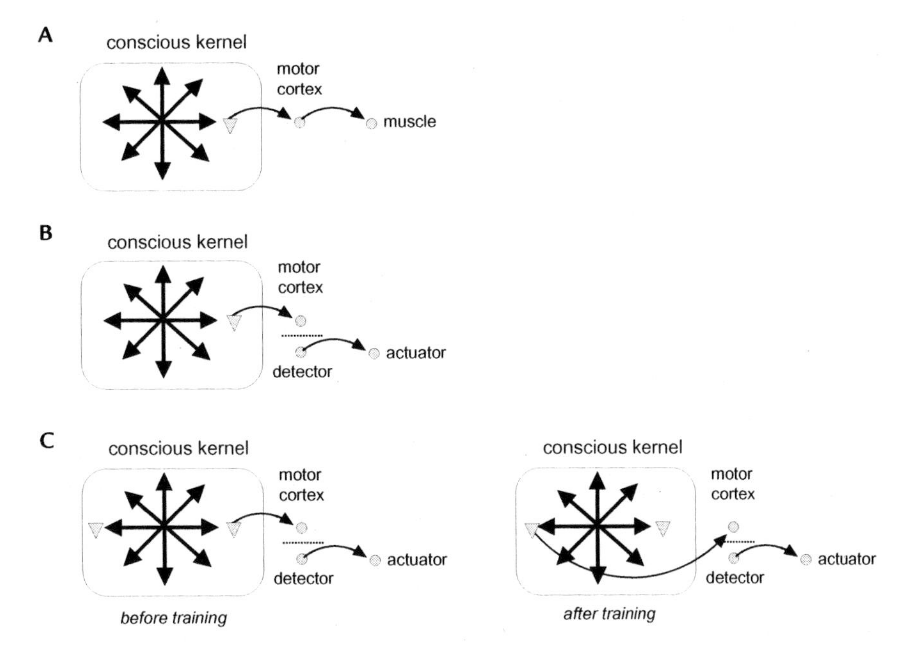

FIGURE 6.5 **A: Under ordinary circumstances, a thought to move in a certain way (represented by the triangle) is translated into a motor action outside of the conscious kernel, and thus one is only aware of the original thought and not the motor sequence producing it. B: Likewise, a detector picking up signals from the motor cortex and driving an actuator represents an analogous situation: one will only be aware of the thought inside the kernel, and it will seem as if this thought drives the actuator directly. C: During training, the effort involved may result in extra conscious thoughts being generated; however, the associative learning will eventually allow a direct connection between the initiating thought and the neural set being detected, and transparency will also be achieved.**

That control of a BMI entails a large degree of transparency, either initially or eventually after training, should not come as a surprise; it is simply a more technologically advanced version of what happens all the time. When we drive a car, for example, after passing through the awkward learning stage, we no longer ordinarily process most control events at a conscious level. If we want to go faster the thought translates into pushing down on the accelerator without us having to notice this action directly. It is as if the acceleration of the engine is driven directly by our minds, rather than our feet. It is only in a difficult situation, such as parallel parking, where the computation is difficult to automate, that the steering wheel, the accelerator, and the brake return to awareness. Likewise, brain-machine interfaces will eventually become seamless extensions of the self, enabling greater physical and intellectual power as if by magic. We will return to this topic in Chapter 8, where cyborgs, true man-machine blends, are discussed in more detail.

READING FROM THE BRAIN

There are two primary means of reading the brain, noninvasive and invasive. The former include a variety of techniques that detect brain signals without direct penetration of the skull; the latter is currently characterized by just a single technique wherein electrodes or arrays of electrodes are implanted in the brain. In general, invasive reading can only be achieved after drilling through the skull and physically inserting an electrode into the brain; as one might expect, this process is beset by a number of other complications. However, the difficulty of this method is offset by the ability to obtain a much higher quality signal. In contrast, all noninvasive techniques suffer from a large degradation of the neural signal before it reaches the recording device, leading to problems with noise, and with spatial and temporal resolution. These two methodologies are discussed first. There is a third possible technique in which electrodes are placed beneath the skull but on the surface of the cortex. As discussed at the close of this section, this semi-invasive procedure may eventually hold the most promise and act as a compromise between the potentially destructive nature of invasive reading and the poor signal quality of noninvasive techniques.

Noninvasive methods

Table 6.2 summarizes the characteristics of the primary noninvasive imaging techniques to be considered here[5], including their spatial and temporal resolutions, and specific limitations with respect to their use in BMI. As with all discussions in this chapter, the primary focus is on this latter capability; no attempt will be made to summarize the extensive literature on both the physics of such devices and clinical and research results. It should also be noted at the outset that currently, none are ideal BMI tools, either because not enough information can be extracted from the brain in any given time period (low bandwidth), or because of difficulty of use and high expense, or both. However, this may change in the future, obviating the need to move to invasive and semi-invasive techniques to produce a viable system.

The first modality we will consider, the EEG, was also the first to be discovered historically. German physiologist Hans Berger began experimenting with the human EEG in the 1920s as a means of better understanding the brain's response to various arousal conditions. Although the EEG signal is now recorded (and analyzed) digitally, the fundamental nature of this device has not altered much. Electrodes are attached to the scalp, ideally after placing a conducting gel between the electrode and the skin to minimize the impedance and to reduce

[5] PET and SPECT, not discussed in this section, are useful in clinical and research settings but difficulty of use make their application to BMIs unlikely.

TABLE 6.2 A summary of the characteristics of the primary noninvasive imaging methods.

Technique	Spatial resolution	Temporal resolution	Limitations wrt BMI
EEG	2–5 cm	Millisecond range	Low bandwidth, largely based on cortical potentials, artifacts
MEG	~1 cm (depending on the number of sensors)	Millisecond range	Expense, poor spatial resolution
fMRI	1–3 mm	1 second	Expense, poor temporal resolution, response latency
fNIR	1 cm	.1 second	Neocortical signals only, response latency

the noise in the signal. Figure 6.6 shows a rather fanciful application of EEG, that of creating music (see Chapter 8 for a fuller discussion of the intersection between neuroengineering and art); typical applications are much more mundane and include the diagnosis of pathologies such as epilepsy, the identification of sleep cycles, measurement of a so-called evoked response to a presented stimulus, and the extraction of fundamental rhythms as a means of identifying the general state of the brain or a portion thereof.

EEG can be characterized by the band in which the dominant frequency of the signal falls; Figure 6.7. illustrates some of these bands. Delta waves (DC to 3 Hz) are present in adults in deep sleep, and also in infants. Theta waves (4 to 7 Hz) indicate drowsiness in adults, and are also present in children when not drowsy. The presence of alpha waves (8 to 12 Hz) is an indication of an awake but relaxed state. Occipital alpha is blocked by visual processing and converts from a slow synchronized (i.e., high-voltage) signal to a faster asynchronized state (i.e., low voltage) upon eye opening. Likewise, the so-called mu-alpha rhythm over the sensorimotor cortex is blocked by contralateral motor movement. In both cases, alpha is replaced by the beta (12 to 30 Hz) rhythm, a catchall category for any state in which brain activity in the measured region is relatively high. There has been recent interest in gamma waves (>30 Hz) as an indicator of long-range coherence in the brain (see the discussion of insight in Chapter 9). However, gamma is difficult to measure with EEG because these high-frequency waves are filtered out by the scalp and skull, and is better assessed with an ECoG, as discussed later.

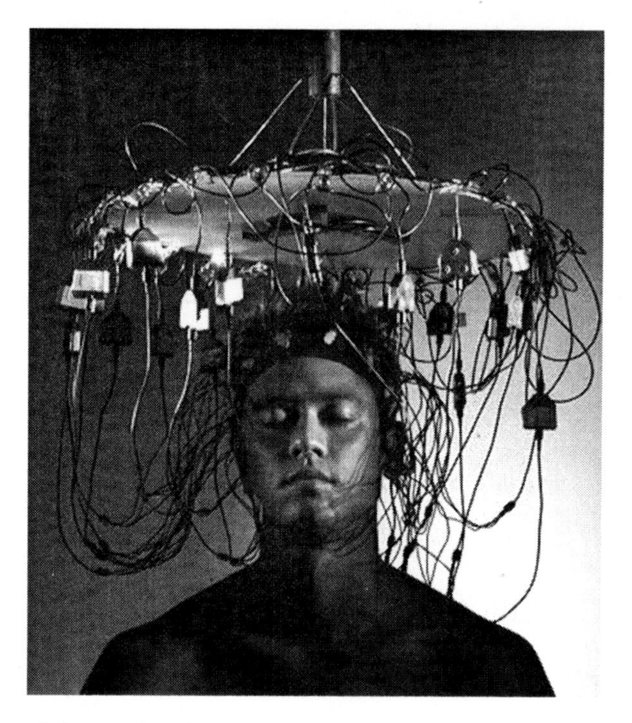

FIGURE 6.6 **EEG being used to drive a device that creates music.**
SOURCE: **http://en.wikipedia.org/wiki/Image:Musical_brainwave_performance_at_deconism_gallery.jpg.**

Thus, the characteristic frequency of EEG may be thought of as an indicator of the overall arousal state of the given cortical area, and cannot easily be modulated for BMI purposes. The best bet for obtaining fine-grained control information from this signal is to look at the variation within a given waveform. Indeed, as shown in Table 6.2 EEG possesses the best temporal resolution of any of the listed techniques with the exception of MEG. Unfortunately, this does not currently translate into a methodology for extracting large amounts of information from the brain. The central problem is that the EEG signal at any given electrode represents the summation of the effects of large numbers of neurons, whereas thoughts and intentions may be encoded at a finer grain, or in signals that EEG cannot pick up. The EEG signal is believed to originate in the summation of the postsynaptic action (i.e., dendritic) potentials of vertically oriented pyramidal neurons; the axon potentials of both these neuron types and other neuron types are randomly oriented and thus these signals tend to cancel out. Thus, EEG represents a mass action effect, and not necessarily a comprehensive one at that. Furthermore, the dendritic potentials that *are* captured represent the sum of two distinct effects, that of thalamocortical inputs, and contralateral cortical

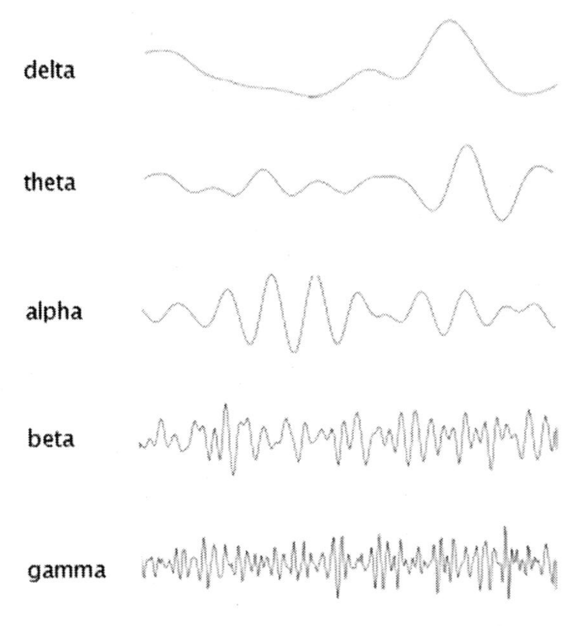

FIGURE 6.7 **Example waveforms in each of five EEG frequency bands.**
SOURCE: **http://en.wikipedia.org/wiki/Electroencephalography.**

afferents.[6] In addition, the EEG signal is smeared by the skull and the scalp, further reducing the amount of usable information. Finally, EEG is beset by numerous muscular and ocular artifacts, which must be filtered from the signal if it is not to be dominated by misleading noise. Taken together, these factors are currently prohibitory in the creation of a high-bandwidth BMI. However, there are some intriguing results on the horizon, including the harnessing of multiple electrodes to increase bandwidth, and the measurement of emotional states with EEG; these are discussed later in this chapter.

Magnetoencephalograpy (MEG), a more recent technique than EEG, measures the magnetic field of the brain instead of the electric field. Its chief advantage with respect to EEG is that the magnetic field is not affected by organic intermediaries; thus the intervening structures between the device and the relevant neurons, the scalp, skull, and cerebrospinal fluid (CSF) do not interfere with this signal. However, the magnetic field produced by the brain,

[6] It is the former that are responsible for the large-scale rhythms found in EEG. When the thalamus is in so-called burst mode, it induces synchronized firing in cortical pyramidal neurons, resulting in slower waves (alpha or below) with higher amplitude. When it is in transmission mode, the idiosyncratic nature of its input allows sensory stimulation to dominate. Thus, burst mode is characterized by drowsiness or sleep, transmission mode by arousal and alertness.

even when large numbers of pyramidal cells are being stimulated in unison is extremely weak and is at approximately 8 orders of magnitude less intense (100 million times weaker) than the magnetic field associated with the earth. To adequately detect these weak fields, superconducting quantum interference devices (SQUIDSs) are necessary. These need to be supercooled, typically by liquid hydrogen, making the cost of a typical MEG device on the order of 2 to 3 million dollars. This makes it prohibitive in a BMI setting, and in fact very little work has been done with MEG in this context. The layered magnetic shielding for a typical such device is shown in Figure 6.8; it need hardly be added that even if expense were not an issue, this is not a device that could be used outside of the laboratory.

In addition, MEG is beset by many of the same difficulties as EEG. First, because the magnetic field produced by the brain is so weak, it is subject to a number of magnetic noise sources. In many cases, though, this can be largely eliminated by the use of dual detection points, in which constant magnetic fields are subtracted from the signal, while larger differences are left remaining; long-range magnetic fields have a relatively constant gradient

FIGURE 6.8 **Magnetic Shielding necessary for MEG. Courtesy of Elekta.**
SOURCE: **http://www.elekta.com/healthcare_international_functional_mapping.php.**

whereas the short-range magnetic field produced by the brain, though weak, falls off rapidly with distance and therefore produces a much higher relative gradient. Second, like EEG, a MEG measures the dendritic activity of a large numbers of cells; it has been estimated that it takes approximately 100,000 similarly aligned and synchronously firing neurons to produce a detectable signal. Thus, like EEG, a MEG is measuring a mass action effect, and one that excludes nonaligned cells and cells with postsynaptic potentials out of phase with the general population.

The final problem that a MEG shares with EEG is source localization. Both of these techniques idealize the production of electric fields and magnetic fields, respectively, as the result of dipole sources, that is, a set of ionic currents that flow in approximately a straight line in the brain. Assuming a set of dipoles one can uniquely calculate either the associated electric or magnetic fields. However, the converse is not the case. Given a set of signals read from a number of detectors, there will, in general, be many nonunique sets of dipoles capable of producing this signal collection. This makes localization of activity on the cortex somewhat problematic, and limits the ability of these techniques to read subcortical activity. To some extent, with the aid of simplifying assumptions, one can make reasonable guesses as to the location of the dipoles, and recent advances have greatly aided this effort (see Michel et al. 2004).

However, this methodology does not produce the degree of certainty with respect to the three-dimensional locus of neural activity as the next method in our noninvasive imaging litany, functional Magnetic Resonance Imaging (fMRI). In many ways, fMRI has become the dominant imaging modality because of its excellent spatial resolution and the ability to unambiguously reveal the workings of the brain beneath the cortical surface.

The full details of the mechanism behind fMRI can be quite complex and are beyond the scope of this book, but can be summarized as follows. The metabolic demands of an increase in neural activity in a given region initially produce a reduction in oxygenated hemoglobin in that area. However, this is quickly followed by an overshoot in which the brain, perhaps anticipating future activity in this region, increases oxygenated hemoglobin, and there is also a corresponding decrease in deoxygenated hemoglobin levels. It is the difference between these quantities, the so-called blood oxygen level dependent (BOLD) contrast, that fMRI measures.

This is accomplished by means of a strong magnet, typically on the order of 3 tesla or greater. Hydrogen atoms become aligned with the field produced by this magnet, and in addition are characterized by a resonance frequency proportional to the strength of the field. An additional oscillating magnetic pulse applied at this frequency creates a precession of the atoms around this field and in a direction orthogonal to it. By measuring this orthogonal field, the rate at which the return to the original orientation can be determined. Crucially, deoxyhemoglobin and oxyhemoglobin possess differing magnetic properties and also affect this rate; thus the relative concentrations of these substances can be detected. In addition, by creating a gradient of the magnetic field across one or more spatial dimensions one can alter the resonance frequency for different points in space, and thus selectively measure the localized BOLD response. By accumulating a number of such responses, a three-dimensional image of approximate neural activation can thus be produced.

Figure 6.9.A shows the activation map on a brain slice for a simple experiment in which a moving image alternates with darkness; this produces "hot

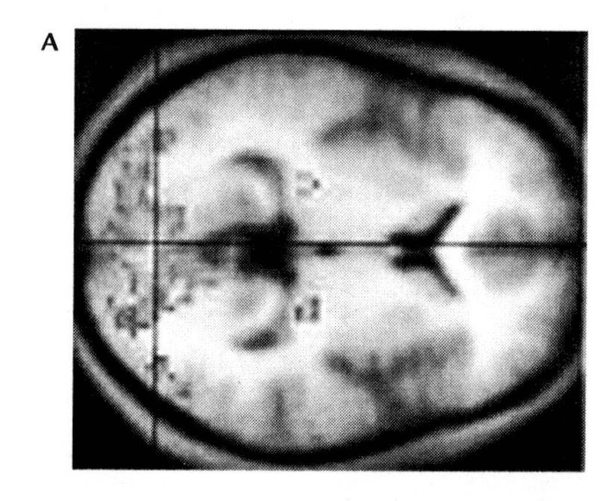

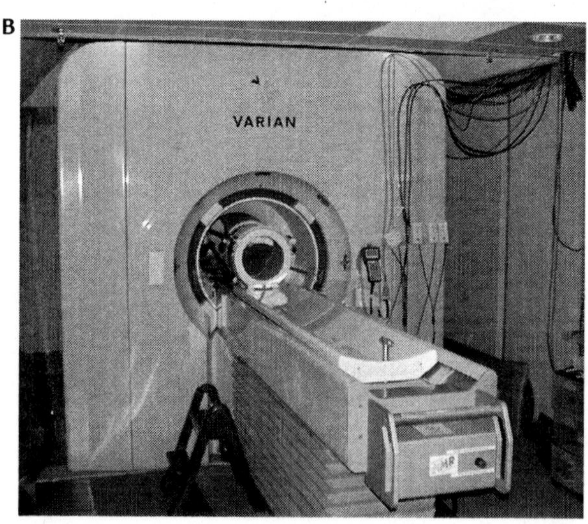

FIGURE 6.9 **A: An activation map for a simple fMRI experiment. B: A typical fMRI scanner.**
SOURCE: **A: http://commons.wikimedia.org/wiki/Image:FMRI.jpg. B: http://commons.wikimedia.org/wiki/Image:Varian4T.jpg.**

spots" (darker areas) indicating activation in the occipital cortex as expected. As in all such images, what is being shown is not absolute levels but differences in the BOLD response between the two conditions. Figure 6.9.B shows a typical fMRI scanner. As is apparent, such devices are extremely bulky, and the expense matches this size. While there are currently plans to develop smaller, more portable, and presumably cheaper fMRI imagers, the sheer cost of the device makes its use prohibitive for BMI in all but research settings. Additional limitations include the low temporal resolution; for the moment, controlling a two-dimensional mouse, for example, is out of the question when one is getting only a single signal per second. Finally, because the technology is based on the BOLD response, which in turn is based on cerebral blood flow, there is of necessity a delay between thought and action of at least a few seconds. This is perhaps the most difficult factor to overcome if fMRI is to be used as a real-time BMI.

Like fMRI, functional Near Infrared Spectroscopy (fNIR) also assesses blood oxygenation. However, fNIR uses a simpler, more direct, and consequently much less costly means of assessing this quantity. The skin, and to a lesser extent bone, are not completely opaque to light. You can see this by shining a flashlight on your hand; some of the light, particularly at longer wavelengths (red light) gets through. This fact can be leveraged to optically sense the levels of oxy- and deoxyhemoglobin. These substances have different absorption coefficients to infrared light, depending on the frequency. Two low-intensity lasers with wavelengths of 780 and 830 nanometers are used; by comparing the relative intensity of the small proportion of light that is reflected back to the surface of the scalp, the relative levels of oxygenation and deoxygenation can be obtained.

A typical fNIR device is shown in Figure 6.10.A, and the path of the light taken from emittance to detection is shown in Figure 6.10.B. As can be seen from the shallow path, it is not possible to measure deep brain structures with fNIR; typical depths are on the order of 3 cm. fNIR also shares many of the difficulties of fMRI with respect to BMI. It is dependent on the BOLD response, which is both an indirect measure of neural activity and also delayed with respect to this activity. However, the low cost and ease of use of fNIR may make this a viable BMI technology, especially if the response latency can be overcome. Work is currently under way that attempts to assess neural activity directly via optical means, and success in this project could provide an alternative to invasive strategies in the long run.

All of the above-mentioned imaging modalities, and especially fMRI, have blossomed in recent years as a means of understanding brain pathologies and also as a means of studying cognitive and affective processing in the normal brain. However, as the previous analyses indicate, all also suffer from considerable deficits in the context of BMI: either they are too expensive for continuous everyday use, do not provide sufficient information bandwidth, produce data

A

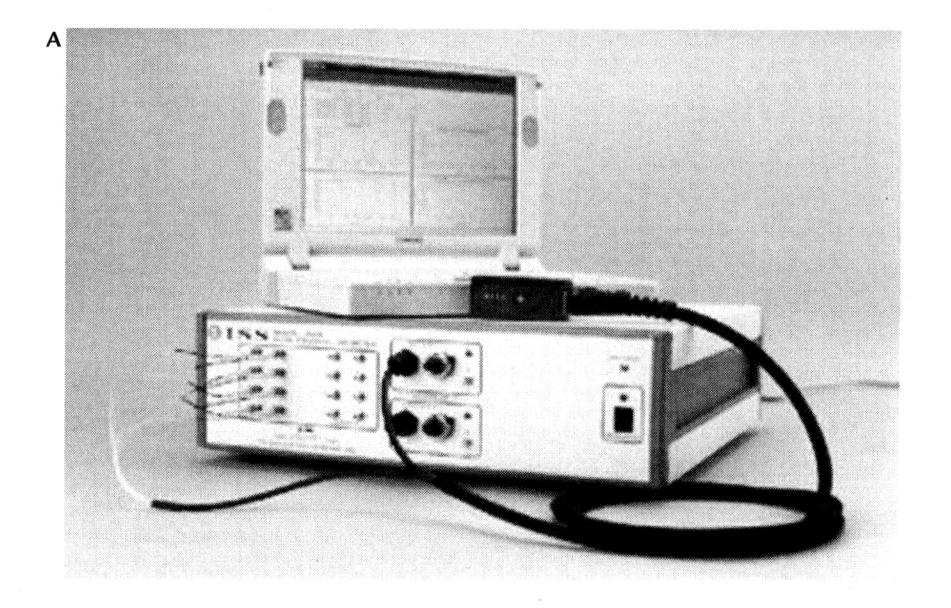

B

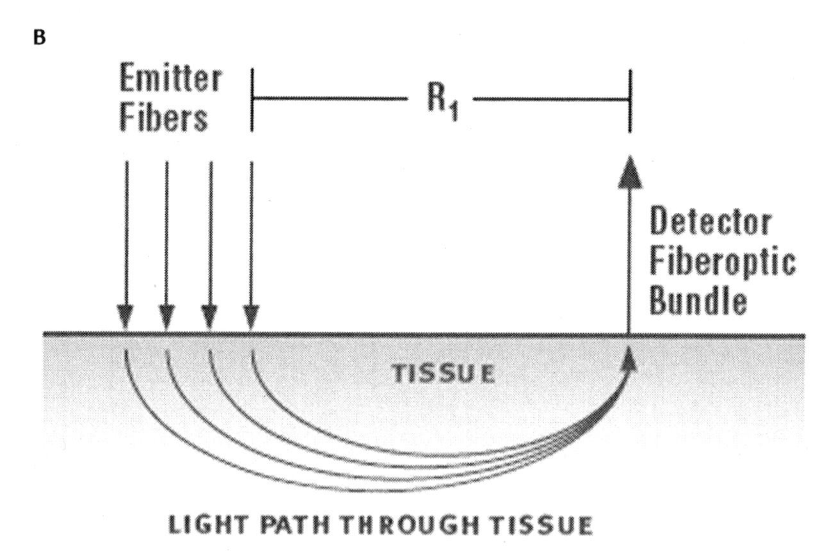

FIGURE 6.10 A: The OxiPlexTS fNIR device, ISS Inc. The black cord contains the optical fibers that guide the lasers to the target; there is also a detector on the emitter array at the end of the cord. B: Typical near-infrared light path.

that is too noisy, provide data that is delayed with respect to neural response, or some combination thereof. Although future advances may change this, the sober reality is that there are no current devices that can reliably and inexpensively indicate the activity of a relatively well-confined population of neurons in realtime. This has led researchers to consider a more radical step, that of placing invasive electrodes into the brain to reveal activity at spatial resolutions as low as that of the individual neuron and with temporal resolutions that reveal individual firings; this possibility is considered in the following section.

Invasive methods

Estimation of what the brain is doing using extracranial devices is a bit like trying to figure out what a computer is doing by measuring the electric field on the outside of the casing. The surest way of obtaining electrical signals is by measuring on the circuit boards themselves; likewise, intracranial measurement with invasive electrodes provides the most direct route to neural activity. A number of related approaches have emerged in recent years; all involve the use of thin electrodes inserted into the brain. While it is possible to record with such electrodes, as described previously the coarse coding of response described earlier usually means that one will want to record from multiple units within a given area. One means of doing so is with a microelectrode array (Figure 6.11). The primary disadvantage of a grouped array such as this is that unlike individual microwires, it cannot measure deep structures in the brain (the maximum insertion depth is 1.5 mm). An alternative that does allow some degree of multiple recording and depth placement is to place multiple recording sites on a single probe. This also permits recording along a vertical column as opposed to the horizontal readings produced by an array.

A single microwire or element in an array can potentially record in three ways: single-unit, multiunit, and local field potential (LFP). Single-unit recording registers the axonal spiking of individual cells. Multiunit recordings register the superimposed spiking activity of multiple cells. Recent evidence has suggested that a better prediction of movement can be obtained with this scheme (Stark & Abeles, 2007), although much more work needs to be done in order to confirm this result and its generality over a range of tasks. An LFP recording registers not axonal spiking but the sum of the dendritic activity in the implant region. Thus, it may better correspond to EEG signal, and it has also been suggested that it better corresponds to the BOLD response obtained with fMRI or fNIR (Scherberger, Jarvis, & Andersen, 2005). In addition, because LFP records afferent influences on a region as opposed to the final efferent decision encoded by spiking, it may better reflect high-level intentions as opposed to actual actions; considerable more research is also required to confirm this claim.

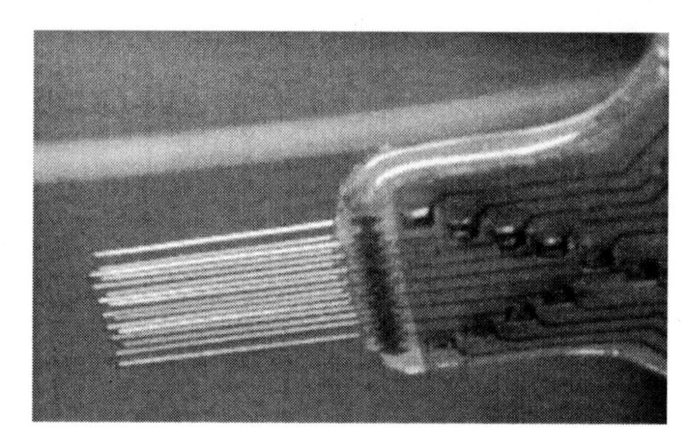

FIGURE 6.11 **A 16-wire microelectrode array.**
SOURCE: **http://en.wikipedia.org/wiki/Image:16wire_electrode_array.jpg.**

Regardless of the signal being detected, one has the choice of both the number of electrodes and the number of distinct sites to record from. The Carmena et al. (2003) results discussed earlier, and *a priori* considerations, suggest that the more the better, especially given that not all electrodes will remain stable over the course of a chronic implant. However, these results also suggest that there are diminishing returns (see Figure 6.4) once the number of electrodes reaches a critical quantity, and it may be desirable to use less in order to minimize trauma. Carmena et al. have additionally suggested that recording from multiple regions in consort can, in principle, provide a more robust signal than from a single region. Once again, much more work is required to show that nonmotor intentions can be read across multiple neural modules; it is likely that the distribution of such is context-dependent, with higher-level cognition more distributed than simpler motor actions.

While it is clear that direct microelectrode implants provide the best means of obtaining a high bandwidth signal with multiple degrees of freedom, there are a number of impediments to the use of chronic electrode implants to record cell activity in humans. First, there is the obvious fact that any such insertion requires brain surgery and its attendant risks. Next, there is the possibility of trauma from the insertion itself. There is some evidence that slow insertion may reduce this risk (Moxon, et al. 2004) but it is unlikely that injury to cells that are on the path that the electrode takes to reach its target is completely avoidable. Finally, glial cells will form around the electrode tip in an attempt to expel this foreign body. This process, known as gliosis, can cut off the electrode from the neuron or neurons it is recording from, degrading the quality of the signal, and also may be dangerous in itself due to the influence of necrotic substances on nearby cells. Ongoing research may alleviate this problem, for example the Kennedy electrode is encased completely in glass, insulating the electrode from gliosis. The case contains neurotrophic chemicals that encourage nearby neurons to move closer to the electrode.

Semi-invasive methods

As the previous two sections indicate, BMI stands on the horns of a dilemma. There is the choice of either acquiring brain signals via noninvasive means and having to live with noisy and low bandwidth signals, or alternatively of implanting electrodes directly into the brain, with the attendant short- and long-term difficulties that this route entails. One possible means of skirting this dilemma is via an electrocorticographic (ECoG) reading. ECoG works in a similar fashion to EEG, but with the electrodes placed directly on the cortex instead of the surface of the scalp. Like EEG, an ECoG registers primarily the synchronized dendritic activity of a large population of pyramidal neurons; however, it provides a considerably cleaner signal than EEG, with better spatial resolution, less filtering of high frequencies, and far less vulnerability to artifacts. In contrast to depth recording, because there is no cortical penetration, it provides less clinical risk and potentially longer-term stability.

Of course, one still needs to remove the skull to place the electrodes, and this is the single greatest barrier to both scientific testing and, ultimately, to practical use. Accordingly, most work that has been done to date has concentrated on patients that need to have their cortex exposed for other reasons, for example, those in preparation for surgery to remove the focus of epileptic activity. Not surprisingly, there are still relatively few volunteers for this surgery who are not compelled to need it. Later in this chapter we will discuss how the improved signal-to-noise ratio afforded by ECoG over EEG has enabled BMI devices with greater degrees of freedom and bandwidth than is currently possible with the latter technology. These results indicate that if a technology were invented which is able to accurately place electrodes beneath the skull but without actually removing it, it would provide a great boon to BMI, and could conceivably form the primary means of obtaining brain signals both for the disabled and for cognitive/physical augmentation in the able-bodied.

SIGNAL INTERPRETATION

Regardless of the means of reading from the brain, considerable signal manipulation must be performed to render these data into a form applicable to the task at hand. A loose analogy can be drawn with a device driver for a computer: the native digital signals must be translated before they can be understood by the device. Here, the problem is considerably more complex, however. The data obtained from the brain will be voluminous, it will be noisy, and it will likely be partially redundant. Two distinct steps are required before it can be reduced into usable form. First, relevant features must be extracted from the data and separated from the noise, and then these features must be translated into the appropriate action.

The elimination of noise and artifacts from the data is largely modality-dependent, and these methods are beyond the scope and interest of this book; the interested reader is referred to (Landini, Positano, & Santarelli, 2005; Sanei & Chambers, 2007) for a complete cataloging of these methods and for the appropriate references for fMRI and EEG respectively. Nor we will attempt to be comprehensive with respect to the feature translation process. Instead we will concentrate on representative algorithms for each of two problems, linear and nonlinear discrimination. We will close this section, however, with an under-represented approach to signal processing, namely, a discussion of the interplay between a priori assumptions about how the brain works and how to interpret the signals it generates.

The linear discrimination problem is represented in schematic form in Figure 6.12.A. In this simplified example, a set of two distinct actions are present; for concreteness, we can think of them as situations in which the subject attempted to move a mouse to the left (positive examples) or the right (negative examples). The examples are placed on the graph according to the values of the features being measured. For example, the two dimensions shown in the figure could be the firing rates of two neurons being recorded with electrodes. The task of the linear classifier is to derive a line that separates the positive from the negative examples. If it is able to do so, it can then predict the intended action merely by looking at the firing rate of the two neurons, and these cells could then be used to drive the mouse directly.

Figure 6.12.B shows a problem in which it is not possible to linearly separate the data. There does not exist a line such that all positive examples are on one side and all negative examples on the other. This nonlinearly separable problem requires a more advanced learning algorithm, and one that will usually take considerably more time than that needed for the simpler case shown in Figure 6.12.A.

The next two sections will describe, in more detail, two possible adaptive algorithms for acquiring the correct classifications. Before proceeding, however, it is worth bearing in mind the difference between an actual BMI learning process

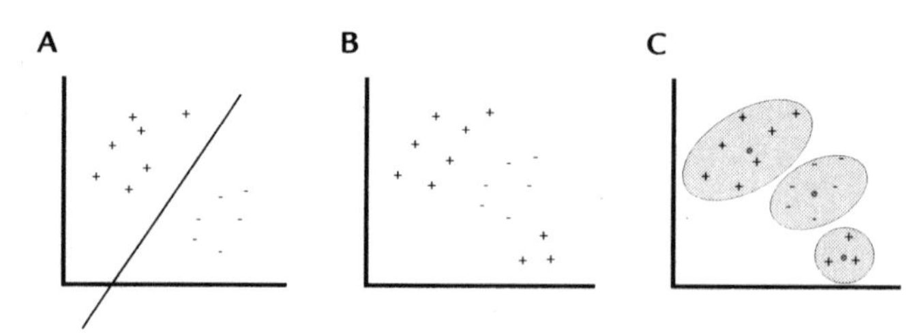

FIGURE 6.12 **A: A linear discrimination problem, with positive and negative examples clearly separable by a straight line. B: A non linear discrimination problem. Here there is no single line that separates the positive from negative examples. C: The formation of example clusters in the LVQ algorithm. Once separated in this way, the clusters can easily be classified.**

and the toy examples shown in Figure 6.12: a) typically, many more dimensions of information will be present (in that case, the line in Figure 6.13.A will become a hyperplane in this multidimensional space), b) many of these dimensions will have little or no information with respect to the separability of the classifications and therefore must be ignored or effectively filtered from the learning procedure, c) the output, depending on the context, may need to be a numerical value rather than a discrete category, or there may be multiple discrete outputs corresponding to multiple classifications, or both; this is inherently a harder learning task, the difficulty of which depends on the granularity of the numerical classification and/or the number of categories, d) noise in the data will often place a particular data point where it should not be (for example, a negative example may end up in the middle of the positive cluster), and finally, and most significantly for BMI purposes, e) learning the correct classification on any given set of prior data does not guarantee the same performance on future data; the brain is a dynamic organ and responses for a fixed task are in a constant state of flux. Any of these factors alone or in combination will serve to produce output errors, regardless of the degree of training. Of course, the subject can then try again to achieve the desired effect, but this will lower the effective bandwidth of the system.

Linear methods

All linear methods, such as the Naïve Bayes Classifier and Linear Discriminant Analysis (Duda, Hart, & Stork, 2000), attempt to separate data such as that shown in Figure 6.12.A by forming lines in two dimensions or hyperplanes in n dimensions. In general, we can say that a given set of feature values x_i are placed in a category j if, and only if,

$$\sum_i w_{ij} x_i > \theta_j,$$

where the w_{ij} are weights and θ_j is the threshold for this category. The task of the classifier is to generate these weights and the threshold such that this inequality holds for all examples of the given class but for none of the examples of the other classes.

One relatively simple algorithm for doing so, and the one we will examine here, is Perceptron learning (Rosenblatt, 1958). As first discussed in Chapter 3, Perceptron learning iterates through the set of examples, both positive and negative, each time adjusting the weights according to the rule

$$\Delta w_{ij} = \lambda(T - x_j)x_i,$$

where Δw_{ij} is the change in the weight between the ith feature and the category j, x_i is as before, x_j is the classification (1 if the weighted sum given earlier is greater than the threshold, and 0 otherwise), T is a teaching signal stating whether the output x_j should or should not be 1 for this example (i.e., if the example is a member of the class), and λ is a parameter that controls the rate of learning. It is easily seen that the weight changes only occur if the teaching signal T does not match the actual output; if these two are the same the classifier is behaving correctly, and it is assumed that there is no need to make any adjustments.

Significantly, if there exist a set of weights that are capable of separating the positive from the negative examples, then the perceptron algorithm is guaranteed to find them, and in a relatively short amount of time. Alternatively, if the problem is not linearly separable, then no amount of training can provide the correct response. As discussed in Chapter 3, this motivated the introduction of backpropagation and other more powerful algorithms. However, in BMI contexts, it will often turn out that a linear classifier, such as the Perceptron, is able to provide good results, or at least as good as the nonlinear methods described in the following section, and with significantly fewer training iterations. The reason for this is the large amount of features typically associated with most imaging modalities. The larger the number of features, the more likely it is that a subset of them will separate the data into linearly separable classes. For example, when recording over a large number of neurons, it will often turn out that a subset of these are encoding for a particular response. They will do so noisily, that is, the output of the classifier will not always be correct; however, because the underlying situation (without the noise) is essentially linear, it may turn out that a nonlinear classifier does not significantly improve performance.

Nonlinear methods

If the learning task is especially difficult, or if there are relatively few dimensions of information, it may turn out that the examples look something more like Figure 6.12.B. In this case, the positive and negative examples are not linearly separable, and an alternative approach is called for. Possibilities including the backpropagation algorithm (Werbos, 1988), Support Vector Machines (SVM) (Cristianini & Shawe-Taylor, 2000), and the algorithm to be described here, Learning Vector Quantization (LVQ) (Somervuo & Kohonen, 1999). LVQ comes in many variants; here we describe the essence of the approach. The algorithm is an adaptation of Kohonen's Self-organizing Maps (SOM), an unsupervised neural network learning technique, to the supervised case.

In SOM, categories are formed on the basis of the Euclidian distance to a prototype, or centroid of the category. This is illustrated in Figure 6.12.C. Here, there are three distinct clusters of examples, with the centroid marked as a dot in each. As each example is presented to the system, a winning unit is chosen among these classifiers by finding the minimal distance among the vectors corresponding to values of the features and the vectors for the weights from the features to each of the classifying units. The weights between the features and this winner are then steered closer in the direction of the feature vector according to the rule

$$\Delta w_{ij} = \lambda(x_i - w_{ij}),$$

where x_i is the value of feature i, w_{ij} is the weight between this feature and the winning unit j, and λ, as before, is the learning rate. It is easily seen that the net effect of this rule is to pull the weight closer to the input (the change is in the direction of the difference between the values of these quantities), and when this is done on all the weights and all the examples the weight vector will come to occupy the centroid of the category.

Furthermore, it is also easy to see that the classification task has been "linearized." All that is needed is a separate classification layer that takes the SOM category outputs as inputs to another layer. In this case, illustrated in Figure 6.12.C, this extra layer will classify any input that belongs in the upper or lower cluster as a positive example, and those that belong in the middle cluster as a negative example. Variants of LVQ allow the classifications of examples to influence the SOM categorization via reinforcement of correct classifications, and other variations in which the classifications attempt to optimally divide the example space, but all follow this basic procedure.

A practical, general comment is in order with respect to linear and nonlinear classification. One can spend a tremendous amount of time playing with various learning schemes in order to optimize performance. However, whenever possible, it is usually far better to transform the problem space than to tweak the classification algorithm. An excellent example of that is provided by the work of (Birbaumer, et al. 1999). These researchers found, consistent with the low-bandwidth characteristics of EEG, that it was extremely difficult to get subjects to move a mouse to a given screen location to select a letter by a series of discrete mouse movements. However, they achieved far better results by trying to read the desired letter directly from the EEG signal. A sequential, time-consuming task was reduced to a single discrimination, yielding far better effective bandwidth. It is unlikely that tweaking classification algorithms alone would have yielded as good as a result. Here, as in many other cases (not just in adaptive routines for BMI but in artificial intelligence (AI) problems in general) a clever representation scheme almost always beats a clever algorithm hands down.

Mind reading vs. brain reading

A characteristic of most approaches to signal manipulation is that they make few assumptions about how the brain works; they treat it as a black box from which to extract the appropriate information. However, it may be possible to leverage a priori assumptions about neural processes in order to enable better performance.

One area in which this may be possible is the separation of conscious strands of thought from the overall neural context in which they are embedded.

The current focus of neuroengineering is on enabling the disabled with motor tasks, and as argued previously, most of the multidimensional information associated with these acts is of necessity beneath awareness. However, as the field progresses, we will see more and more effort devoted to reading the conscious signals only. *Pace* Freud and the psychoanalytic tradition, there is no modern neurological evidence for repressed demons or other items of interest dwelling beneath the surface of consciousness. What lies beneath awareness consists of: a) neural activity not in coordination with that subserving the current conscious stream, and b) synaptic efficacies not currently being accessed. The former are as external to our identities as any other aspect of the universe not in causal interaction with our brains, and the latter cannot be imaged in any case—they are outside of the dynamic chain entirely. It is the conscious stream that contains our hopes, wishes, and desires, both immediate and long-range, and a focus on this stream could act as a filter that removes unwanted noise from the read signal. Furthermore, the ability to build a mind-reading device, i.e., a device that reveals just what a subject is thinking or experiencing at any given moment (see Chapter 8), depends critically on this separability.

The Tong et al. (1998) study was the first published attempt to perform this separation; they looked at the BOLD response as revealed by fMRI in the two areas, the fusiform face area (FFA) and the parahippocampal place area (PPA) during binocular rivalry. They found that the former area "lit up" when the subjects attended to the eye receiving a picture of the face, while the latter area showed evidence of greater activity when a picture of a house was attended to. A fascinating and potentially more meaningful study, in that it demonstrated that one can separate the conscious from unconscious signal within the same neural module, was recently carried out by out by Kamitami and Tong (2005). They first trained a linear classifier on sets of voxels (three-dimensional pixel volumes) to correctly classify lines of differing orientation. They then presented subjects with two sets of lines oriented perpendicularly to each other, but had them only attend to one set of orientations. They found that the same classifier accurately predicted which of the two orientations the subjects actually saw, despite the fact that both line sets struck the retina with equal intensity. This study is notable in that: a) consistent with the predictions made in Chapter 4, the primary visual cortex contributes to the construction of the conscious perceptual field, and b) conscious information is there for the taking, even with an imaging modality that is relatively poor at producing fine-grained temporal information. A similar technique was used by Haxby et al. (2001) to detect the category of presented objects (e.g., shoes, bottles, faces) by processing activity in the ventral temporal cortex, although no attempt was made to separate attended from unattended stimuli.

These experiments, however impressive, still use purely data-driven methods to separate the conscious wheat from the subconscious chaff, and it is likely that there will be strong limitations that will arise with these techniques when we try to obtain more precise information regarding conscious content. If the proposal put forward in Chapter 4 with regard to consciousness has any validity, however, then we ought to be able to apply theoretical means to enhance this process. That chapter argued, among other things, that consciousness is most likely a function of the cell populations in the kernel, i.e., those populations in mutually causative relationships during any given time slice. This will manifest itself in differing ways depending upon the imaging technology, but assuming that it has a fine enough temporal grain it should be possible to separate out those signals that are in synchrony with each other and those that are not. Furthermore, we can also assume that if this synchrony extends over a sufficiently long period of time (say, >50 ms), then it is unlikely that this temporal alignment is a coincidence. We will then be able to build up an image of the brain that consists not merely of cells that are active, but those that are part of the kernel. If we were to apply visualization software to this object, it will look something like waves of activity that course through the brain, with differing modules moving in and out of the kernel as they come "online" in order to contribute to the construction of conscious content.

In order to accomplish this, the imaging modality must have sufficiently good temporal resolution to be able to detect these synchronies. Currently, EEG and MEG are the only candidates for such a procedure, and indeed, Tononi and Edelman (1998) have proposed methods for extracting what we are calling the conscious kernel from the overall electrical signal. As other imaging techniques improve, it may also be possible to apply a similar methodology to the signal. Regardless of the source of the signal, it cannot be emphasized enough that in many cases, and especially for high-level applications, we are interested not so much in reading from the brain, but from the mind, i.e., the subset of neural activity that subserves conscious thought. A full-fledged mind-reading device, however, will require not only identification of the conscious kernel but a means of translating the accompanying signals into specific conscious contents. This procedure still eludes us, but if the mental supervenes on the physical, then such a procedure must exist, and if the way in which it is does so it law-like, then it should be possible in principle to discover it (see Chapter 8).

REPRESENTATIVE EXAMPLES

Complex engineering feats, such as BMIs, are often best understood by reference to working systems, rather than general principles, because the final product can only be understood as the sum of the interaction between its parts.

Accordingly, we consider representative examples in three categories in this section, invasive animal, noninvasive human, and invasive human. No doubt many of these systems will seem quaint in construction and weak in performance a mere few years hence, but future systems will only arise once the limitations in current ones are well understood.

Animal, invasive

Invasive animal BMIs have been most closely associated with the pioneering work of Miguel Nicolelis and colleagues at Duke. Nicolelis's early work concentrated on controlling simple actions. For example, a rat was trained to press a lever for a reward of water. The activity of motor neurons was monitored during these pressings, and a model was built that predicted when the lever would be pressed. The output of the model was then used to control the lever, and at the same time mechanical control was suspended. At first the rat was perplexed that its reward was cut off, and then it turned its attention to other matters. Eventually, however, it found out that it could operate the lever by its thoughts alone, and it was able to generate its reward without any motion.

These researchers also tested the transfer from mechanical to purely thought-driven control in the context of a considerably more complex control task with monkeys (Carmena et al. 2003). Figure 6.13 illustrates the experimental setup. The macaque holds a pole that moves a robot arm; the pressure of its grip also controls the gripper on the arm. The animal receives feedback regarding the position of the arm in two dimensions on a display, with the size of the circle indicating the strength of the grip. This setup allows three separate tasks to be tested, movement of the arm alone, gripping alone, and the two in succession, simulating a real motion task. While the animal controlled the pole to achieve these goals, multiple electrodes at multiple sites in frontal and parietal regions were recorded from. Using linear and artificial neural network (ANN) models, a prediction of movement and gripping strength was made from these signals. The pole was then decoupled from the robot arm, and the animal's ability to control this arm via neural signals alone was measured.

This experiment produced a number of significant findings:

i) First, and most significantly, as with the rat and the lever, the monkey was able to make the transition from mechanical to thought control without undue difficulty. There was a small decrement in performance immediately after the transfer, but plasticity within the monkey's brain quickly closed the gap between this reduced performance and earlier performance levels.
ii) Visual feedback was absolutely critical; otherwise the monkey would have no idea what its thoughts "meant." Augmenting this feedback with

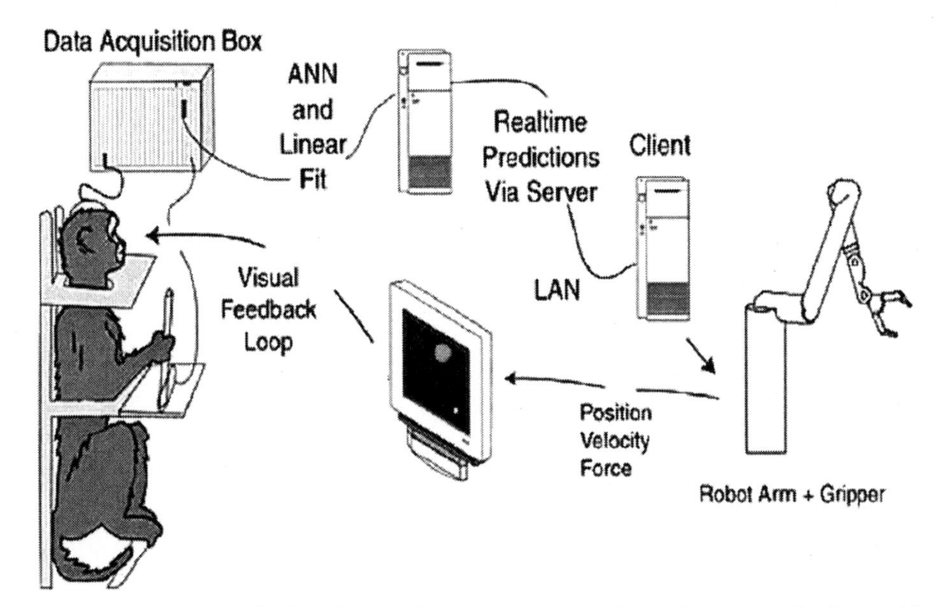

FIGURE 6.13 **Monkey think, robot do. The macaque controls a robot arm and gripper with a pole, and receives visual feedback. The animal also learns to control the arm with thought alone when the pole no longer drives the arm.**
SOURCE: **Carmena, JM Learning to Control a Brain–Machine Interface for Reaching and Grasping by Primates. PLOS Biology 1:2003.**

more direct proprioceptive input may enhance this process (see the following text).

iii) Multiple tasks were capable of being carried out (in this case reaching and grasping), without the signals for each interfering with each other.

iv) As discussed earlier in the context of coarse coding, the more electrodes included in the prediction, the more accurate it was, although adding more electrodes after a certain point produced diminishing returns.

v) Individual neurons tended to be more predictive of movement and grasping over time, indicating the existence of an adaptive process aided by the visual feedback. Nicolelis speculates that the robot arm (or at least the display corresponding to the arms position) became partially incorporated into the animal's body image, in much the same way that any tool or external device (a pencil, tennis racquet, or even a whole car) does the same with practice.

vi) Although as one might expect measurements from the primary motor cortex proved the most predictive of behaviors, neurons from other areas were also correlated with these actions, including those from dorsal premotor cortex, supplementary motor area, posterior parietal cortex, and primary somatosensory cortex.

An even more dramatic example of the same sort was carried out recently by the Duke lab in conjunction with the Computational Brain Project of the Japan Science and Technology Agency. Microwire electrodes at multiple sites recorded from a monkey walking a treadmill and the BMI was trained to predict the animal's movements from these signals. These predictions were sent to the Japanese lab over a high-speed connection that drove a robot to perform nearly identical movements on another treadmill (see the video at *http://www.youtube.com/watch?v=SSaBOd4pQpM*). Remarkably, the monkey was also able to continue to control the robot for a few minutes after it stopped walking itself simply by watching the robot.

As impressive as these experiments are, much work remains to be done in this area. Current focus is in two main directions. The first is to incorporate haptic (peripheral touch) and proprioceptive feedback into the system. Visual feedback is critical to drive the adaptive process, but it is likely that BMIs would benefit to a much larger degree from additional touch-based feedback. Visual feedback typically provides only partial information regarding all the degrees of freedom of a multijointed effector, such as an arm, whereas proprioceptive feedback provides immediate and full feedback (twirl your arm behind your head, for example; you have a very good idea of the position of your joints in space at all times without being able to see them). The other focus is to finally carry out what was the original goal of this research all along, the transfer of this technology to motor-impaired people. In this case, the output of the system could drive an external device, such as a wheelchair or a Segway, it could actuate an exoskeleton or other brace-like device, or use what is known as functional electrical stimulation (FES) to activate existing musculature. In the next few sections, we review the prospects of driving such devices with human BMIs.

Human, noninvasive

Human noninvasive BMIs may be divided into two categories: those using EEG, and those using fMRI. We can further divide the first category into techniques in which the subject attempts to modify some aspect of the signal in order to achieve control of a device, those that rely on the event-related potential (ERP), and those that attempt to extract the appropriate information from a number of aspects of the EEG signal.

Exemplary of the first technique is the Karim et al. (2006) work with the nearly locked-in patient HPS. HPS worked as a lawyer prior to contracting ALS in 1989, and since 1993 he has been almost completely paralyzed. To give HPS some semblance of a normal life, the researchers developed a Thought

Translation Device (TTD) that examines the slow cortical potential (SCP) for change relative to a baseline. The SCP is believed to reflect a mechanism that regulates cortical excitation or inhibition; the former results in a negative potential and latter a positive potential. The signal itself is obtained from the EEG by putting it through a 1 Hz low-pass filter. Significantly, people can learn to control the SCP via operant conditioning, for example, by watching a visual representation of the signal.

In the case of the TTD, the subject first learns to control his SCP through feedback in this manner. Once mastery has been obtained, the polarity of the SCP relative to a baseline measurement drives the cursor upward or downward. HPS achieved a mean accuracy of 80% with this method. This was sufficient to accomplish a number of tasks, including Web surfing and writing email. HPS is able to view 5 to 9 pages per hour, and he produces 2 characters per minute when composing text. While this performance relative to normal manipulation of the mouse and keyboard is poor, it is safe to say that the life of HPS has been immeasurably enriched by the device; he reads newspapers online, he sends email to his friends, and he is able to order law books online.

A completely different approach than having the subject learn to modulate his EEG is to modulate the stimulus instead and look for a characteristic indication that this has occurred in a set of brain signals; this is the basis of the P300 approach. The *P300* is an ERP characterized by positive upturn in the EEG signal approximately 300 ms after the appearance of an unexpected or odd event. This fact can be leveraged in the context of a BMI by selectively altering a target stimulus in a set of such stimuli. If the subject is attending to the altered target, one then sees a positive upswing in the signal 300 ms later.

For example, in the Donchin et al. (2000) study, the letters of the alphabet and the numbers were placed in a six-by-six grid. Rows and columns in this grid then flashed at random, thus creating a continuous series of unpredictable events. If the subject were attending to a letter in the flashed row or the flashed column, then 300 ms after the flash, the P300 signal would be produced. By noting the time course of these signals, one can thereby identify the flashed row and column, and thus uniquely identify the letter. Using this methodology, Donchin et al. was able to show that a typing rate of 7.8 characters/min could be obtained with 80% accuracy, and 4.8 characters/min at 90% accuracy. This compares favorably with methods based on EEG feedback, although the latter have more potential for improvement. The P300 methodology can only improve so much, as it requires that all the potential targets be altered in succession[7], and also relies on a signal with a

[7]One could also perform a binary search on the targets, reducing the number of alterations to approximately $\log_2 n$, where n is the target number.

latency relative to the identification of the target. This technique may prove useful in some environments, though. For example, Bayliss (2003) reports using a similar methodology that allows a user navigating through a virtual reality environment to control devices in this world by merely looking at them (lights flashing on the virtual device allows the P300 detector to determine which device is attended to).

An alternative to both of these methods relies on adaptation both within the brain and within the device to optimized performance. Such was the case with the Wolpaw and McFarland (2004) study. They had four subjects (two paralyzed, two control) attempt to control cursor movement in two dimensions via mu (8–12 Hz) or beta (12–26 Hz) signals over left and right sensorimotor cortices. The frequency chosen for each subject was based on performance. In addition, vertical movement was determined by a weighted sum of the left and right signals, and horizontal movement by the difference in these signals as in an earlier study by the same researchers. However, in this case the weights were adaptive, and were adjusted so as to produce optimal performance for a given user.

Results were quite good, and comparable to the same task given to monkeys with invasive measurements. Movement time from the center of the screen to one of eight targets took 1.9 s on the average, and accuracy was 92%. The authors also point out that these results were obtained with a relatively small subset of electrodes and with much of the EEG signal within an electrode ignored; there is also the possibility of higher quality signals being obtained with ECoG (see the next section). Whatever medium is used, harnessing the power of both the brain and the system to jointly adapt to each other will likely lead to the best results in the future.

As previously argued, fMRI is an impractical imaging modality both because of the expense and the poor temporal resolution of the signal. Nevertheless, a system constructed along these lines is of theoretical interest because of the potentially large amount of information about the brain that can be generated, and a future system that resolves these impediments may thus be a viable alternative to current methods. Yoo et al. (2004) describe one such experiment. They had subjects attempt to navigate through a maze by an fMRI-driven thought translation device. To achieve upward motion of the cursor the subjects were instructed to carry out a mental calculation, to move down subjects were told to produce internal speech, and to move right and left they imagined moving their right hand and left hands. These were usually accompanied by BOLD increases in the anterior cingulate gyrus, the posterior aspect of the superior temporal gyrus, and left and right somatosensory cortices, respectively.

The data indicate the high accuracy of this technique: two of three subjects were able to navigate the maze without error, and the remaining subject had an 8% error rate. There was, however, one huge drawback with this methodology: determination of a single up-down-right-left movement took over two full minutes! This is much slower than the slowest EEG method, and between 100 and 1000 times slower than manual mouse movements. Prospects for greater performance do exist, however. One possibility is the joint adaptation between man and machine discussed previously. Another is to more fully exploit the information across the entire brain, a technique known as multivoxel pattern analysis. However, neither of these methods is likely to make fMRI the preferred medium for BMIs in the near future; this awaits a much faster means of accomplishing this type of imaging, if this is forthcoming.

Human, semi-invasive and invasive

Work on invasive BMIs for humans is, of necessity, much more limited than noninvasive studies. Here we consider two studies, the first using ECoG on patients suffering from epilepsy, and in the second more ambitious experiment, the implantation of a multi electrode array in a patient with quadriplegia.

As previously discussed, ECoG represents a compromise between the noisy signals obtained by noninvasive means, and the dangers and complications accompanying electrode implantation. As such, one would expect performance improvement with ECoG relative to EEG and that is exactly what the Schalk et al. (2007) study found. Five subjects had a subdural grid of up to 64 electrodes placed in their temporal, parietal, and frontal regions for one week for the primary purpose of monitoring the foci of epileptic seizures. This opportunity was also used to study the ability of information obtained from these electrodes to control a cursor in two dimensions. As with any BMI, considerable signal processing is required to extract the information present from these multiple sites and convert these into the appropriate response. However, these researchers also found an unexpected result: at least some of the electrode sites coded for the movement more directly. That is, the untransformed direct signal could correspond directly to the movement of the cursor in a the x or y direction. The researchers labeled this signal the local motor potential (LMP).

By combining the LMPs from the various electrodes with a linear model they were also able to show that cursor control was comparable to that of animal studies with invasive electrodes. This provides partial confirmation of the claim that under some circumstances the information obtained with ECoG may rival that of this latter method without the attendant risks. Another finding of significance is that although much of the information used to control the cursor was obtained from hand and arm areas of the motor cortex, as one might expect, a significant amount was also obtained from other brain areas; these were also highly subject dependent. This is consistent with the idea that we should necessarily limit ourselves to given regions of interest (ROI) on an *a priori* basis with BMIs, but allow for the possibility that the brain often encodes useful information regarding a particular task in a more distributed manner.

We now turn to a more radical technique of obtaining this information, although one that may yield a richer vein of information about the mind's intent. Matthew Nagle was an amiable former high school football start when he stepped in to help a friend during a fight after the Independence Day fireworks in Wessagussett Beach, MA. He was stabbed in the left side of his neck and the blade severed his spinal cord between the third and fourth vertebrae. He was almost completely paralyzed from the neck down and could not breathe without a ventilator, although he did maintain some shoulder movement. Three years later he agreed to have the Cyerkinetics BrainGate interface implanted in his primary motor cortex for one year under the supervision of John Donoghue, the founder of this firm, and the director of the Brain Science program at Brown. Matthew has since died from complications due to his injuries, but the sacrifice of this brave young man has taught us much about the feasibility and efficacy of human invasive BMIs.

The implanted device was the 10 × 10 microelectrode array pictured in Figure 6.11. Figure 6.14 shows a clearly ebullient Nagle attempting to control a cursor with his thoughts alone. The question, as with all BMIs, is how to

FIGURE 6.14 **Matthew Nagle moving a cursor with his thoughts (from Abbot, 2006).**
SOURCE: **Abbot, Neuroprosthetics: In search of the sixth sense, Nature, 442, (2006). Reprinted by permission from MacMillan Publishers Ltd: *Nature*, 442, 2006.**

transform the signal received into the appropriate action. Figure 6.15.A shows that in at least some cases, single channels are indicative of individual actions (Hochberg, et al. 2006). For this experiment, Matthew was asked to think about moving his body a certain way. As shown at the top of the figure, channel 38 unambiguously codes for hand movements (the vertical lines show the spiking activity, and the graph is the integral of these spikes). The situation is somewhat more cloudy for channel 16, which appears to code for both wrist and shoulder movement. It is important to note that these results are significant in themselves, apart from any further transformations one might perform on these channels. These results confirm that the motor cortex continues to encode for movement well after it has ceased to actually produce such movement (although this was also suspected from fMRI studies).

A linear filter was used to classify the activity vector as produced by the array, so as to produce the optimal result for a given task. The most studied such task was cursor control. Performance on this task was equal to or exceeded those of monkeys with similar implants and similar signal processing (and thus also superior to the just-discussed ECoG study). Figure 6.15.B shows a line drawn by a technician and Matthew's attempt to match that line. Figure 6.15.C shows a task in which the aim is to hit all the green circles and none of the red squares. As these diagrams indicate, performance is impressive given the nature of the device, but is significantly weaker than normal mouse control. Matthew was able to control a robot hand, however, and to produce rudimentary movements with a robot arm. He was also able to play a very good game of Pong (this requires movement in one dimension only, and is therefore a simpler task than cursor control).

Perhaps as significant as the formal results were the comments Matthew made about the system (we of course do not get comments in animal experiments). One quote in particular captures two items of interest: "I can't put it into words. It's just—I use my brain. I just thought it. I said, 'Cursor go up and to the right.' And it did, and now I can control it all over the screen. It will give me independence." His words echo the principle of transparency discussed earlier. Once training has been completed, the feeling of a direct correspondence between one's thoughts and the desired actions will arise, in much the same way that we directly control our limbs under normal circumstances without the need for intermediary cognition. The other aspect of importance is Matthew's invocation of independence. This, after all, is one of the main goals of such research, and it is gratifying to see that some small progress has been made toward this end.

This completes our study of BMIs. Although these devices are at a relatively primitive stage, initial results are encouraging, and we can expect to see rapid gains in the coming years. It may very well be, for example, that bypassing the injured spinal cord entirely with a BMI will prove more efficacious in the short run than biological attempts to repair the injury with stem cells or other means.

As previously mentioned, one underexplored area that would greatly increase the ability of BMIs to work effectively is to simulate proprioceptive feedback via invasive brain stimulation. In the next chapter, we turn to the study of stimulating the brain for this and other means.

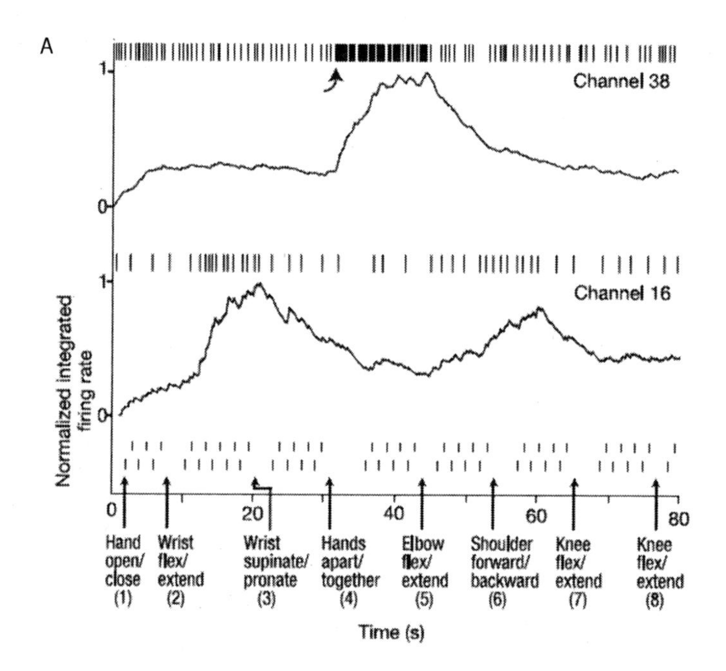

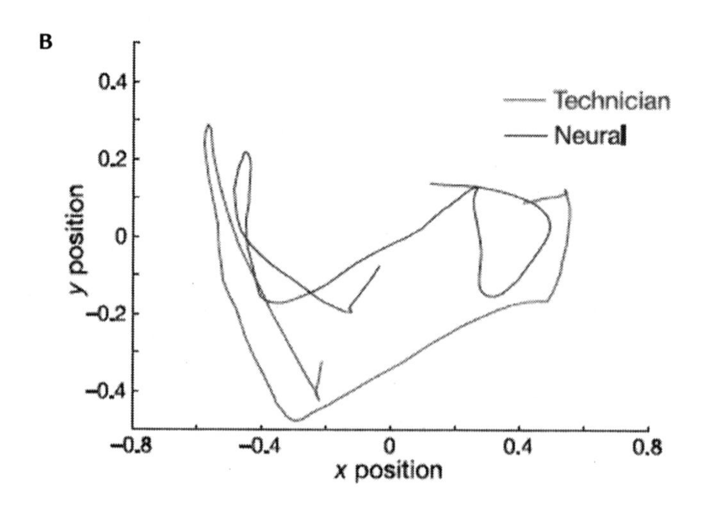

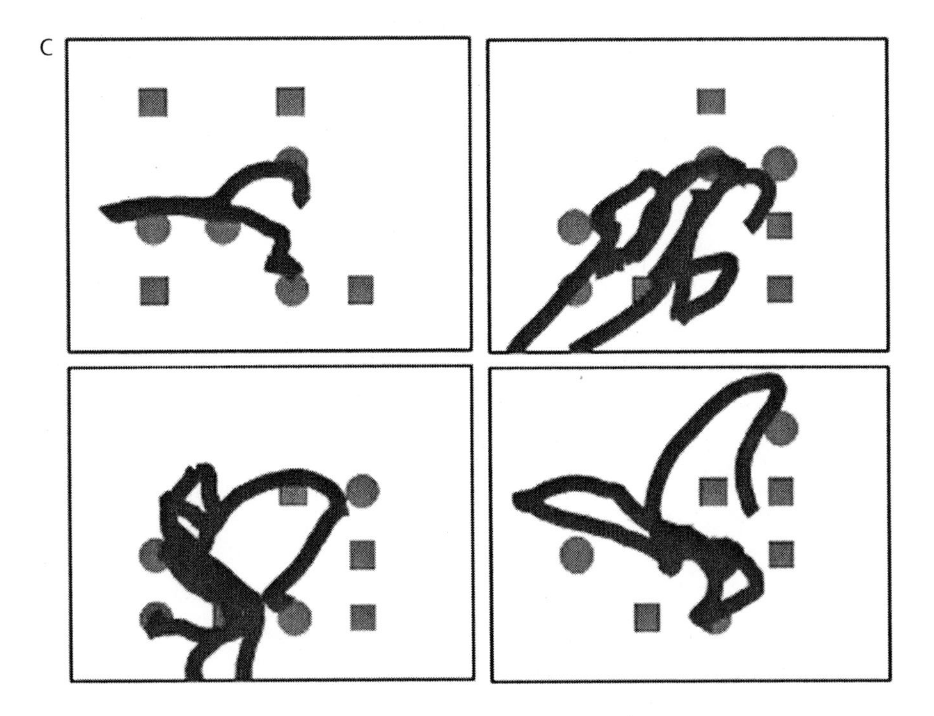

FIGURE 6.15 **Results for the implant array (adapted from Hochbert et al., 2006). A: The matching of mouse-drawn contours with the device. B: Attempt to touch the green circles without hitting the red squares. C: Recorded activity for two electrodes as a function of imagined body movement.**
SOURCE: **A, B, C: Hochberg et al., Neuronal ensemble control of prosthetic devices by a human with tetraplegia Nature, 442 (2006).**
Reprinted by permission from MacMillan Publishers Ltd: *Nature*, 442, 2006.

REFERENCES

Abbot, A. (2006). In search of the sixth sense. *Nature, 442*, 125–127.

Bayliss, J. (2003). Use of the evoked potential P3 component for control in a virtual apartment. *IEEE Transactions on Rehabilitation Engineering*, 113–116.

Birbaumer, N., Ghanayim, N., Hinterberger, T., Iversen, I., Kotchoubey, B., Kübler, A., et al. (1999). A spelling device for the paralysed. *Nature*, 297–298.

Carmena, J. L., Crist, R., O'Doherty, J., Santucci, D., Dimitrov, D., Patil, P., et al. (2003). Learning to control a brain-machine interface for reaching and grasping by primates. *PLoS Biology, 1*.

Cristianini, N., & Shawe-Taylor, J. (2000). *An Introduction to Support Vector Machines and other kernel-based learning methods.* Cambridge: Cambridge University Press.

Donchin, E., Spencer, K., & Wijesinghe, R. (2000). The mental prosthesis: Assesing the speed of a P300-based brain-computer interface. *IEEE Transactions on Rehabilitation Engineering, 8*, 174–179..

Duda, R., Hart, P., & Stork, D. (2000). *Pattern Classification.* Hoboken: Wiley-Interscience.

Georgopoulos, A. P., Schwartz A. B., & E., K. (1986). Neuronal population coding of movement direction. *Science, 233*, 1416–1419.

Haxby, J., Gobbini, M., Furey, M., Ishai, A., Schouten, J., & Pietrini, P. (2001). Distributed and overlapping representations of faces and objects in ventral temporal cortex. *Science, 293*, 2425–2430.

Hochberg, L., Serruya, M., Friehs, G., Mukand, J., Saleh, M., Caplan, A., et al. (2006). Neuronal ensemble control of prosthetic devices by a human with tetraplegia. *Nature, 442*, 164–171.

Kamitani, Y., & Tong, F. (2005). Decoding the visual and subjective contents of the human brain. *Nature Neuroscience, 8*, 679–685.

Karim, A., Hinterberger, T., Richter, J., Mellinger, J., Neumann, N., Flor, H., et al. (2006). Neural Internet: Web surfing with brain potentials for the completely paralyzed. *Neurorehabilation and Neural Repair*, 508–515.

Kennedy, P. (1989). he cone electrode: a long-term electrode that records from neurites grown onto its recording surface. *ournal of Neuroscience Methods, 29*, 181–193.

Landini, L., Positano, V., & Santarelli, M. (2005). *Advanced Image Processing in Magnetic Resonance Imaging.* Boca Raton: CRC Press.

Michel, C., Murray M.M, .., G., L., S., G., L., S., & G. d. (2004). EEG source imaging. *Clinical neurophysiology, 115*, 2195–2222.

Miller, G. (1956). he Magical Number Seven, Plus or Minus Two: Some Limits on Our Capacity for Processing Information. *63*, 81–97.

Moxon, K., Leiser, S., Gerhardt, G., Barbee, K., & Chapin, J. (2004). Ceramic-based multisite electrode arrays for chronic single-neuron recording. *IEEE Transactions in Biomedical Engineering*, 647–656.

Rosenblatt, F. (1958). The perceptron: a probabilistic model for information storage and retrieval in the brain. *Psychological Review, 65*, 368–408.

Sanei, S., & Chambers, J. (2007). *EEG Signal Processing.* Hoboken: Wiley-Interscience.

Schalk, G., Kubanek, J., Miller, K., Anderson, N., Leuthardt, E., Ojemann, J., et al. (2007). Decoding two-dimensional movement trajectories using electrocorticographic signals in humans. *Journal of Neural Engineering, 4*, 264–275.

Scherberger, H., Jarvis, M., & Andersen R.A. (2005). Cortical local field potential encodes movement intentions in the posterior parietal cortex. *Neuron, 46*, 347–354.

Somervuo, P., & Kohonen, T. (1999). Self-organizing maps and learning vector quantization for feature sequences. *Neural Processing Letters*, 151–159.

Stark, E., & Abeles, M. (2007). redicting movement from multiunit activity. *The Journal of Neuroscience, 27*, 8387–8394.

Taylor, D., Tillery, S., & Schwartz, A. (2002). Direct Cortical Control of 3D Neuroprosthetic Devices. *Science, 296*, 1829–1832.

Tong, F., Nakayama, K., J. Vaughan, J., & Kanwisher, N. (1998). Binocular rivalry and visual awareness in human extrastriate cortex. *Neuron*, 753–759.

Tononi, G., & Edelman, G. (1998). Consciousness and the integration of information in the brain. *Advances in Neurology, 77*, 245–279.

Waggenspack, W. (2007, August). Mind Game. *Duke Magazine.*

Werbos, P. (1988). Backpropagation: Past and future. *Neural Networks*, 343–353.

Wolpaw, J., & McFarland, D. (2004). Control of a two-dimensional movement signal by a noninvasive brain-computer interface in humans. *Proceedings of the National Academy of Sciences*, 17849–17854.

Yoo, S., Fairneny, T., Chen, N., Choo, S., Panych, L. P., Park, H., et al. (2004). Brain-computer interface using fMRI: spatial navigation by thoughts. *Clinical Neuroscience and Neuropathology*, 1591–1595.

Chapter 7 Stimulating the Brain

Stimulating, or writing to the brain, is the converse operation to brain reading, the subject of the previous two chapters. The contrast between the two operations, both in the nature of the technology, and the capabilities of current systems, could not be greater. The kinds of reading devices that we have looked at yield, depending on the technology used, an indication of the brain's activity at a fairly precise location over a limited time span, and moreover at a number of such locations simultaneously. There is no analog to this process in the reverse direction. With deep brain stimulation (DBS), it is true, we can precisely pinpoint an area to be stimulated, but because this involves the implantation of electrodes, it can only be done in a few places at once. Transcranial magnetic stimulation (TMS) with multiple magnets may hold some hope of reaching arbitrary structures in the brain, but this is still experimental, and in any case, the spatial resolution of this process is many orders of magnitude less than functional magnetic resonance imaging (fMRI). In short, there are no devices, invasive or otherwise, that can write an advertising jingle directly into your consciousness, that can make you smell strawberries, or that can project a video directly into your visual cortex.

On the other hand, a number of recent developments, especially with DBS, are nothing short of astounding. In many instances, it suffices to stimulate only a single area of the brain, and leverage the fact that the downstream effects of such stimulation will suffice for the task at hand. For example, not only is treatment for Parkinson's disease (PD) viable with DBS, since 1997 it has been approved by the Food and Drug Administration (FDA), for clinical practice. DBS for refractory depression is still in the experimental stage, but has shown very good results, with minimal side effects relative to pharmacological therapy. Finally, and most remarkably, a limited number of results have indicated that in at least some cases, minimally conscious patients may be revived from their quasi-coma-like state and return to something approaching normalcy.

In summary, if DBS were the only form of neuroengineering in existence, then the main thesis of this book would still be justified, namely, that neurotechnology will eventually transform not only how we live, but also what it means to be human. The future can only improve on these early results, and in later chapters, we speculate on what it will be like to be happy on demand, or to achieve a eureka-like moment when desired. Let us first examine, however, what can be done today.

SENSORY PROSTHESES

Strictly speaking, a sensory prosthetic is simply a replacement of an already-existing means of providing input to the brain rather than a novel means of stimulating it. Still, the successes and difficulties in doing so tell us much about the nature of building an interface between hardware and wetware, and thus this is an excellent place to begin our discussion of writing to the brain. We will consider two prosthetics for sensory modalities in order of degree of development: cochlear and retinal.

Cochlear implants (CI) are by far the most common prosthetic, and are also a good example of a neurotechnology that has passed into the mainstream. Well over 100,000 people have acquired this implant, and public figures as diverse as the congenitally deaf Miss America of 1995, Heather Whitestone, and Rush Limbaugh are recent recipients. In the postlingually deaf CIs (after language acquisition) deaf, success is difficult to assess before the operation has taken place, although results are often quite good; in Limbaugh's case, he was able to return to his radio-hosting work soon after the implant. In the case of those deaf from birth, the earlier they receive the implant, the better the results (Manrique et al., 2004). There is also tremendous individual variability in performance depending on the nature of the deafness, the preservation of auditory neurons, and other nonperipheral factors, but it is safe to say at this point that almost everyone benefits to some extent from this device.[1]

A CI consists of five components, a microphone, that picks up the sound; a speech processor, transforming the sound into a form appropriate for the auditory neurons; a transmitter, sending the transformed signal to the cochlea; a receiver to

---

[1] By this we mean that they can almost always process sounds better than before the implant (any latent auditory facility is destroyed by the implant, so postoperation the device provides the sole means of input). However, CIs remain controversial among some in the deaf community, especially for children, as they encourage reliance on what may be a degraded sound signal rather than the relatively noise-free signal of sign language. Hence, the decision to undergo this operation may governed by other, nontechnological factors.

pick up the transmitted signal; and the electrode array, implanted in the cochlea itself. These parts can be seen in Figure 7.1. The transmitter and receiver are located above the ear, and these send the processed signal to the electrode array in the cochlea (the spiral-like structure).

The working premise of CI technology is the place theory of audition, in which hair cells in the cochlea respond selectively to differing frequencies, with the apex (center of the cochlear spiral) responding selectively to low frequencies, the base (outer part of the spiral) responding to high frequencies, and with other frequencies arranged tonotopically between these extremes. In the hearing impaired, it is usually these hair cells that are damaged; however, the auditory neurons are usually intact and it is these that receive the processed signals from the electrodes. Among other things, this processing separates the sound into frequency bands such that each electrode receives only the intensity of the signal in a single band, and this electrode then targets the nerve fibers that would respond to this band in normal circumstances. A key variable in a CI is the number of such electrodes, with more accurate sound perception associated with a greater quantity. Modern CIs have between 16 and 22 points of stimulation, falling far

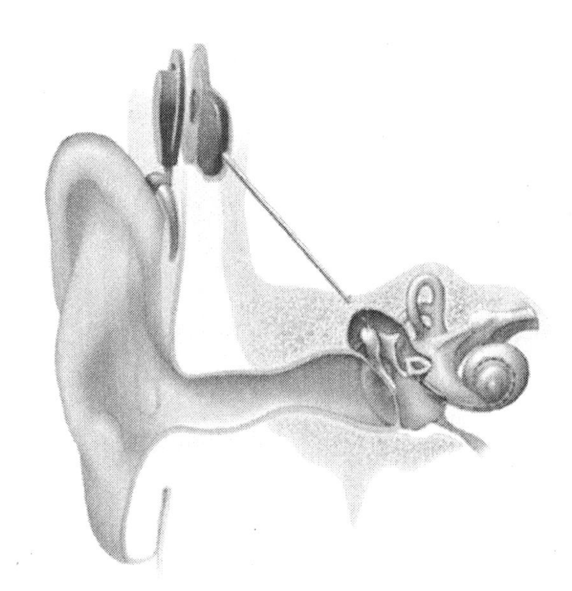

FIGURE 7.1 **The typical CI setup. Transformed sound signals are transmitted through the skull and onto the electrode implanted in the array, which stimulates the auditory nerves and thereby allows audition.**
SOURCE: **http://commons.wikimedia.org/wiki/Image:Cochlear_implant.jpg.**

short of the approximately 3,000 in those with normal hearing, but sufficient for speech recognition.

Two factors make this reduction possible. First, speech recognition is a highly context-dependent mechanism, leveraging not only the surrounding phonemes and the surrounding words, but also expectations. As anyone learning a foreign language knows, it is always easier to pretend to speak a language in a constrained context, such as ordering in a restaurant, or buying something at a store, than in an environment in which unanticipated speech arises. The second factor is neural plasticity, which was important enough to be elevated to a fundamental principle in the previous chapter. It is no less important in this instance. Typically, when the device is first turned on, it sounds like mangled noise. It is only after tuning of the device parameters by the technician, and a period of adaptation on the part of the recipient, that a degree of effortless listening and speech recognition becomes possible.

The ultimate goal, however, of CI technology is not merely to hear but to hear well, and this includes listening to music as well as making out speech. Here, the technology falls somewhat short, although good progress is being made toward this end. In a 2005 *Wired* article, Michael Chorost describes his quest to find the right processing to hear Ravel's *Bolero*, his favorite piece from the time before full deafness set in. With earlier technologies he had no trouble recognizing that it was *Bolero* that was being played; the problem was that it didn't sound like *Bolero*, or at least enough like it to be enjoyable. The central difficulty was that conventional CIs are constructed to maximize performance for speech recognition. As such they concentrate on a relatively compressed frequency range (the range of speech is much narrower than that for music), and have relatively few frequency bands in this range. This means that notes as far as an octave apart may sound more or less identical. After many false starts, Chorost's quest finally brought him to an Advanced Bionics device that makes use of current steering to turn 15 electrodes into 120 virtual spectral bands. Current steering makes use of the fact that one can interpolate between two electrode positions by applying currents to these electrodes proportional to the position (Bosco et al., 2005). For example, suppose electrode 1 is centered on nerve cells for 440 Hz and electrode 2 on nerves cells for 600 Hz; to achieve an approximation to 560 Hz, or 3/4 of the way to the second frequency, one would apply a pulse 3 times as large to electrode 2 as to electrode 1.

After Chorost had the software to accomplish this uploaded into his CI processor, the unexpected happened. Tears came to his eyes as this device was switched on as his favorite piece was being played. For the first time in many years he could hear the passion in the music. Current steering, however, is not the end of the road for CI, and this device as yet does not provide anything like a normal auditory experience, but it is likely that this young technology will make rapid progress in the coming years.

The same can be said of retinal implants (RI), although here the difficulties are much greater, befitting the greater complexity of vision, and as such the current state of the art is still relatively primitive. Despite the fact that there are four companies actively pursuing a commercial RI device, the technology is currently stalled at the clinical testing stage. The need, however, as in the case of CI, is substantial. Approximately 900,000 Americans have profound vision loss, and another 2.5 million have some degree of impairment (Winter, Cogan, & Rizzo, 2007); this is likely to increase as the population ages. Although vision deficiencies can be caused anywhere in the chain from light striking the retina to the visual cortex, the chief cause of age-related vision loss is macular degeneration, or degeneration of the photoreceptor cells (rods and cones) in the central retina. Likewise, retinitis pigmentosa, and more debilitating diseases with earlier onsets affect primarily these detectors.

Crucially, in both conditions ganglion and bipolar cells (see Chapter 2) are substantially preserved, making retinal implants that stimulate these remaining cells a viable technology. Other means of restoring vision include stimulation of the optic nerve and direct stimulation of the occipital cortex. These prosthetic forms would cover a wider range of deficits, but both are beset by technical as well as theoretical difficulties. Because the retina performs considerable pre-processing of the image, this would need to be replaced in both cases; in addition, occipital stimulation would need to account for processing in the lateral geniculate nucleus, and thalamo-cortical feedback interactions.

Thus, most effort has focused on the two kinds of RI, epi- and subretinal (Zrenner, 2002). Epiretinal implants are driven by an external camera and sit on the surface of the retina stimulating the retinal ganglion cells. Subretinal implants contain photodiodes that are directly driven by incoming light and attempt to replace lost photoreceptors directly. The advantages of epiretinal relative to subretinal implants include a reduced amount of microelectronics, as the signal processing is external to the device, and the fact that the vitreous cavity can act as a heat sink. The chief advantage of the subretinal implant is that it stimulates cells earlier in the processing chain (recall from Chapter 2 that retinal processing is inverted with respect to the light path); however, current photodiodes are not as efficient as native rods and cones and do not provide as much stimulation as the latter. Active groups are currently pursuing both types of RI, and it is unclear which one will become the dominant technology.

Obstacles to current progress in both epiretinal and subretinal progress include trauma related to surgery and damage related to chronic implantation, gliosis (see the discussion of DBS in the next section), reducing the effective strength of the signal, and most critically, a tradeoff between electrode size and resolution. The larger the stimulation site, the stronger the effect, and the more likely that retinal cells will be stimulated. However, the larger the electrode, the smaller the resolution of the resultant image. In addition, vision is an information-rich modality, requiring substantially more inputs than audition, for example, before adequate perceptual levels are achieved.

This is illustrated in Figure 7.2. Panel A is a picture of Einstein from the 1950's. 7.2.B shows the resolution reduced to the approximate theoretical limit for object recognition, $25 \times 30 = 750$ pixels (compared to a minimum of approximately 8 stimulation sites before speech recognition can be achieved). Note that all fine detail is lost, such as the wrinkles and the famously unruly hair, but the image is still vaguely recognizable as that of an older man (squinting helps). Figure 7.2.C shows the upper limits of current implant technology, an $8 \times 10 = 80$ pixel image. Recognition except for relatively simply objects is not possible at this resolution. Furthermore, it is unlikely that the image would appear in this form to the implant patient. Mitigating against this appearance are three factors: a) although the retina is topographically organized, the regular grid may not be preserved depending on the site and spacing of the stimulated cells, b) the target cells may not be preserved, either because of the original illness, or because of damage caused by the implant itself, causing some pixels to be lost, and c) the effective number of shades per pixel may not be as high as indicated in the figure. However, there are a few factors that may help the perceptual process: a) as always, there will be some degree of plasticity in higher cortical areas, depending on the nature of the signal and age at the time of the implant and the age at which blindness occurred, b) less obviously, a moving image may be able to be recognized better than a static image, and c) actively exploring the image with eye movements may yield better recognition than simply passively registering movement (this would give the subretinal implant an additional advantage over the epiretinal implant because the latter requires an external camera that does not respond to eye movements).

A

B

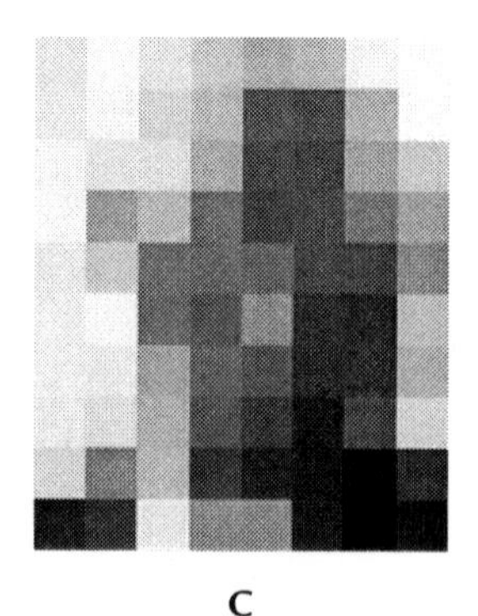

C

FIGURE 7.2 **A: A pensive Einstein, from his later years. B: Resolution reduced to the theoretical limit. A older man's face is still vaguely recognizable, but fine detail has been completely lost. C: The limits of current technology. Object recognition is not possible.** *SOURCE:* **http://commons.wikimedia.org/wiki/Image:Albert_Einstein_in_later_years.jpg.**

Future progress in RI is likely to be relatively rapid, despite the numerous obstacles that need to be overcome. We are not quite at the point of creating a bionic man or woman (but see the next chapter), although object recognition quality should not be out of the question in the next decade, and detailed vision another decade further. Olfactory, gustatory, and tactile/kinesthetic prostheses are even further down the road, as the demand for them is correspondingly less. The latter, however, may receive a considerable boost as BMI systems become more common, because of the need for proprioceptive feedback in fine-grained motor movements.

DEEP BRAIN STIMULATION (DBS)

In essence, DBS is an extremely simple idea. One or more thin electrodes (~1 mm in diameter) are placed in various positions in the brain. An external signal generator with a power supply is also implanted somewhere in the body, typically in the chest cavity. An external remote control sends signals to the generator varying the parameters of the stimulation, including the amount of current, the frequency of the current (typically around 100 Hz), and the duration and frequency of the pulses. Thus, the patient does not have any direct control over the stimulation[2], and this, among many other things, separates modern treatment paradigms from the earlier self-stimulation studies of questionable scientific merit, such as those conducted by Robert Heath at Tulane in the 1950s and 1960s (see Chapter 10).

Despite increasing use, the precise mechanism by which DBS works is still a matter of considerable debate. It is generally agreed that the dominant effect of DBS is on axons rather than on cell bodies (soma), as the former exhibit a significantly greater excitability than the latter. It is also known that to a first approximation DBS acts in an inhibitory manner, as the result of this stimulation is comparable to ablation of the target area, an earlier treatment for PD. In fact, one of the chief advantages of DBS over this prior method is that it achieves the same effects but is reversible, as it does not create any permanent damage. However, whether DBS reduces neural output directly or operates through a more complex mechanism is a matter of dispute. McIntyre et al. (2004) summarize four possible mechanisms of action: depolarization blockade, synaptic

---

[2] At least not yet, but build it and they will hack. There can be little doubt that as neural implants become more common, so will "off-label" uses of these devices. This is a dangerous game; it is one thing to unlock an iPhone, quite another to hack into the extended circuitry of your own consciousness. Still, it is likely that new discoveries will be made through this unconventional practice.

inhibition, synaptic depression, and stimulation-induced disruption of pathological network activity. They argue that the first three are inconsistent with the independent effect that DBS has on axonal firing, leaving the last possibility as the most viable candidate. Supporting this notion is the hypothesis that the symptoms of PD are caused by pathological synchronous oscillations between structures in the basal ganglia; DBS may disturb this pattern (Kringelback et al., 2007). Further study is needed to confirm this idea, and also to see how well it can be extended to the mechanism by which DBS works for other pathologies.

DBS has a long history in animals, starting with Delgado's controversial experiments in the sixties controlling the fighting, playing, and sleeping behavior of monkeys and cats with remote-controlled stimulation. In the most famous example of his work, and the one that received worldwide attention, Delgado stepped into the ring with a charging bull fitted with a remote stimulation device; he was able to stop the bull in its tracks. In a much more recent, and equally well-publicized result, rats were guided through a three-dimensional obstacle course by remote stimulation (Talwar, 2002; see also http://www.youtube.com/watch?v=PpfjmzZ4NTw for a short BBC report). The rats were fitted with three stimulators, one that projected to the left whisker cortical representation, one to the right, and one to the reward center in the medial forebrain bundle (MFB). They were first trained in a figure eight course, with the appropriate whisker cued for turning, and a reward signal presented to the MFB when the rat made the correct turn. They were then placed in a three-dimensional obstacle course, such as that in Figure 7.3. Significantly, the signal to the MFB acted not only as a reward, but also a motivator, urging the rat forward regardless of the obstacles ahead. With such a device, for example, they were able to encourage the rat to explore brightly lit open spaces, an environment this cautious animal would naturally tend to avoid.

In humans, DBS is used exclusively for beneficial purposes, not control of course.[3] Table 7.1 presents the major pathologies associated with DBS treatment, the major sites of stimulation, and other pertinent information. Although the data is still scanty in many areas, a wide variety of illnesses are represented, and it is likely that this table will need to be augmented in just a few short years. Note, however, that all current targets are subcortical. This is in keeping with the notion introduced at the start of this chapter that this form of stimulation is not intended to introduce particular contents into the conscious stream; rather, the goal is a global modulation of either motor or affective responses. Thus, DBS as

[3] It is easy to envision science-fiction scenarios where humans are controlled remotely to carry out the bidding of an evil genius. Whether this would work, however, is an open question. People are more resistant to operant conditions than animals, and may be able to override such signals by force of will. One cannot preclude, however, the possibility that the appropriate combination of reward and punishing signals could result in at least partial control of human behavior.

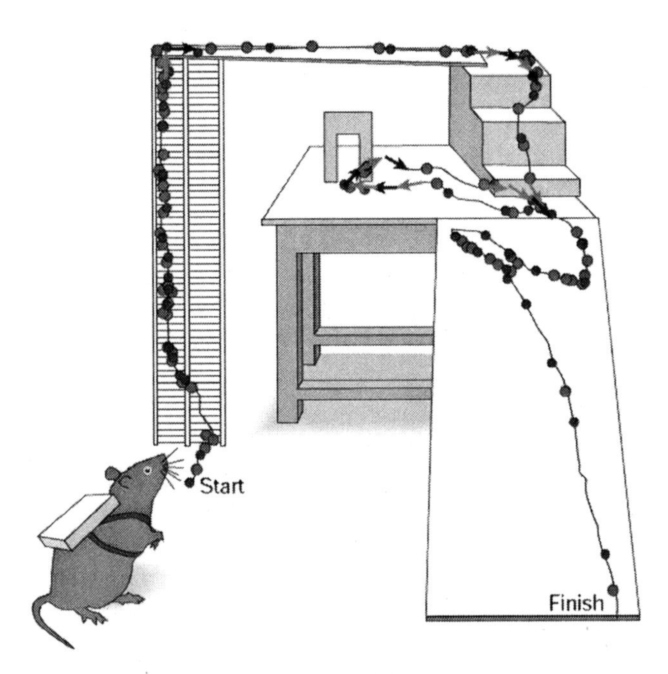

FIGURE 7.3 **A remotely controlled rat running through an obstacle course. Dots represent motivating MFB stimulation and also turning cues (adapted from Talwar et al., 2002).**
SOURCE: **Talwar et al., 2002.**
Used with permission of the Nature Publishing Group; *Nature*, vol. 417, p. 37–38, 2002.

currently practiced is best thought of as a corrective mechanism that acts on one or more critical components in a larger regulatory circuit.

The clearest illustration of this can be found in the treatment of PD. PD is a progressive movement disorder characterized by tremors, slowing of movements, and difficulty initiating action (bradykinesia), stooping, shuffling gait, speech softening, fatigue, and a number of associated motor symptoms including depression, often preceding motor difficulty, and possibly dementia in later stages. PD has a prevalence of approximately 1 in 100 people and primarily affects those over the age of 50. The putative cause is lowering of dopamine production in the pars compacta region of the substantia nigra. Figure 7.4 shows a simplified diagram of the motor circuitry implicated in PD; the left of the diagram shows normal functioning and the right pathological functioning associated with lowered SN output. As indicated in Table 7.1, three targets in this pathway can be useful in ameliorating some PD symptoms, the globus pallidus internus (GPi), the ventral intermediate nucleus (VIM) (the middle third of the ventral nucleus of the thalamus, not shown, receiving projections from the cerebellum, also not shown), and the subthalamic nuclei (STN) (Benabid, 2003). The latter is preferred as it apparently enables a reduction in levodopa or other drugs, although

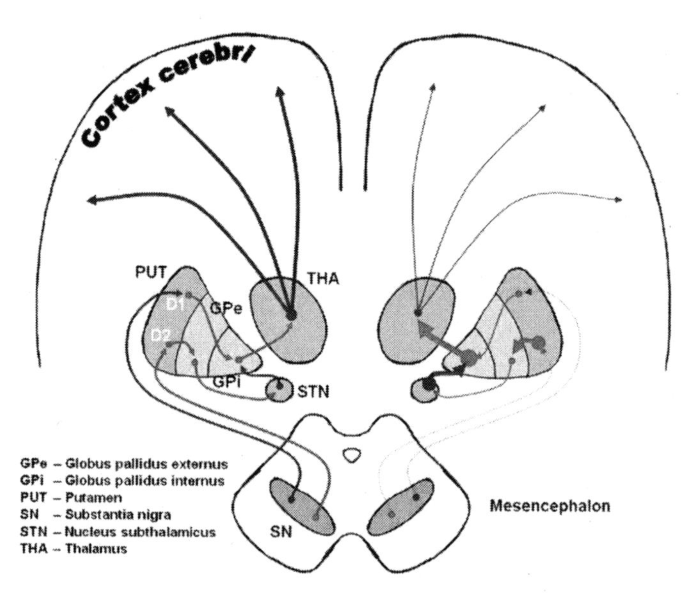

FIGURE 7.4 **A simplified picture of the circuitry involved in movement. Reduced domaminergic input from the substantia nigra in PD (right) results in reduced input to the motor cortex and less control over the activity in this region.**
SOURCE: **http://commons.wikimedia.org/wiki/Image:DA-loops_in_PD.jpg.**

more study is needed to confirm that this is not also the case for the GPi (MB Stern, personal communication). In at least some cases, DBS can provide dramatic relief of PD symptoms. Simple tasks, like drinking a glass of water, made difficult because of tremor become possible; the ability to walk can also be recovered. However, DBS of the STN does have side effects, including possible cognitive effects, and is thus contraindicated in the case of dementia.

Equally dramatic results have been shown in the two main studies conducted on depression and DBS, although the data in this case is unfortunately much more sparse reflecting the recency of this treatment. Two brain centers have been targeted for this affliction. The first, and most natural place to stimulate is the nucleus accumbens (NAcc), a limbic system structure in the ventral striatum discussed at more length in Chapter 8. Here we simply note that the NAcc is a critical part of the reward circuitry, and that rats allowed to self-stimulate this structure will do so until exhaustion, ignoring even food and water. In a recent study, (Schlaepfer, et al., 2008) implanted electrodes in this area for three patients suffering from refractory depression; they were resistant to improvement with any of the standard treatments, including drug and talk therapies. All three showed significant improvement on subjective ratings of depression symptoms as well as PET imaging data indicating a renormalization of their metabolism. The

TABLE 7.1. Stimulation targets for the primary pathologies treatable with DBS.

Condition	Stimulation target	Comments
Parkinson's disease	Ventral intermediate nucleus (VIM)	*Reduces essential tremor (ET) only*
	globus pallidus internus (GPi)	*Reduces most motor symptoms but does not permit drug reduction*
	subthalamic nucleus (STN)	*Similar reduction in motor symptoms, allows drug reduction, possible side effects*
Depression	nucleus accumbens (NA) subgenal cingulate	*Limited but encouraging studies*
Tourette's	various	*Limited studies*
Obsessive-compulsive disorder	anterior limb internal capsule nucleus accumbens (NA)	*Limited studies*
Intractable pain	periventricular/periaqueductal gray (PVG)	*Indicated in nociceptive pain*
	sensory thalamic nuclei	*Indicated in neuropathic pain*
Minimally conscious state	central thalamus	*Single subject only, but promising*

Mayberg et al. (2005) study targeted the subgenal cingulate white matter based on prior evidence that this region is overactive in depressives. They obtained good results with 4 of 6 subjects. As remarkable as the "objective" improvement, as revealed by imaging and depression inventory scores, were the spontaneous reports of the subjects as the current was switched on. These included a "disappearance of the void," a feeling of "connectedness," a brightening of the room and intensification of colors.[4] This is consistent with results from the NAcc study, in which a previously lethargic patient, upon receiving stimulation, suddenly realized that he was in Cologne and had not yet visited the famous Cologne Cathedral; he did so the next day.

Ultimately, the use of DBS for depression and other affective disorders could far surpass that for PD. There are a number of reasons for this. First, depression is so common it has been dubbed the "common cold" of mental illnesses. It afflicts approximately 5% of the Western population at any given time and up to 1 in 12 will have a major depressive incident in their lifetime. The second reason is that current pharmacological treatments, such as serotonin selective reuptake inhibitors (SSRI), are marred by a combination of decreasing responsiveness with time, possible serious side effects, and little or no efficacy in 10–15% of the population. It is also instructive to compare the mechanisms by which drug therapies and DBS work. The former target-specific neurotransmitters, which are active in many structures throughout the brain, often serving multiple purposes. In contrast, DBS acts on a specific area only. Thus, in principle at least, this latter methodology should be able to more precisely target the part of the brain that is malfunctioning with minimal effects on other non-targeted areas.

DBS has also met with some success in the treatment of symptoms for Tourrette's syndrome (Wichmann & DeLong, 2006) and in obsessive-compulsive disorder, in which the target was also the NAcc (Sturm, et al., 2003). The other major pathology that may be amenable to invasive stimulation is intractable pain. As with depression, drug treatments are often ineffective or have too many side effects because of their global action on the brain. In a paper that describes 15 years of clinical experience, Kumar et al. (1997) describe the results of DBS on 68 subjects with chronic pain. Of these, 62% achieved greater than 50% reduction in their pain levels. Those that responded well include patients with lower back pain, deafferentation-type pains (pain due to the damage to afferent nerve fibers, presumably leading to a lack of inhibition of the pain response), and trigeminal neuropathy. Thalamic pain syndromes, postherpetic neuralgia, and lesions of the spinal cord did not fare as well.

Finally, there is the dramatic case of the bringing "back to life" of a minimally conscious 38-year-old male patient (Schiff, et al., 2008). Significantly, PET and other functional imaging studies indicated a largely reduced global brain metabolism but otherwise intact structures. This made him an excellent candidate for thalamic DBS. Recall from Chapter 2 that the thalamus modulates arousal signals from the brainstem and acts as a sensory gateway for the neocortex; when firing in burst mode the brain is awake and receiving external input, and when synchronized firing occurs, this gateway closes, and the brain is drowsy or asleep. Indeed, the patient showed significant recovery in all areas of normal conscious behavior upon stimulation. He can eat without a feeding tube, he can watch movies, he cries and laughs, and perhaps most significantly, he can interact, to a limited extent, with other people. How well this can be extended to other similar patients and perhaps others with more severe impairments, such as persistent vegetative state and coma, remains to be seen. Certainly, we can say at a minimum that DBS can work only to the extent that the rest of the brain is preserved in spite of the trauma; it cannot miraculously restore brain function where none exists. What it can do is replace or mimic thalamic regulation, and in many cases, this may be extremely beneficial.

[4] It is tempting to speculate that the feelings accompanying being in love are generated by a kind of natural DBS operating on the same circuitry.

Future developments will no doubt improve the efficacy of the treatments shown in Table 7.1, as well as deepen it to include other diseases. One noteworthy aspect of this table is that only pathologies are present. There is no attempt with DBS to improve cognitive or affective functioning in otherwise normal subjects. This is understandable; brain surgery is not something to be taken lightly, and although the side effects can be minimized, they can be severe. The set of possible enhancements, many of which are explored in later chapters, are also potentially dramatic, however. Many may wish to make this wager if the circumstances allow. There can be no doubt that it will require as much an attitudinal revolution as well as a technological revolution before nonpathological neural enhancement becomes widely accepted. To see that this is the case, one need only point to the current controversy swirling around the used of steroids and other additives in sports. By comparison, purely elective brain surgery is a much more radical step, and accordingly will require a more radical change in the way body (and mind) alteration is thought of.

VAGUS NERVE STIMULATION (VNS)

The *vagus* (from the Latin word for wandering) nerve, the longest of the cranial nerves, was originally believed to contain only efferent fibers for the purpose of parasympathetic regulation of the heart, larynx, lungs, and other organs, but it is now known that approximately 80% of this nerve is dedicated to the processing of afferent signals. It terminates in the nucleus of the solitary tract (NTS) of the medulla, which in turn projects to a number of brain structures, including the hypothalamus, anterior cingulate, and other brainstem nuclei. Stimulation of this nerve has the potential, in principle, to affect brain dynamics as a whole and indeed it has been found that weak stimulation causes EEG synchronization whereas stronger stimulation (>2 mA of current) results in desynchronization (Duncan and Groves, 2005).

The latter effect may be responsible for the primary use of VNS, the control of epileptic seizures that are the result of pathological recruitment of a relatively large body of cells into a synchronized subnetwork. As in DBS, the pulse generator for VNS is placed in the chest cavity, but in this instance the device is connected directly to the vagus nerve (only the left vagus is stimulated, as the right innervates the heart). Also as in DBS, the parameters of stimulation can be controlled from outside the body with a wand or remote. VNS has proved efficacious in the prevention of seizures, and unlike medication for this disorder, it seems to show a small improvement over time (Morris & Mueller, 1999); to date tens of thousands of patients have been implanted with the Cyberonics, Inc. prosthesis for this purpose. Adverse effects are few and are usually limited to hoarseness and cough, and these can be reduced with proper parameter settings.

Early on, physicians also noted a curious adjunct to VNS therapy. Many epileptics also suffer from depression for an unknown reason. In some of these sufferers, VNS caused a marked improvement in mood. By itself, of course, this is only suggestive evidence; it could very well be that patients were simply happier that they were having fewer seizures. Accordingly, systematic research into VNS treatment for nonepileptic depressives has been carried out. In general, the findings are suggestive of the benefit although not nearly as dramatic as the initial results with DBS; this has led Medicare to deny coverage for this treatment in the case of refractory depression. The Sackeim et al. (2001) study generated a typical result profile. Approximately 35% of patients showed a greater than 50% reduction in scores on standard depression inventories. It is also not known how long benefits for this procedure last when it is effective, a critical factor in any procedure requiring surgery. VNS has also been investigated for its use in the reduction of anxiety, in cognitive enhancement in Alzheimer's, and also as a possible treatment for migraines.

Although VNS may have potential clinical benefits for these and other maladies, it is, in the long run, uninteresting from the point of view of neuroengineering. Its primary advantage, providing a line into the brain that bypasses the spinal cord[5] that is relatively easy for modern surgical methods to access, is also its downfall. Because its innervation of the brain is fixed, under the best of circumstances it is unlikely that we will be able to obtain the fine-grained control over neural dynamics that we are ultimately seeking. We turn next to a device with possibly no more clinical success than VMS but with much greater potential to achieve such control.

TRANSCRANIAL MAGNETIC STIMULATION (TMS)

The history of TMS is both varied and colorful; at the start of the twentieth century the physicist Silvanus P. Thompson attempted unsuccessfully to stimulate his brain with large magnets. A number of later informal experiments showed that magnetic stimulation near the retina could produce phosphenes, or flashes of colored light in the visual field, but magnets were generally too weak at this time to produce cortical effects. True progress in the field awaited the development of a strong and reliable magnet by the Sheffield group headed by Anthony Barker. With such a device they were able to show for the first time in 1985

[5] It has recently been suggested, for example, that the reason some women with severe spinal injuries (and no feeling below the neck) can experience orgasm is due to the afferent information carried by the vagus nerve to the brainstem (see Chapter 10).

that stimulation of the motor cortex could lead to muscle movement (Walsh and Cowey, 2000). Since this time, a number of speculative claims have swirled around this technology, including the ability of suitably situated magnets to induce out-of-body and religious experiences. Uses on firmer scientific footing include diagnosis of motor diseases, the creation of temporary lesions for investigative purposes, and the treatment of depression and epilepsy.

TMS works as follows. An electric current is induced through a circular coil as in Figure 7.5.A (or a coil with a figure eight configuration for greater focus). This produces a magnetic field orthogonal to the direction of the current, which in turn will produce an electric field proportional to the time derivative of the magnetic field and in the opposite direction to that field on the surface of the scalp, by Faraday's law of electromagnetic induction. Thus, in order produce a large electric field it is desirable that the magnetic field switches on rapidly; capacitors that store charge and release it quickly are used for this purpose. The generated intracranial electric field, pictured in Figure 7.5.B, in turn affects the polarization of cortical axons, but only if there is a gradient in the ionic distribution along the axon. This will be achieved if either the axon is straight, and the electric field curves through it, or if the field is linear, and the axon curves, as in the pyramidal axon in Figure 7.5.C.

TMS can be performed in two modes, single pulse and repetitive TMS (rTMS), in which multiple pulses in succession are generated. The effect of the former is believed to be short lived. The latter, in contrast, can lead to long-term

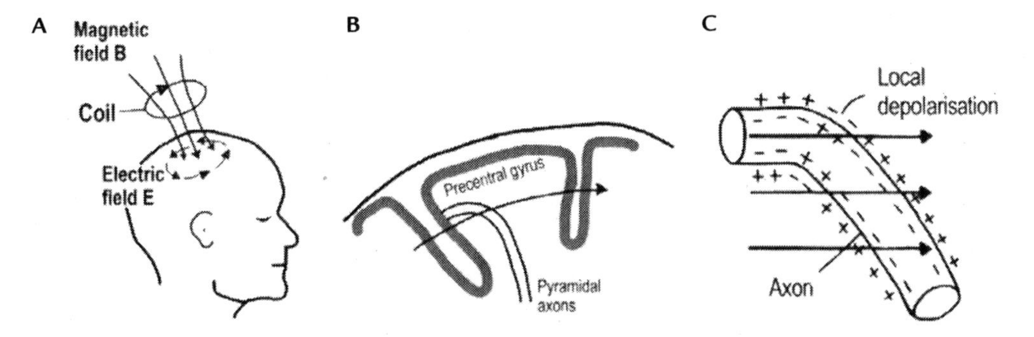

FIGURE 7.5 **The mechanism behind TMS. A: A magnetic field will induce an electric field at a distance that travels in the opposite direction to the coil current. B: This penetrates the skull and affects cortical axonal behavior. C: The degree of effect will be proportional to the curvature of the axon relative to the electric field.**
SOURCE: **http://www.sciencedirect.com/science?_ob=ArticleURL&_udi=B6SYS-48Y0DD2-1&_user=10&_coverDate=09%2F30%2F2003&_alid=713111463&_rdoc=1&_fmt=summary&_orig=search&_cdi=4842&_sort=d&_docanchor=&view=c&_ct=1&_acct=C000050221&_version=1&_urlVersion=0&_userid=10&md5=397584ecc2f4f97a0926e843b412f9e3.**
Reprinted from *Brain Research Reviews*, vol. 43, Sack et al., "Combining Transcranial Magnetic Stimulation...," p. 16, Sept. 2003, with permission from Elsevier Ltd.

potentiation (LTP) or long-term depression (LTD), with high (3–20 Hz) and low frequencies (< 1 Hz) respectively. One mechanism for this action is that alteration of synaptic efficacy, although this is still not well understood (George et al., 2003). In any case, because of its continuing effect on neural excitability, rTMS is the preferred mode of operation for therapeutic uses. High-frequency rTMS can lead to seizure, however, so caution must be exercised with its use.

Applications of TMS may be divided into three categories: research, diagnosis, and therapy. The first often takes advantage of the fact that TMS disrupts neural activity and thus can create a "virtual lesion" in the targeted area. In an notable use of TMS in this manner, Cohen et al. (1997) showed that congenitally blind subjects relied on the primary visual cortex to read Braille, indicating that this area had been recruited to accomplish this spatial task. Note that in this case, and in similar instances, TMS data complement and reinforce imaging data. For example, the fact that the visual cortex "lights up" when Braille is read by these subjects in not necessarily an indication that it is involved in this activity. It could be that tactile signals are merely routed to the occipital cortex, but that the computations accomplished there are not used in letter and word disambiguation. The fact that disruption of this area also disrupts reading provides additional confirmation that this is not the case, assuming that the final effect of TMS is not due to alteration of downstream processes.

TMS can also be used directly in combination with imaging methods to reveal connectivity in the brain. An especially apropos study in light of this book's emphasis on consciousness is the Massimini et al. (2005) study of connectivity during the conscious and unconscious states. They looked at how the activation spread after stimulation of the right frontal cortex in subjects who were awake and subjects in non-REM sleep. In the former, EEG indicated that activation migrated to the parietal cortex and corresponding areas in the left hemisphere, but in the latter activity was confined to the original site of stimulation. These results are consistent with the notion put forth in Chapter 4 that anesthesia works by reducing the coordination between interacting brain modules, and the larger theory also advanced in that chapter that for a brain area to participate in the generation of the conscious content, it must be in reciprocal communication with other areas in the conscious kernel.

TMS has also proved useful in diagnosis and treatment. Stimulation of the motor cortex can provide information about its excitability and the propagation time of neural signals, and may prove useful in the diagnosis and prognosis of multiple sclerosis, Bell's palsy, stroke, and other maladies with motor components. Most interest in TMS therapy has centered on treatment for depression, especially for treatment-resistant forms of this illness (Gershon, Dannon, & Grunhaus, 2003). The typical regime involves the application of high-frequency (>5 Hz) rTMS to the left dorsolateral prefrontal cortex (DLPFC), in an attempt to normalize the putative deficit in activity in the left relative to the right prefrontal cortex in depressives. Using a similar justification, low-frequency (~1 Hz) rTMS is sometimes applied to the right DLPFC in an attempt to lower metabolism in this region. In the standard procedure, the area of stimulation is found by first locating the motor cortex, and then moving 5 cm forward from this area. A key difficulty in assessing the effectiveness of this treatment is a potentially larger than average placebo effect. Latent notions about the magical powers of magnets combined with the elaborate apparatus needed to achieve the desired effect could easily influence the suggestibility of subjects, although this is less likely to occur in cases of severe depression. Accordingly, most studies are careful to include a realistic sham condition, in which the magnet is fed the identical current as in the therapeutic condition, but is held at an angle that is unlikely to stimulate the cortex. With this blinding in place, results of TMS on depression are mixed, despite some initially encouraging results. Two metastudies from 2003 and 2004 (Martin et al., and Couturier, J.L., respectively) show little or no difference in efficacy between sham and actual treatment. At best TMS for depression appears to occupy the scientifically uncomfortable region above statistical significance but below consistent clinical effectiveness.

Unlike, VNS, though, future prospects for TMS are considerably brighter. There are two reasons for this. One of the major obstacles to successful treatment for depression and other disorders with TMS, if not *the* major obstacle, is its inability to penetrate into deep brain structures. Ordinary coils, such as the figure eight, would require high-intensity stimulation to reach these depths, possibly leading to seizure as well as other effects associated with nontarget stimulation, including severe facial pain; this limits conventional stimulation to a depth of approximately 1.5 cm. However, new designs, such as the Hesed coil (Zangen et al., 2005), are designed so that the magnetic field falls off more slowly as a function of distance. They do so by including multiple magnetic sources, each of which is designed to reduce the nontangential component of the electric field, thereby reducing the accumulation of electric charge on the surface of the brain. Combined with multiple sources and possibly other alternative coil geometries it may be possible to reach deeper into the brain without generating undesirable effects along the way.

The other innovation on the horizon is the combination of TMS with neuroimaging methods. Currently, this is done in order to determine the precise effects of magnetic stimulation (Sack and Linden, 2003) on the brain, but there is no reason that the stimulation parameters cannot be dynamically adjusted to achieve a desired goal. This topic will be taken up in more detail at the start of the next chapter.

Deeply penetrating magnets in combination with neuroimaging may ultimately result in what may aptly be described as the holy grail of TMS: localized noninvasive deep-brain stimulation. The ability to provide precise targeting of deep structures without the need for surgery would provide us with unprecedented powers to alter both the cognitive and affective contents of consciousness. The next few decades are likely to show potent advances in this direction, and as they do occur, neuroengineering will rapidly transit from the research laboratory into everyday life.

CRANIAL ELECTROTHERAPY STIMULATION (CES)

The therapeutic use of electricity has a long history. The Roman physician Scribonius Largus recommended standing on a live torpedo fish to alleviate gout and headache. Fish tend to be difficult to control in both research and clinical settings, and it was the invention of electricity that initiated the modern era of this form of treatment. In fact, up until Flexner's 1910 report throwing cold water on this treatment paradigm, there were up to 10,000 physicians using some form of electrotherapy in their practice. Transcutaneous Electrical Nerve Stimulation (TENS) is probably the most common form of electrotherapy in common use today. TENS is used for pain relief and involves application of current on the order of 60 mA near the affected region. CES, in contrast, involves the application of small currents (<1 mA) to the ear lobes.

CES has been suggested for treatment of a wide variety of disorders including anxiety, depression, sleep disorders (when CES arrived in America from Europe it was called "electrosleep"), headache, dementia, and insomnia (Gilula & Barach, 2004). The mechanism by which CES works is not known; however, it is hypothesized that current flows to the brainstem from the skin, stimulating the production of 5-HT (serotonin). This in turn inhibits cholinergic and noradrenergic inputs to the thalamus, lowering the overall arousal signal to this gateway. Reduced firing in the thalamus reduces tonic activity in the cortex, and also induces alpha wave synchronization according to this account; in fact, there is direct evidence to show an increase in alpha activity with CES (Kennerly, 2006). Thus, the putative net effect of this stimulation is to lower cortical arousal, lowering anxiety in the over-aroused patient, and encouraging sleep otherwise. Curiously, though, the device does not result in immediate sleep, and its proponents claim that if it is used anytime during the day it results in better sleep at night.

Results with CES are a mixed bag and difficult to interpret at this point. A number of studies in the primary areas of application, anxiety, sleep disorders, and depression show good results. However, most of these have been carried out directly by device manufacturers and/or are not properly blinded. A comprehensive meta-analysis did show that CES significantly outperformed sham treatments for anxiety (Klawansky et al., 1995), although the authors of this study suggest caution because, for the most part, these studies they did not include therapist binding. Hence, more research needs to be carried out before definitive conclusions can be drawn; the literature on CES, though promising, is orders of magnitude smaller than VNS and TMS, for example. It should also be noted that treatment is always a matter of balancing cost and benefit, and the side effects of CES appear to be mild, especially when compared to current drug treatments.

CES, as it currently stands, is more of a medical curiosity than anything else, but it is included here for the sake of completeness; there is always the possibility that future developments may more clearly demonstrate its effectiveness or may yield better means of delivering stimulation. It is also conceivable the dominant medical paradigm for psychic illnesses, which focuses on neurotransmitters and means of the pharmaceutical regulation of such, are implicitly or actively suppressing alternative methods such as CES, although this does not seem to be the case with TMS and VNS. However, as a neuroengineering method CES must be counted as only mildly interesting, for many of the same reasons given for VNS. Regardless of future clinical success, CES is unlikely to be converted into a general means of stimulating specific brain areas. The smearing of current by the scalp and skull makes its effects too diffuse for such a use. Hence, if CES is to prove more general, it will probably be in the form of a stimulation equivalent of ECoG, in which electrodes on the surface of the cortex deliver a more targeted signal to the brain.

REFERENCES

Benabid, A. (2003). Deep brain stimulation for Parkinson's Disease. *Current Opinion in Neurobiology, 13*, 698–706.

Bosco, E., D'Agosta, L., Mancini, P., Traisci, G., D'Elia, C., & Filipo, R. (2005). Speech perception results in children implanted with Clarion devices: Hi-Resolution and Standard Resolution modes. *Acta Otolaryngologica, 125*, 148–158.

Chorost, M. (2005, November). My Bionic Quest for Bolero. *Wired*.

Cohen, L.G., Celnik, P., Pascual-Leone, A., Corwell, B., Falz, L., et al. (1997). Functional relevance of cross-modal plasticity in blind humans. *Nature, 389*, 180–183.

Couturier, J. (2005). Efficacy of rapid-rate repetitive transcranial magnetic stimulation in the treatment of depression: a systematic review and meta-analysis. *Journal of Psychiatry and Neuroscience, 30*, 83–90.

George, M., Nahas, Z., Lisanby, S., Schlaepfer, T., Kozel, F., & Greenberg, B. (2003). Transcranial magnetic stimulation. *Neurosurgery Clinics of North America, 14*, 283–301.

Gershon, A., Dannon, P., & Grunhaus, L. (2003). Transcranial magnetic stimulation in the treatment of depression. *American Journal of Psychiatry, 160*, 835–845.

Gilula, M., & Barach, P. (2004). Cranial electrotherapy stimulation: A safe neuromedical treatment for anxiety, depression, or insomnia. *Southern Medical Journal, 12*, 1269–1270.

Groves, D., & Brown, V. (2005). Vagal nerve stimulation: a review of its applications and potential mechanisms that mediate its clinical effects. *Neuroscience and Biobehavioral Review, 29*, 493–500.

Kennerly, R. (2006). *Changes in quantitative EEG and low resolutin tomograpy following cranial electrotherapy stimulation*. PhD dissertation, University of North Texas.

Klawansky, S., Yeung, A., Berkey, C., Shah, N., Phan, H., & Chalmers, T. (1995). Meta-analysis of randomized controlled trials of cranial electrostimulation. Efficacy in treating selected psychological and physiological conditions. *Journal of Nervous and Mental Disease, 183*, 478–484.

Kringlebach, M., Jenkinson, N., Owen, S., & Aziz, T. (2007). Translational principles of deep brain stimuation. *Nature Reviews Neuroscience*, 623–635.

Kumar, K., Toth, C., & Nath, R. (1997). Deep brain stimulation for intractable pain: a 15-year experience. *Neurosurgery, 40*, 736–746.

Manrique, M., Cervera-Paz, F., Huarte, A., & Molina, M. (2004). Advantages of Cochlear Implantation in Prelingual Deaf Children before 2 Years of Age when Compared with Later Implantation. Advantages of Cochlear Implantation in Prelingual Deaf Children before 2 Years of Age when Compared with Later Implantation. *The Laryngoscope, 114*, 1462–1469.

Martin, J., Barbanoj, M., Schlaepfer, T., Thompson, E., Perez, V., & Kulisevksky, J. (2003). Repetitive transcranial magnetic stimulation for the treatment of depression. *British Journal of Psychiatry, 182*, 480–491.

Massimini, M., Ferrarelli, F., Huber, R., Esser, S., Singh, H., & Tononi, G. (2005). Breakdown of cortical effective connectivity during sleep. *Science, 309*, 2228–2232.

Mayberg, H., Lozano, A., Voon, V., McNeely, H., Seminowicz, D., Hamani, C., et al. (2005). Deep brain stimulation for treatment-resistant depression. *Neuron, 45*, 651–660.

McIntyre, C., Savasta, M., Kerkerian-Le Goff, L., & Vitek, J. (2004). Uncovering the mechanisms of action of deep brain stimulation: activation, inhibition, or both. *Clinical Neurophysiology, 115*, 1239–1248.

Morris III, G., & Mueller, W. (1999). Long-term treatment with vagus nerve stimulation in patients with refractory epilepsy. *Neurology, 53*, 1731–1735.

Sack, A., & Linden, D. (2003). Combining transcranial magnetic stimulation and functional imaging in cognitive brain research: possibilities and limitations. *Brain Research Reviews, 43*, 41–56.

Sackheim, H., Rush, A., George, M., Marangell, L., Husain, M., Nahas, Z., et al. (2001). Vagus nerve stimulation (VNS) for treatment-resistant depression: Efficacy, side effects, and predictors of outcome. *Neuropsychopharmacology, 25*, 713–728.

Schiff, N., Giacino, J., Kalmar, K., Victor, J., Baker, K., Gerber, M., et al. (2008). Behavioural improvements with thalamic stimulation after severe traumatic brain injury. *Nature, 448*, 600–603.

Schlaepfer, T., Cohen, M., Frick, C., Kosel, M. B., Axmacher, N., Joe, A., et al. (2008). Deep brain stimulation to reward circuitry alleviats anhedonia in refractory majory depression. *Neuropsychopharmacology, 33*, 368–377.

Sturm, V., Lenartz, D., Koulousakis, Treuer, H., Herholz, K., Klein, J., et al. (2003). The nucleus accumbens: a target for deep brain stimulation in obsessive-compulsive and anxiety-disorders. *Journal of Chemical Neuroanatomy*, 293–299.

Talwar, S., Xu, S., Hawley, E., Weiss, S., Moxon, K. A., & Chapin, J. Rat navigation guided by remote control. *Nature, 417*, 37–38.

Walsh, V., & Cowey, A. (2000). Transcranial magetic stimulation and cognitive neuroscience. *Nature Review Neuroscience, 1*, 73–79.

Wichmann, T., & DeLong, M. (2006). Deep Brain Stimulation for Neurologic and Neuropsychiatric Disorders. *Neuron, 52*, 197–204.

Winter, J., Cogan, S., & Rizzo, J. (2007). Retinal prostheses: current challenges and future outlook. *ournal of Biomaterials Science, Polymer Edition, 15*, 1–25.

Zangen, A., Roth, Y., Voller, B., & Hallett, M. (2005). Transcranial magnetic stimulation of deep brain regions: evidence for efficacy of the H-Coil. *Clinical Neurophysiology*, 775–779.

Zrenner, E. (2002). Will retinal implants restore vision? *Science, 295*, 1022–1025.

Chapter 8 Near-term Trends

By near-term we do not necessarily mean these trends are right around the technological corner, but rather that these are things that most of us are likely to see, eventually, in our lifetimes. No doubt some predictions may seem far-fetched if not downright impossible, but change in this field, when it does arrive, will be rapid and will build on itself—success will lead to investment, leading to more successes, etc. If the Internet is any guide, we are likely to see an initial period of great exuberance leading into a period of over-exuberance, followed shortly by a more sober but also more far-reaching set of changes. Also, as in the case of this earlier technology, many of the changes will not be so much direct byproducts of the technology itself, but the influence that the technology has on the structure of society and interaction within that society. However, before we consider these issues, let us turn to a more immediate consequence of current developments.

CLOSED LOOP BRAIN STIMULATION (CLBS)

CLBS does not yet exist except in theoretical form, but it is perhaps the item on our agenda that is closest to fruition—certainly no more than ten years away.[1] It may be defined as not only monitoring the effects of brain stimulation, however this is achieved (deep brain stimulation (DBS), transcranial magnetic stimulation (TMS), or otherwise), but continuously adjusting the parameters of the stimulation

[1] A similar idea is found in many Augmented Cognition systems, but in these cases the stimulation is sensory and not directly to the brain (see Chapter 9). Likewise, closed loop systems and BMI systems also use sensory feedback.

to achieve a desired end. Thus, the goal of CLBS is to achieve a given brain state not by delivering a stimulation burst and hoping that things go well but rather to ensure that the stimulation works by adjusting it until this end is achieved.

Figure 8.1 illustrates a hypothetical CLBS setup. As previously stated, the goal is not to merely measure the effect of the stimulation but to alter it. This is accomplished with a control system that determines the distance between the goal state and the actual state of the brain by some suitable metric. If this distance is large, it will attempt to reduce it by searching for alternative stimulation parameters, and if the distance is small, it will attempt to hold the brain in that state. However, as in the case of a much more primitive control system, a thermostat and its environs, it may be necessary to initiate action even when the desired state has been achieved.

The need for CLBS follows from the fact that it can be extremely difficult to determine the effect of stimulation on the brain, and therefore it is not, in general, possible to produce a given brain state with open loop stimulation. There are four basic obstacles to accurate achievement of this goal:

a) Determination of the generated electric field
In the case of DBS and TMS, it is a relatively straightforward matter to apply standard electromagnetic wave theory to determine the resultant field in a homogeneous medium. But the brain is not homogeneous after the application of the field, and the accumulation of charges can affect the resultant field. The Hesed coil, discussed in the previous chapter, for example, is designed to reduce charge build-up on the surface of the cortex and to thereby allow the magnetic signal to penetrate further into the brain.

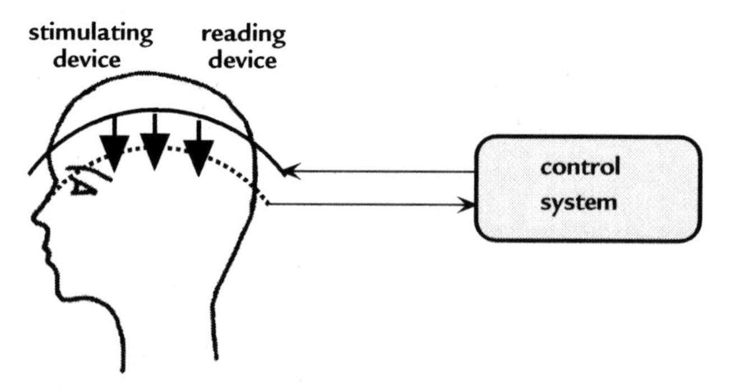

FIGURE 8.1 **A hypothetical CLBS system. The solid arc represents the stimulation component and the dotted arc the imaging component; both are connected to a control system.**

b) The influence of the field on neurons
There are a number of unknowns in these calculations. While it is possible to model the influence of an external electric field on axonal ionic transport, other influences are less understood. For example, there is currently no accepted theory to explain why low-frequency TMS leads to a long-term decrease in excitability of cells and high-frequency stimulation leads to the opposite result. Moreover, any proposed model must contend with the geometry of the components of the affected neurons, which do not oblige us by falling conveniently into fixed angles with respect to the stimulation.

c) The effect on the cell network
Ultimately, we are not stimulating individual cells but large networks of such. To take a trivial case, if we increase the excitability of a cell population, which in turn has an inhibitory effect on another structure, the net effect of our "positive" stimulation may be to inhibit cerebral activity. Of course, the situation can be much more complex than this, and in general, we will only be able to predict the precise effect of stimulation if synaptic efficacy data is included in our calculations.

d) Individual differences
Finally, and far from trivially, there exist considerable individual differences in both the gross anatomy of the brain and also the connectivity within that anatomy. To take two examples of the former from the literature on TMS, the standard procedure for locating the DLPFC is to move forward from the motor cortex. However, serious questions about this procedure have been recently raised because of individual variations in cortical anatomy (Herwig et al., 2001). In another example, one theory for the reduced efficacy of TMS for elderly depression centers on the claim that brain shrinkage increases the distance between the stimulation coil and the targeted cortex. Even when these anatomical differences are accounted for, there remains considerable difference in neural connectivity. This is not surprising, as these sum the net effect of all of our life experiences.

The beauty of CLBS is that while it helps to have some of these factors under one's command, it is not absolutely necessary to pinpoint precisely the effect of the stimulation; one can adjust the parameters until this is achieved. Moreover, the optimal set of parameters for a given patient can be stored, and then these can be used as the starting point for the next session. As a paradigmatic example of the synergy made possible by simultaneous reading from and writing to the brain, let us examine stimulation of the nucleus accumbens (NAcc) with TMS.

As we saw in the previous chapter, stimulation of this reward center may be beneficial for the treatment of depression; we will also suggest in Chapter 10 that direct stimulation of this structure in the nondepressed subject may help achieve a feeling of well-being.

However, currently TMS cannot penetrate deep enough to stimulate the NAcc without adversely affecting other parts of the brain. Suppose, however, we use multiple magnets spread over the surface of the skull. There is the possibility (but not the certainty) that *some* set of stimulation parameters for these magnets will initiate a dynamic effect that results in net excitatory activity to the NAcc. For example, this area is fed mainly by the ventral tegmentum area, which in turn is fed by the hypothalamus, the substantia nigra, and the amygdala, and each of these structures in turn have numerous afferents. While it may not be possible to reach the NAcc directly, or its immediate neurophysiological antecedents, this does not mean it cannot be reached by the appropriate stimulation of cortical structures (or in the case of deep TMS, white matter bundles). Moreover, because CLBS is not wedded to a particular theory of cerebral circuitry or geometry, it may be able to adjust for individual idiosyncrasies in brain anatomy and physiology. In summary, "blind" stimulation of the brain may soon very well be seen as pointless as blind surgery; it is only by first having clear neurophysiological goals, and then by monitoring how close we are to achieving these goals, that we will be able to make reasonable progress in this field.

THE EXPANDED SENSORIUM

The neuroengineering community currently focuses on restoring lost faculties whether these are of sensory, motor, or cognitive in origin. This is only natural, as this is the most pressing need; it also doesn't hurt that both private foundations and government funding is funneled almost exclusively toward this effort. However, as the field progresses, more and more emphasis will be on the extension of existing capacities. One area that may see rapid and relatively early growth is an expansion and enhancement of sensory input. In order of most difficult to understand theoretically and to realize in practice, we may distinguish among three types of modification: tweaking existing sensory modalities, adding new qualitative experiences within a modality, and adding entirely new modalities.

The simplest possible alteration we might want to make is to expand the sensory range. It is not generally possible to hear below 30 Hz and above 20 KHz, and in practice adult audition is considerably more constricted than this. Most over the age of 30 have little hearing above 14 KHz; clever children have been known to download ringtones above this value so that their parents cannot hear their offsprings' phones ring. Using a cochlear implant or suitably enhanced future device it would be a relatively trivial matter to enable audition in the very low frequency range, to enable it in the ultrasonic range, or to restore adult hearing to what it was before he or she attended rock concerts. Other effects are also possible, including harmonization of musical sounds, melodization of non-musical ones, and timbre alteration. Crickets pulsing at night could sound like lush strings; the sound of water dripping from the tap could form the substrate for a Glassian composition if one was inclined in this direction. Likewise, visual prostheses need not be limited to "conventional" vision. Apart from the obvious abilities, such as night vision, and "Bionic Woman" capabilities, such as the ability to zoom in on a distant target, a host of Photoshop or psychedelic effects are also conceivable.

It is important to realize that these transformations work by mapping sensory signals to currently existing qualia; in principle, this could be done with a suitable souped-up iPod, or a camera with a special effects processor and a personal media viewer and headphones. By tapping deeper into sensory processing in the brain, however, it may be possible to evoke entirely novel sensations. To motivate this idea, let us reexamine a fascinating experiment on color perception, now a quarter of a century old. In a 1983 *Science* paper, Crane and Piantanida suggested that it was possible to create a scenario in which novel illusory colors were perceived. They presented retinally fixed adjacent red and green or blue and yellow stripes to the eye. Any object that is fixed on the retina quickly fades, and the resulting perception is thought to be a function of the border colors. They claimed that as the stripes faded, the intermediate zone was filled in with both opponent colors, in violation of the theory of color opponency (see Chapter 2). Previously impossible bluish-yellows and red-greens, which under ordinary circumstances would be perceived as shades of gray, were supposedly seen by their subjects.

This result has recently been questioned by Hsieh and Tse (2006). They claimed that careful measurement showed that the actual colors were well within the range of ordinary vision, and were variants of brown and other muddied hues. It may very well be the case that it is not possible to create novel sensations via actual sensory stimuli, but this does not mean that it would not be possible via DBS or some other brain stimulation means. Let us take theory rF (developed in Chapter 4) as a working hypothesis, for example, in which qualia are a function of the resulting graph of causal interactions (the following discussion may still apply, *mutatis mutandis,* if computational functionalism or something resembling strong AI turns out to be valid). Only certain graphs will naturally arise from ordinary input. For example, we would not expect to see both blue and yellow mutually active within a given phenomenal pixel, because ordinarily they would inhibit each other. However, if we could clamp them with a current sufficient to overcome this interaction, then this would indeed be possible. And given that qualitative perception is a function of this causal interaction, one would also expect novel sensations to arise.

Just what these sensations would "look" like, would be hard to describe, as we have never seen them. It is that much harder to describe what a novel sense would be like, whether it was some amalgam of existing senses, or something entirely new. We also have no idea of how it might be created, because we currently do not know precisely why the visual brain creates visual feels and why the auditory brain creates sounds and likewise for the rest of the senses. If theory rF is valid, however, we may speculate that vision has the quality that it does because of a high-level property of the causal graph associated with this type of property. Speculating further, we could posit that this property depends on the lattice-like structure that results from lateral interaction between processing of adjacent phosphenes. Auditory processing, alternatively, has a more linear one-dimensional structure, with notes processed along a single continuum, but with some interaction between the same note in the different octaves and perhaps notes that stand in the relation of a perfect fifth to each other. In other words, the processing of given sense produces a unique topology, and the overall quality of this sense is a function of this topology. Finally, it follows that if we were to artificially induce a novel topology, then we would also produce a novel sense.

There are admittedly many links in this speculative chain, and it is easy to be skeptical of the final conclusion. However, we may, as always, rely on our old fallback, supervenience, to produce a more minimal conclusion. In this view, the quality of a modality is functionally dependent on how the brain treats the inputs for this sense. Unless our current senses are complete (and there is no reason to believe that they are), that is, they cover all the possible ways in which inputs may be acted on, then there will be alternative ways of processing a given set of stimuli, or alternative ways of stimulating the brain to achieve the same effect. Thus, there is a very good chance that our rich sensory world may be made even richer still by the addition of entirely new ways of perceiving in the not too distant future.

IMPLANTABLE CHIPS, EXTENSIBLE BODIES

Chronically implanted functional chips are not likely to become a reality overnight. Even if we were to magically solve the numerous physical biocompatibility problems the informational compatibility problem would remain; the brain is a radically different type of processor than the Boolean devices that dominate the digital world. Still, halting but steady progress will be made and it will be possible within the next quarter century to make implantable chips small enough, with a sufficient power supply, and in such a way as to provide cognitive enhancement without the need for an external processor.

In a much-publicized result, for example, considerable progress has already been made in constructing an artificial hippocampus, the part of the brain involved in the laying down of new memories (patients with significant hippocampus damage live in the continuous present, with each moment seeming to arise *ex nihilo*). Belying the notion that scientific revolutions invariably involve a lone wolf operating against the current orthodoxy, but consistent with the complexity of the task, this project comprises six distinct research groups operating at different universities (Graham-Rowe, 2003). They share a common set of goals: a) to construct an artificial hippocampus through reverse engineering, b) to test the device on animals, and c) to test the device on humans. While much progress has been made on the first goal—they claim 95% accuracy in mimicking the rat hippocampus—the latter two goals have proved more elusive. They also estimate that an effective clinical device for humans is approximately 15 years away. This may be overly optimistic, but even if we multiply this by a factor of 2, this means that within approximately three decades we will begin to replace brain parts in the same way that we replace knees and hips today. It should also be realized that this project represents an ambitious attempt to reproduce *in toto* an existing brain structure. A "simple" device such as an ECoG-based reader, chronically implanted on the surface of the cortex, will prove considerably easier to construct and to install.

Thus, we can conservatively estimate that within a quarter century implantable brain chips will be possible in practice. Let us now turn our attention to a more theoretical topic, that of the difference between chips that merely extend cognitive capabilities and those that will be truly integrated into the consciousness of the user. In Chapter 4 we labeled these C^- and C^+ devices respectively. The former, by theory rF, or more broadly, any functionalist account in which the chain of causality determines what is and what is not incorporated into the phenomenal field, include devices in which the interaction with the brain is on a time scale so as to preclude such interaction. The latter are highly interactive with existing wetware and thus can affect the contents of consciousness and also have properties similar to "naturally" generated qualia including cerebral celebrity and its corollary, a strong effect on what is and is not remembered.

Let us look at an example of the difference from the point of view of the implantee.[2] Suppose that a C^- device provides mathematical results of some sort, the formulas for indefinite integrals, for example. This could work by a "mind-reading" mechanism (see the next section) that detects the thought "I need the result of integrating such and such," in combination with a virtual "heads up" display that writes the result back into the visual field. Here, we have bidirectional interaction between man and machine but the afferent and efferent influences

[2] One must also keep in mind for future discussions on "mind uploading" and similar topics that strictly speaking, a C^+ device and, to a lesser extent, a C^- device can become incorporated into the identity of the pre-op subject. Thus the distinction between implant and implantee is blurry, but we will ignore this complication for now.

are one-time only. Thus, this case is merely a more convenient means of obtaining the same result by MATLAB, or by a Web search.

A C^+ interaction, although more difficult to realize, is also more interesting. In this case the implant acts as a true extension of the mathematical abilities of the recipient. This person is aware of all intermediate steps in the process of generating the solution, and moreover, it is the will of this thinker that guides the solution process. Why should the C^+ process be superior to a C^- process? After all, the former heavily loads consciousness, and is more time-consuming and more distracting. The answer is that C^+ solutions are generalizable to other problems, whereas C^- are not. If the answer is not within the capability of the C^- device, it just spits out "no solution found" and that is that. If, on the other hand, the solution process is an intimate part of one's psyche, the elements of this process can be entered into one's mathematical bag of tricks, and these can be applied to a host of other problems, both similar and, in some cases, novel.

In summary, the difference between a C^- and C^+ device is similar to the difference between a good math student and a bad one. It may be hard to believe, but students have been known to use computational shortcuts to solve problems they should be working out by hand, and thereby get to spend less time on their homework than their more diligent counterparts. The difference between the two, however, becomes apparent at crunch time, at a midterm or a final, in which a novel problem is presented, or in which the solution to a given type of problem needs to be incorporated into a larger theoretical framework.

One can make the same distinction between physical devices. One can merely have the thought "I want to do a triple somersault with a half-pike off of this diving board and not embarrass myself," and it is done, or one can be given the flexibility of a gymnast and allowed to achieve this routine over time. As in the purely mental case, the C^+ process is more likely to lead to the ability to perform other sorts of dives outside the canon of the physical device, or perhaps generalize to other sports. Here, the advantages of the C^+ device over the C^- device are not quite so dramatic, however. Generally speaking, unless we are athletes, we are satisfied with physical results, no matter how they are achieved. In addition, as has been previously stressed, to be an athlete is to be *unaware* of how one's body achieves its ends; the conscious mind is too slow and clumsy to hit a great passing shot, or to find the path through three defenders that leads to the layup.

As to the nature of the physical extensions that will become popular, whether of the C^- or C^+ variety these are too varied and numerous to be treated within the confines of this work. Suffice it to say that we should not expect, except at the initial stages, that these will be merely more powerful or more graceful variants of the human form. Once we allow control via BMIs over the external world, anything that could be controllable likely will be controllable, and the human animal will no longer be defined by its default evolutionarily provided anatomy. Further possibilities along these lines are explored at the close of Chapter 11.

MIND READING

The ability to peer into the cognitive and affective contents of consciousness of a sentient entity would represent both a considerable technological and theoretical advance. In strictly practical terms, this capacity would enable or advance lie detection, in which the degree of truthfulness of a statement is assessed (or for that matter, a concealed thought is revealed, as in Figure 8.2); the burgeoning field of neuromarketing, in which the hidden desires of the consumer are revealed; a dream detector, in which one's dreams could be played out on an external screen like a movie; and a host of other applications, sinister and otherwise.[3]

This device would also represent an unprecedented advance in our conception of mind. A cornerstone of numerous philosophical positions is that we have access only to the contents of our own consciousness, and indeed, part of the lack

FIGURE 8.2 **Thoughts cannot always be inferred from behavior.**
SOURCE: **http://www.wjh.harvard.edu/~wegner/pdfs/ConsCausBehav.pdf.**
ARTIST: **Joe Sutliff**

[3] Unwanted probing of one's mind is in many ways the ultimate violation of privacy. We are already seeing a considerable erosion of privacy due to online data; this would be the next step. Ideally, how to balance the considerable gains that such a device would enable with these sorts of issues is something that should be worked out now, before these devices become common; this may be sooner than we think.

of progress in consciousness studies is that without firm data about such contents it is difficult to build theoretical traction. If mind reading were a possibility, this would not be the case. Of course, there is a circularity hidden in this argument. We cannot build a mind reader until we have a theory of consciousness, and it is difficult to construct a *full* theory of consciousness without one. However, this does not preclude a bootstrapping process whereby each element slowly builds on the other until this is achieved, with the data of verbal reports serving as the initial substrate on which a theory may be built.

The key to building such a device is to: a) acquire sufficient data about the brain from some suitable imaging device, and b) to separate the unconscious chaff from the conscious wheat. We do not care, in general, about a complex motor routine initiated by the cerebellum; what we want is to know what the person is trying to do. Likewise, we do not wish to know the contents of the entire visual field; this can be picked up with an eye-movement-directed camera. What we want to know is what aspects of the visual field the person is paying attention to, and what it looks like for them.[4]

Requirement *a* we will not comment on here except to say that there will be slow but steady improvements in resolution, cost, and ease of use of neuroimaging devices, and we will gradually come to a state where there is enough information in the signal to decode at least a portion of the conscious content. Requirement *b* is more interesting, and there are two basic approaches. We have already discussed in Chapter 4 the groundbreaking research of Kamitami and Tong (Kamitani and Tong, 2005) in which they extracted from the fMRI signal which of two orientations of lines were being consciously perceived. However, this was accomplished in a theoretical vacuum; raw machine learning techniques were used to separate out the two cases. The complexity of both the phenomenal field and the underlying neural substrate for that field prohibit this as a general approach; there is simply too much to learn.

Therefore, true progress will only come when some filtering can be done *a priori*, on the basis of theoretical considerations. One possibility is suggested by theory rF and its explanatory antecedents; namely, that conscious content is a function of the cell population subset in a state of quasi-simultaneous mutual interaction, and this set only (Katz, 2008). Previously, we referred to this as the conscious kernel. However, there may be extra constraints. For example, it may turn out that the kernel must also adequately engage the frontal lobes for an item to reach consciousness. Crick and Koch (1990) have suggested as much from a

[4] There may be exceptions to this rule, especially in the case of research. However, let us designate a device that captures everything about the neural state of a brain-reading device, and one that concentrates only on conscious content as a mind-reading device; here we are concentrating on the latter.

number of empirical and theoretical considerations, including the fact that this cortical area would be the natural place to collate both sensory afferents and also higher-level cognitive and affective processes. If this does turn out to be the case, however, it will likely be so because of the dynamic integrative properties of this region, rather than because of any special static properties at the cellular or subcellular region. We await a more dynamic, not-locality-based approach, but this cannot be too terribly long in coming, and when this is combined with a suitably advanced imaging device we will indeed be able to listen in on anyone's thoughts, and to hear them as clearly as if the appropriate signals were routed to the vocal chords instead of being silently voiced.

NEUROART

The Internet, of course, is not merely a better way to buy books; it has opened up a host of artistic and entertainment possibilities that were barely conceivable ten years ago. Likewise, it would be a mistake to think that neurotechnology will confine itself to pragmatic ends. As the brain-machine interface becomes more advanced, we should expect to see a rash of experimental efforts to extend existing artistic modalities and to create entirely new ones. Many of these will no doubt fail, because they will focus too tightly on the gee-whiz aspects of the technology, and not on the aesthetic nature of the endeavor. The ones that *will* succeed will be those that harness the new machines to do what art has always done, that is, to communicate an abstraction from one mind to another that cannot be transmitted in a more direct fashion. This will be made possible by the increased expressiveness afforded by bypassing the tools for creation that evolution has provided us—hands, legs, and bodies—and tapping more directly into the original thought stream. Here, we consider three such possibilities, two of which are extensions of existing artistic media, music and dance, and a third that does not yet exist, but may one day prove to be the dominant artistic endeavor, that of the direct projections of the thoughts of the artistic "cognizer" into the minds of the audience.

Musical generation via EEG is perhaps the most natural place to begin a discussion of neuroart, and in fact there have been many attempts to harness brainwaves in this manner (see Figure 6.6). However, without exception, all violate the principle discussed earlier, namely, that the technology should be at the service of art and not vice versa. We generally are not interested in listening to someone's brainwaves; we are interested in the musical thoughts that generate these waves. If neuroimaging is to be useful, it must act as a means of increasing the harmonic, melodic, and rhythmic expressiveness of those thoughts.

One way to do this is to cut out the middleman in the musical generation process, the body, and route the output of the brain directly to a synthesizer.

As things stand now, the bandwidth and degrees of freedom that can be routed through the body far exceeds that which can be garnered through imaging. However, the body is an extra step in the chain and therefore it can only act as a bottleneck. In principle, there are more musical thoughts inside a musician's head than can be expressed through any instrument. As neuroimaging methods improve, this will become apparent. In at least some gifted musicians, we will see not only control of harmony and rhythm, as when the guitar accompanies singing, or when Dylan plays the harmonica, but also an added rhythmic component. Furthermore, in the case of musical geniuses, we may be able to hear the output of a small chamber orchestra, or perhaps even a symphony. We can be fairly certain that the Mozart of Amadeus would relish the experience of being wired into such a device, and we can also be certain that the real Mozart, whether he resembled this whimsy-driven savant or not, would be sure to delight us if so connected.

As with music, so with its artistic close cousin, dance. The most fundamental constraint on dance is imposed by gravity, and while it is true that art often succeeds not in spite of constraints, but because of them—when we watch Nuryev leap it is the *apparent* suspension of gravity that thrills—it is unlikely that all balletic inspiration must be subservient to the severe restriction that the gravitational constant $g = 9.8\ m/s^2$. By means of a BMI that reads the dancer's thoughts, and a suitable set of guy wires, the dancer could suspend or partially suspend the action of gravity, bringing the third vertical dimension into full play. Note that this is not the same thing physically or artistically as a circus-like routine in which the performer floats through the air. These are predefined routines designed to provide the illusion of gravity suspension. In the proposed mechanism, the dancer is continuously and actively invoking vertically by her thoughts, so as to maximize the intended expressive effect. Taking BMIs one step further, it need not be the human body that does the dance. Anything that can be controlled can be controlled artistically. In the future, we will see robots, anthropomorphic and otherwise on the stage, and perhaps a host of other unusual objects such as wall-climbers, hover-bots, and even liquids in this cirque d'esprit.

Finally, and somewhat farther down the engineering pike, but well worth contemplating while we are on the subject, is the notion of an art without form, in which the thoughts of the artist are broadcast directly to the audience. This would be, bar none, the purest form of artistic expression, as there would no be no intervening medium, simply the baring of one's soul for others to experience. If we are willing to pay dearly for concerts now, how more desirable would the dreams, reveries, and streams of consciousness of especially talented thinkers be? These deep thinkers and deep feelers could not fake their offerings, because their audience would be inside their heads with them. Very few would likely be able to stand the intensity of having to deliver on a regular basis in this manner, but the few "thoughmeisters" that could truly deliver would have the capacity not only to produce unprecedented phenomenal visions in their constituents, but also to effect a more permanent, and one hopes salutary effect on them. This, the ushering in of a new, and better way of thinking, is the topic of the next section.

A NEW "NEW AGE"?

Every twenty years or so Western culture experiences a minor crisis of confidence and turns its collective gaze Eastward, upward, and, in some cases, backward toward an earlier and supposedly more golden age. Invariably, however, these brief flirtations with alternative modes of being are brought rapidly down to earth, and usually with a vengeance. The contingencies of life remain the same despite our best efforts, and it is these that, in the end, dictate the shape of civilization. Going forth and proclaiming the freedom of the soul from the rooftops is all very well, but it doesn't put tofu on the table, let alone pay the mortgage.

Spiritual and other movements fail because human nature is an approximate constant. But with the sorts of the changes discussed in this chapter, and certainly with the kinds of changes that are previewed in coming chapters, that will no longer be the case. It would be as serious an error to conclude that culture will not change as a result of advances in neuroengineering as to say that individual minds will not change, as the former is simply the result of the mass action of the latter. What is much more difficult to predict is the direction that culture will take when these advances begin to take hold.

This book has not shied away from going out on a conceptual limb, however, and it would be inconsistent to do so now. Let us assume the following comes to pass as a result of neuroengineering advances, perhaps not as a result of the first wave of this technology, but in the not too distant future:

a) The average IQ is raised to about 150. This does not mean that everyone will necessarily become genius—see the discussion on creativity in the next chapter—but it does mean that all of us will have the intellectual firepower now reserved for all but the lucky few, and
b) The neurotic has been almost completely purged from the psyche. This does not imply that we will be without passion in the appropriate circumstances. It does mean that we will be free from unfruitful emotional episodes, that we will take a deliberate and balanced view of things, that we will behave calmly and rationally, and in the ideal case harness the now excess emotional energy into what truly matters: art, science, love, and faith.

We have no evidence that: a) acting in unison, or b) alone, would make a better human being, but the two taken together could have powerful consequences, both at the individual and cultural levels. In particular, the following, in rough order from the quotidian to the sublime, could be seen:

a) Fashion and style
If there is single area of everyday life more hampered by conservative thought, it is difficult to know what it would be. The key difficulty is the tripartite function of fashion: it has a practical component, that of protection and warmth; it has a component of swagger, in the case of the men, and more purely aesthetic in the case of women; and it has a final and more nebulous, but equally important component in which it acts as a signifier of status or of identity. It is this last component that acts on a drag on innovation. If the past's view of the future had any validity, by now instead of the business, pantsuits, and tracksuits we should be seeing flowing robes, intricately woven fabrics, and all manner of headdresses and jewelry.

But we do not see this. If anyone came into an office with this rigging they would find themselves at best politely ignored, at worst an object of ridicule. No one, quite understandably, wants to walk into the modern corporate arena looking like Whoopi Goldberg's character on *Star Trek: The Next Generation*. Instead, fashion moves forward incrementally, and then cycles back on itself, leaving us no more exotically clothed then our distant ancestors on the dance floor from the 1970s. If we are to break through the collective neuroticism that puts a brake on all statements of individuality that deviate from convention more than a slightly outré cravat, we must first quell individual insecurity and neuroticism. In short, we must move to the other side of the stylistic tipping point, and once a few people jump on that side of the seesaw, more will surely follow.

b) Hopes for better governance
It is tempting to believe that the quest for power as an end in itself is at the root of social evil. Certainly the twentieth century was painfully afflicted by this disease, and ravages of holocaust, war, and other ailments are testimony to this fact. It is also tempting to believe that a reduction of this desire, at the level of the individual, would be of benefit to society as a whole. Had Hitler, Stalin, and Pol Pot alternative means of expression, through art, say, or knowledge, would we also have suffered from their delusions? For the educated, intelligent, and neurosis-free man it brings as much satisfaction, if not more, to prove a theorem or to cultivate a bonsai tree as to have the masses plaster posters with one's picture throughout the land.

Is it not, however a Pollyannaish and painfully naïve notion to think that evil could be eradicated by technology? Certainly, nothing that has come down the pike in the last 2,000 years of innovation would lead us to expect that this would be case; if anything, man's power of destruction is magnified by technology. There is a short retort to this, one that this work has made many times before, and one that we turn to once again at the risk of repetition. Neurotechnology cannot be thought of as an ordinary technology, because it aims to change not what people do but what they are. In principle, there is nothing preventing us from injecting a soupcon of Ghandi into everyone's soul—not so much to render everyone passive in the face of evil but to make them principally impassioned, activated by justice and not by power. Although it may be difficult to put the blame for aggression on a single neurotransmitter, it is not likely to have an overly complex explanation in terms of neural circuitry, because of its roots in the fight-or-flight complex.

Certainly there is hope that it can be controlled at the individual level without undue unwanted consequences. This is not the problem. The difficulty is that a few bad evolutionarily designed apples could spoil the technologically enhanced bunch. How much evil does there have to be in a society in order to make it fully dysfunctional is a question that neither history nor sociology has a firm grip on, but the answer cannot be high, at most on the order of 10%. What if these holdouts refuse to submit to an otherwise salutary process? There are all sorts of reasons to avoid "the treatment," legitimate and otherwise, from general nonconformity to the belief that if one holds out, one will have that much less competition in the political arena (which may be valid).

The only thing we can say with certainty at this point is that, genetically, through neural means, or some combination of both, we will one day no doubt succeed in deleting the unabashed will to power, and importantly, without compromising the enterprising spirit. But we have no guarantee that it will be within the ensuing years, and we have no ability as yet to predict the effect of the inconsistent application of such a technology on the dynamics of society as a whole.

c) Science and scientific progress

The illusion of the dispassionate scientist working steadily and patiently for the good of mankind, immune to the foibles of human nature, has long ago been punctured, starting with Watson's 1968 book *Double Helix*, and continuing in the writings of numerous scientists to this day. It is not that the members of the scientific community are any worse than the rest of mankind; it is that frequently they are no better. In particular, this community suffers from an emotional, borderline neurotic attachment to existing theories and methods that places a drag on the search for new knowledge. It is perhaps only natural that the vast majority of funding and effort is funneled into existing theory and methods, but the history of science shows that this overemphasis on the current orthodoxies is wrong. New, and very often good, ideas spring out of the blue from unexpected sources, and some mechanism should be at play to at least throw these ideas onto the scientific stage, even if they are to be ultimately rejected. Neurotechnology has the potential both to relax the conservative streak running through science as a whole and also to encourage creativity at the individual level, as the next chapter will suggest.

But there is another, perhaps more important point. If the intelligence and diligence of the average working scientist could be increased even slightly, this would naturally include those working in the neural and computational sciences. This in turn would enable greater increases in intelligence, which would lead to more advances, and the process in principle would snowball into an exponential rate of increase; we may already be on that path and not realize it. While barriers will undoubtedly arise to impede this progress, these will also be overcome, as in the case of the development of computational devices, and it is perhaps not too farfetched to suggest that within a half century a colony of super geniuses will exist solely to toil away on problems both abstruse and fundamental, including that of building the next generation of thinkers.

d) Dopamine dens and consciousness clubs

If neuroenhancement seems a solitary, almost solipsistic affair, this is only because it is natural from both a scientific and engineering standpoint to treat each brain as a stand-alone entity ripe for modulation. Yet we have already hinted at how brains may be eventually connected in networks that are the direct analog of today's Internet groups, and in the last chapter of this book, we will introduce the notion of a full meld between consciousnesses. Before these are fully operational, however, neurotechnology will, one hopes, serve to enhance ordinary human interaction. It is eminently worth stressing that this technology will not, at least initially, change basic needs, among the foremost of which is social interaction and confirmation.

Rather it is the way in which this interaction takes place that will find new expression. We can conceive of a variety of new venues for the enhanced mind, including what may be termed a dopamine den. These places may escape the undoubtedly deserved louche reputation of their predecessor, the opium den, in which a neuromodulator was delivered indiscriminately to the entire brain, by stimulating only selected areas, and providing just enough kick to bring goodwill to the fore. We may also conceive (perhaps dream is more apt) of a kind of salon where the goal is not so much to upstage others as to entice, entertain, and enthrall with music, poetry, and the higher arts. More intrepid neuronavigators will wish to take this one stage further, and explore altered states of consciousness in the company of fellow enthusiasts. Here, the goal would be to explore similar parameter settings of a stimulation device in the company of others.

e) Religion and temples of the neuroritualistic
Finally, is it going too far to suggest that when mutually satisfactory subspaces of consciousness are discovered, that practitioners will wish to build a church around these modes of thought? The indigenous people of the Peruvian Amazon, the Urarina, would never conceive of imbibing ayahuasca, a drink prepared from a macerated vine native to the rainforest, outside a ritualistic context, nor would the Navajo peyote. As has been frequently noted by anthropologists, the ritual confines and the appropriate social setting provide a degree of comfort to what may otherwise be a terrifying experience; there is no reason to suspect that is should be otherwise for technologically induced psychonautical adventures.

The intersection between religion and neurotechnology in general is a fascinating topic, and one that could expand into an additional volume or two, but there is one important point that can be made relatively succinctly within our earlier framework describing how the mental arises from the physical. A key argument revolves around the variable X in the following sentence: Religious faith X brain states. If X = "is nothing but a byproduct of," with the emphasis on the "nothing but," then one is in the territory of Dennett, Dawkins, and paleo-atheists such as Hume. Alternatively, if X = "is a correlate of but not reducible to" one is in the territory of neo-neuro-dualists, such as Newberg, D'Aquili and Rause (2002), who are attempting to gain a richer understanding of religious experience through neural dynamics but are not necessarily dismissing the intrinsic authenticity of faith.

While the chasm between these two views is wide, some would say irreconcilable, strictly speaking what is required from the point of view of neuroengineering is only that X be equal to something like "is supervenient on." As before, this produces a weaker claim than either extreme but one that both sides of the aisle can embrace intellectually, although the committed atheist would naturally be dismissive of the faux spiritual state that was induced. At the very least, though, this induction would have the means of generating the concomitants of religious experience such as goodwill, charity, and compassion. Less obviously, supervenience also implies that if religious experience turns out

> to be something over and above a material state of the brain then true prayer and worship can still be technologically induced. The neuroengineering device would not be doing the actual spiritual work, it is true, but by virtue of placing the mind in the appropriate state it will trigger the corresponding metaphysical events that actually generate these states. Expect to see less squirming in the pews and less impatience for the afternoon's football match as souls are raised upward on virtual biers with the congregation looking on in awe.

This completes our survey of what can be done now, and what will be done over the next few decades. In the remaining four chapters, we turn our vision to the long-term, in which man bypasses evolution, and becomes, for good or ill, the designer of his own destiny.

REFERENCES

Crane, H., & Piantanida. (1983). On seeing reddish green and yellowish blue. *Science, 221,* 1078–1080.

Crick, F., & Koch, C. (1990). Toward a neurobiological theory of consciousness. *Seminars in Neuroscience, 2,* 263–275.

Graham-Rowe, D. (203, March 12). World's first brain prosthesis revealed. *New Scientist.*

Herwig, U., Padberg, F., Unger, J., Spitzer, M., & Schonfeldt-Lecouna, C. (2001). Transcranial magnetic stimulation in therapy studies; examination of the reliability of "standard" coil positioning by neuronavigation. *Biological Psychiatry, 50,* 58–61.

Hsieh, P., & Tse, P. (2006). Illusory color mixing upon perceptual fading and filling-in does not result in 'forbidden colors'. *Vision Research,* 2251–2258.

Kamitani, Y., & Tong, F. (2005). Decoding the visual and subjective contents of the human brain. *Nature Neuroscience, 8,* 679–685.

Newberg, A., D'Aquili, E., & Rause, V. (2002). *Why God Won't Go Away: Brain Science and the Biology of Belief.* New York: Ballantine Books.

Watson, J. D. (1968). *The Double Helix: A Personal Account of the Discovery of the Structure of DNA.* New York: Atheneum.

Part 3 The Future of Neuroengineering

Chapter 9 From Thinking Caps to Eureka Caps

Did your first-grade teacher ever tell you to put your thinking cap on? It is unlikely that he thought you would take the idea literally; he was simply trying to get you to pay attention. Somehow, over the years, the concept became reified to the point where many now take it seriously, and in fact in this chapter a device along these very lines will be proposed. In addition, we will examine the possibility of a device to enhance creativity. These two devices operating in unison could be then combined synergistically to produce what could be called a genius cap. Although we are far from producing such a machine currently, there are important hints in both the empirical and theoretical literature that at the very least we can improve on what nature, and teachers, gave us, and eventually that something along these lines could genuinely be built.

THE AMPLIFICATION OF INTELLIGENCE

The first step in so doing is to augment the purely intellectual capacity of the human mind. There clearly are huge discrepancies in sagacity from person to person, and within an individual, from one time to another. Hence, the natural question is, what is the difference between them, or between those moments? We will thus first examine the nature of intelligence, and how this seemingly nebulous concept may be given some structure. We will then consider *Augmented Cognition*, (AugCog) the Defense Department's ambitious plan to enhance the intellectual capacities of its soldiers and pilots. We will then discuss why this project, while not without interest, falls short of what we are looking for, and how true intellectual enhancement may be achieved via other means.

The nature of intelligence

Intelligence is one of those ideas like jazz, good food, and love that are almost impossible to define but can be spotted from miles away. In the case of intelligence, it seems to come haltingly out of some, but from others it positively oozes forth; think of the difference between your college professors, or between your relatives if you must. To drive this home further, imagine that an alien descends to Earth. He quickly masters Chinese, English, and Urdu, in a few short months he is able to beat all but the best chess and Go players, and to add insult to injury he anonymously enters architecture, poetry, and music composition competitions and bests the most accomplished human practitioners of these arts almost every time. He clearly is in possession of a concentrated form of some ability or quantity that is only loosely and haphazardly distributed among the human species.

How can we define this quantity, and how, for example could we recognize whether an alien or purely silicon-based machine was intelligent? An early but still influential view, perhaps undeservedly so, was put forth by the father of computer science, Alan Turing, well known for his work in decoding the German submarine code during World War II, and also for work on abstract computing devices, offered the following suggestion. Assume that a human judge is communicating both with the unknown entity and a human via a terminal or other mechanism that does not allow a direct view of the participants. If the judge cannot tell the difference between the participants, then the entity must be at least as intelligent as a human being.

As a number of researchers have pointed out (e.g., Russell & Norvig, 2003), the Turing test is tied too closely to an anthropomorphic view of intelligence to be of general use. In order to fool the judge the entity would need to exhibit a number of human traits not necessarily tied to intelligence, including emotion, cultural biases, linguistic tics, etc., and may also need to show traits actually inversely correlated with intelligence such as over-perseverance in belief, irrational associations, and the like. Take our superintelligent alien from the previous example. He would not necessarily pass because he would not have human quirks, or because he had extra traits, such emitting a squeaky noise every 12.7 seconds. Furthermore, if he did pass, all we would know is that he was at least as intelligent as the average human being, but have little feel for his intellectual prowess on an absolute scale.

The Turing test fails precisely because it makes no attempt to relate intelligence to other concepts or theories, but this is exactly what we need if we are to manipulate this quantity. The next natural place to look is IQ tests, and the quantity they are meant to capture, Spearman's quantity g, or a general intelligence independent of the domain of thought. One sometimes hears that intelligence is what IQ tests measure, but on its own, this is a vacuous statement. It doesn't tell us what ability or capacity that these tests measure. For example, is it pattern recognition, is it the number of items that can be held in memory, or is the ability to think quickly, or some combination thereof? Given that IQ tests are correlated with success in other intellectual endeavors, such as schooling and work, that they have a strong genetic component, and that they are relatively stable within a person over time, we have some confidence that they capture at least a portion of g. Furthermore, although intelligence is a catch-all term housing many different faculties, verbal, mathematical, musical, spatial, mechanical, etc., these are generally moderately to well-correlated with each other and therefore it makes sense to talk about a generalized intelligence following Spearman.

The question remains as to what g is, and what IQ tests actually measure. To see this, let us examine a few representative test questions:[1]

Q1) Even the most ______ rose has thorns:
- a) ugly
- b) weathered
- c) elusive
- d) noxious
- e) tempting

Q2) NASA received three messages in a strange language from a distant planet. The scientists studied the messages and found that "Necor Buldon Slock" means "Danger Rocket Explosion" and "Edwan Mynor Necor" means "Danger Spaceship Fire" and "Buldon Gimilzor Gondor" means "Bad Gas Explosion." What does "Slock" mean?
- a) danger
- b) explosion
- c) nothing
- d) rocket
- e) gas

[1] Answers: e,d,d

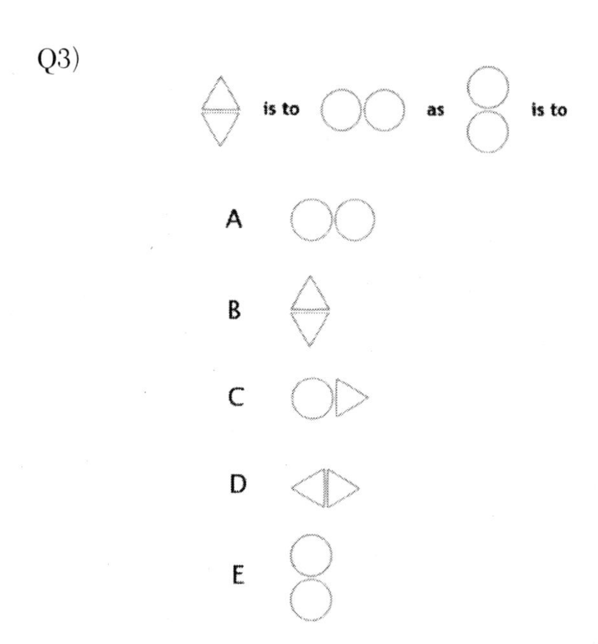

Let us begin with the last visual analogy. The solution can be found by translating the pictures into concepts: two vertically stacked identical objects (except for inversion) are to two identical horizontally ordered objects, but different from the original objects, as two vertically stacked objects are to what? Obviously, two identical horizontally oriented objects, different from the triangles, which leaves only D as the answer. In other words, the key to this problem and many others of this sort is to abstract away the details, the particular shapes, their distance, and other concrete facts and leave only the generalized representation. Abstraction is also at work in the solution to the other two examples. The essence of the first problem is that a desirable object (a rose) has undesirable properties (thorns); the only word that emphasizes the positive aspects of the rose is the last choice. The second and most difficult problem also involves abstraction in the sense that one must ignore word order in the strange language but also realize that there is a one-to-one correspondence between this language and English (rarely true with real let alone extraterrestrial languages, but this is obviously simple expressions in the alien language). Once this is realized, it is simply a matter of matching up the commonalities to determine what word goes with what.

In summary, we can conclude that IQ tests are principally tests of abstraction—the ability to derive and manipulate, in a contextually meaningful way, the high-level properties of the objects, and to ignore the literal properties of these objects. Of course, this is not an accident; IQ tests are designed not as tests of fixed knowledge and associations but as tests of high-level, knowledge-free reasoning. We can also see the relevance of abstraction to the discussion of our prefrontally challenged friend from Chapter 3, the monkfish. The reason the monkfish can only trap fish that happen to go by, and would be useless on land even if it could breathe and walk is that it has not produced an abstract representation of what it is doing. It doesn't know that its spine extension is there to lure fish, and if it did, it would still not have the capacity to understand the entire bait and trap cycle, from luring an animal by presenting something attractive, to restraining or eating it directly.

IQ tests also highlight another aspect of intelligence, that of speed. These tests are typically timed, with about 30 seconds given per question, depending on difficulty. Although it is conceivable that someone could be a slow genius, thinking great thoughts but at glacial speed, and certainly there are different cognitive styles, intelligence has been historically linked to rapid thought. This is born out in the EEG evidence to some extent. The original motivation for exploring neurophysiological measures of intelligence was to provide a culture-free measure of this quantity (Deary and Caryl, 1997); this has since been shown to be an overly ambitious goal but these studies still provide useful information about brain functioning and IQ. In particular, the latency of event-related potentials (ERPs) in the EEG in response to a stimulus have been weakly inversely correlated with IQ; this factor appears to be largely independent of head size, another weak correlate of intelligence, and thus the two can be used in conjunction to produce a moderate predictor of this trait (Wallhovd et al 2004).

The notion of neural efficiency has fared somewhat better as a predictor of intelligence than ERP latency and head size. Neural efficiency means that less effort is required to achieve the same or often better results. Anecdotal evidence suggests, for example, that a good chess or Go player will paradoxically spend less effort to find a *better* move than a weaker player in the same situation. A number of neuroimaging studies ranging from PET to fMRI (see, e.g, Reichle, Carpenter, and Just, 2000) have confirmed this finding. Interestingly, a study that inferred cortical activity from EEG showed that on the training phase of a reasoning task the more intelligent subjects showed greater frontal activity, indicating greater effort, but that after training they exhibited significantly less activity to achieve slightly better results (Neubauer et al., 2004). Presumably, after an initial effort, they required less work to produce the same results as their less-intelligent counterparts.

A number of resting state (i.e., when no explicit cognitive task was given) EEG measures of intelligence have also been proposed. A recent comprehensive study indicated that the following measures are positively correlated with IQ: shorter EEG frontal phase delay (the difference in phase between signals at differing electrodes); longer EEG posterior phase delay; lower EEG coherence (a measure of the degree of match between signals at different electrodes); and EEG amplitude (the total strength of the signal) (Thatcher, North, and Biver, 2005). The third result with regard to lowered coherence is particularly interesting in that it is consistent with the idea that intelligence is related to functional complexity, with the intellectually superior brain performing more independent calculations in parallel.

Our final clue to the puzzle of intelligence lies in the differences along the continuum from human to ape to mammal to monkfish to insect to inanimate objects. The obvious neurological difference as we climb the phylogenetic ladder is the development of the frontal lobes. But what precisely is it about these recent evolutionary developments that result in intelligence? We can look at this from a different angle by asking what is different behaviorally as we go up the ladder. If we start with a rock, presumably a very dumb object, we see that it does move and reacts to its environment, but strictly due to the dictates of the laws of nature. To a lesser extent, this is true of the lower animals. It is true that they transform inputs into outputs, but this is done in a stereotypical fashion. Moreover, and this is the key difference, a reactive transformation does not add information about its environment, or if it does, it does so in a very limited fashion. The behavioral output, though it usually benefits the organism, is no more informationally rich than the sensory input.

To make this more concrete, let us look at an example. You see your Uncle Joe walking on the opposite side of the street, and at the same time, a beetle walks by your uncle. Let us look at this from the beetle's point of view and then yours. The beetle sees something big approaching and then scampers away. It is unlikely that anything is going through the conscious mind of the beetle other than pure fear, but let us be generous and say that the beetle, considered as a system, "knows" that big objects like Uncle Joe are harmful. It may also know something about the way Joe is walking and how much danger it is in. But what do you know from the same sensory information? You know that Joe is going somewhere, you know that he is not at home, you know that he is married to your aunt, that he fathered your cousins, that by virtue of being a person he has two kidneys, a heart, liver, and gall bladder, that he has desires, wishes, and dreams, that he owns a '64 Caddy, and an almost endless list of things that you don't even know that know until you think about them. In other words, you are able to add a large body of knowledge, driven by both specific and general considerations, to your database of information over and above what is provided directly by your senses.[2]

What does this have to with abstraction, which we earlier identified as the key to successfully solving IQ problems? Again, we turn to the monkfish,

[2] This example is slightly unfair to the beetle, because Joe is not his uncle. But it remains the case that we know much more about our world and especially about people, even when they are anonymous, than this insect.

and his limitations. In purely behavioral terms, the monkfish is confined to his niche. He cannot successfully venture outside of his habitat because he has no means of catching his food there. To a first approximation, he is a purely reactive system, with a clockwork-like mechanism that transforms sensory impulses into feeding behavior. If this mechanism receives novel input outside the boundaries of what it expects, it will have no idea of what to do. In other words, the monkfish lacks generality of response. The abstraction "if you want to catch something, dangle an object it wants in front of it" is powerful precisely because it gives us information not just about a particular type of situation, but an almost unlimited range of situations, including, as we discussed in chapter 3, catching sharks, trapping foxes, and phishing in addition to fishing. The nature of abstraction is thus closely tied to the ability to make conclusions about what will and will not work in novel environments in which only limited initial information is available.

These considerations motivate the following working definition of intelligence:

> *An organism is intelligent to the extent that it is able, within a fixed period of time, to produce new information over and above that provided by its senses, but is consistent with this sensory input.*

In other words, we can conceive of an intelligent system, like the human brain, as a kind of information amplifier that produces more knowledge than what it is given. The faster it is able to perform this operation, the more it knows about its environment and the more it will be able to adapt if this environment is altered. The notion of efficiency is intimately related to this definition: an efficient brain is one that in the same amount of time and for the same amount of effort is able to produce more and better quality information about its environment.

This is not a completely unproblematic conception of intelligence (although none is). First, what we really want to say is that truly novel types of information are produced. Once you know a little math you can add 1 to anything and get a new number, as in the child's game, but this does not mean that you have infinite intelligence, because of the limited scope of this type of knowledge. Second, there is no notion of utility in this definition; you could be generating vast quantities of knowledge that are completely useless. Third, and most subtly, it pays to produce new information that may occasionally be wrong, and to not do so is to be far too conservative; in Artificial Intelligence (AI) this is known as default or commonsense reasoning. Speculation, as long as it is done in a principled manner, is not only valuable, it is absolutely necessary because the vast majority of things that we know about are contingent and could always turn out not to be the case.

Despite these limitations, the notion of intelligence as information amplification will prove important in our proposal for the thinking cap. But before describing this device, let us turn to another avenue of research on cognition that is currently being hotly pursued.

Augmented cognition

AugCog is currently identified with DARPA's plan to increase the cognitive capabilities of its soldiers and pilots, but the idea is neither new in principle nor practice. As early as 1960, J.R. Licklider wrote "The hope is that, in not too many years, human brains and computing machines will be coupled together very tightly, and the resulting partnership will think as no human brain has ever thought ..." (Licklider, 1960). In a very real sense, we are already significantly augmenting our native intellectual capabilities with external tools. For example, a pencil and paper augments memory and increases visualization, and the personal computer enables all manner of calculations from tax forms to numerical integration that could not easily be done in the head. The key difference between AugCog and these examples can be found in how tightly bound the man-machine interface is, as the Licklider quote suggests.

In AugCog, the emphasis is on machines quickly sensing the cognitive and affective states of their human masters and presenting information in such a way as to maximize throughput. A human operator is continuously assessed with physiological monitors. These have included EEG (event-related and background), ECG, body posture and position, video pupillometry, fNIR, and various combinations thereof (Reeves and Schmorrow, 2007). The overall cognitive work load of the user as well as other variables are then derived from these monitors. On the basis of these measures, the system adjusts the information presented to the user so as to be maximally consistent with his cognitive/emotional state.

In the prototypical application of AugCog, and the one that most frequently appears in the literature, a subject in an informationally rich environment, such as a pilot, is monitored for cognitive load. Because of the restricted bandwidth imposed by consciousness on thought, only a few displays can be genuinely attended to at any one time, and only then if they can be combined into a single Gestalt. Moreover, if stimulus-rich events occur at the periphery of attention, they will detract from the current stream of consciousness, without offering up much in the way of new information—the worst of all possible worlds. Because the degree of cognitive overload cannot be predicted in advance, as it depends on the user's level of fatigue, his ability to process information in parallel, and the precise nature of the stimulus, the best way of handling this situation is to reduce the amount of information when an overload is detected in real time. This could also be applied in reverse; an understimulated user could be supplied with extra information, or if drowsy, the system could suggest a cup of coffee or a rest.

In what is probably the highest ratio of public relations bravado to actual tangible results in the history of science, a professionally produced twenty-minute movie has been made by DARPA to show how this scenario would work in a war room–type environment in the year 2030 (available online at http://www.augmentedcognition.org/video.htm). The film depicts a scenario that may be both overly ambitious in one sense and in another not ambitious enough. It shows a cognitive load detector that works nearly flawlessly, but does not treat many of the other possibilities that have at least a chance of emerging over the next twenty-five years including memory add-ons, creativity enhancement, specific thought detection and control (as opposed to general state detection), not to mention computational systems that will be considerably smarter.[3] Nevertheless, the research goals shown in the movie as well as the goals of AugCog as a whole are reasonable enough and probably will be achieved in the not too distant future. This is to increase the throughput of the human information processing system. By our previous definition, this should also count as an increase in intelligence. In a given time frame, more information will be treated, and more new and better quality information will be generated when there is an impedance match between the two interacting complex systems, such as a human and a computer.

Toward a thinking cap

The two central questions we must now ask, given our working definition of intelligence, and given the reviewed evidence are the following: 1) are there any means of increasing the information amplification ability of the human mind by modifying its behavior without fundamentally changing its structure, and 2) could this be done by the addition of external devices. Let us take these separately, as they are really two fundamentally different issues: the first is more in the way of a "classical" thinking cap, if we can say this about something that has yet to exist, and the second entails the integration of artificially intelligent devices into the brain and ideally into the consciousness of the user.

The first and most obvious way of enhancing the existing brain is to somehow increase plasticity. Learning in the brain (reviewed in chapters 2 and 3) is by no means completely understood, but it is generally agreed that two factors mediate this process, the alteration of existing synaptic efficacies, and regeneration via

[3] Many have speculated that we may end up being the add-on to the computational system rather than vice versa, performing highly specialized tasks that AI still has not mastered, such as those involving insight or representational restructuring (see also the discussion on the eureka device). But this ignores that, in consort with these developments, technology will emerge to more tightly bind computational systems to the brain of the sort that we have considered throughout this book. In fact, it is unlikely that there will be an identifiable client-server distinction in the future; as we have argued, external and internal computation will seamlessly blend to produce a new hybrid system drawing on the strengths of both.

the formation of new synapses and/or entirely new cells. In the infant both contribute to adaptation, and in the adult the former predominates, although the precise amount of true neural regeneration in the adult brain is in dispute, and may be much more prominent than we previously thought (Eriksson, et al., 1998). Whatever set of mechanisms is at play, it is clear that plasticity is critical to intelligence. Without this, not only will we not be able to lay down new memories, will be not be able to think properly about situations with which we may already be familiar. Although the formation of fast temporary connections is less understood than long-term learning, there must be some mechanism of this sort to adapt flexibly to problems that deviate in any way from what has previously been encountered.

However, it is unlikely that a mere increase in plasticity alone in the normal brain will lead to a significant enhancement of intelligence. There are at least two reasons for this. First, it would be very easy to overshoot the ideal setting of this parameter. When this happens, old memories are erased at as fast a clip as new ones were formed. In other words, at best, a constancy of knowledge and wisdom could be maintained. This may be good if fundamental retraining is needed, but it is hardly a prescription for increasing intelligence. The second reason is more fundamental. In decades getting machines to learn one lesson arises again and again. This is that what cannot be represented or generated cannot be learned. Plasticity in itself will not allow us to think great thoughts, although it may be a necessary condition for doing so, and a necessary condition for remembering them and slotting them into our existing knowledge base once they have been formed.

For this we need to turn to another clue from the reviewed literature. This is that intelligent brains are complex brains, ones that do not settle for the obvious thought but nimbly move among possibilities until the solution presents itself. There is an alternative and instructive way of looking at this process. A useful way of thinking about a very large system, such as the brain, is as a device that takes input and settles into an attractor, or relatively stable state that represents the processing of this input. It then moves on to process the next set of inputs, possibly influenced by this previously generated state. Figure 9.1 illustrates two (simplified) extremes of attractor-based processing. The square represents the space of possible thoughts, the small shaded boxes represent possible inputs, and the lines direction of thought. The left box depicts a simple mind, the right a complex mind.

On the left, in which so-called "hardening of the categories" has taken place, a number of different inputs lead to the same thought. The characteristic of the inflexible thinker is to reduce the complexity of the world to a manageable set of inputs; for example, if the subject of sports arises in any way, he will invariably lament the Dodgers move to L.A., or if gardening arises, he will dwell on his champion acorn squash from the '84 judging. The right of the diagram shows

A

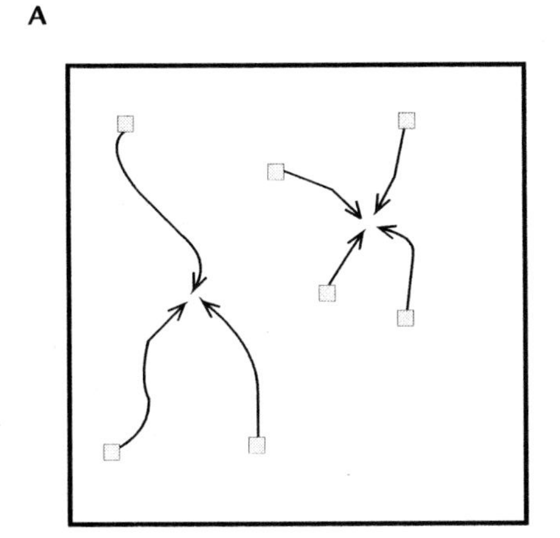

B

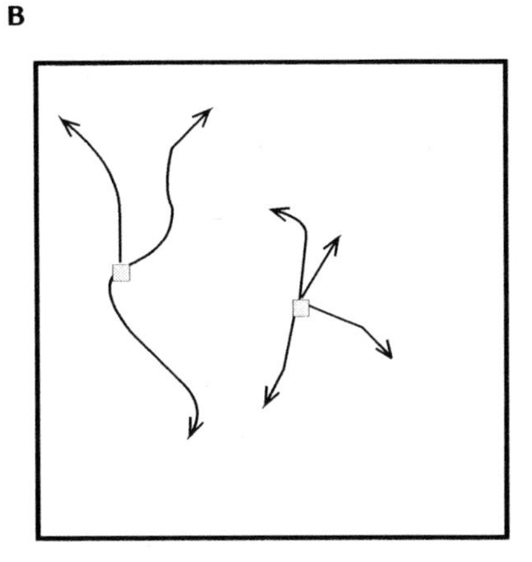

FIGURE 9.1 **Processing in two types of minds. On the left, the simple mind takes a number of different inputs and produces the same thought. The converse is true on the right, it takes single inputs, and produces multiple thoughts.**

the opposite extreme. Not only will each input lead to a new state, it may lead to many states (these are shown being triggered in parallel for simplicity, but this may also be a serial process). In other words, the complex mind rarely says "in other words" to itself (oops); it takes each input at face value, and treats it as a unique occurrence. It savors the perceptual and conceptual richness of its external and internal environments, and is not constantly attempting to reduce this content to something familiar. This is the cognitive counterpart to the intelligence measures discussed earlier that suggest that IQ is correlated with low EEG coherence.

How does this relate to our definition of intelligence in general, and the notion of abstraction in particular? We can see this by reconsidering Q3 from the IQ test given earlier. As previously discussed, the problem can be retranslated into the following analogical form: two identical vertically oriented objects are to two identical horizontally oriented objects as two vertically oriented objects are to what? Of course, once translated into this form, the answer is trivial. The key is to be able to entertain enough abstractions so that the first and third elements in this sequence match. The second element is then derivable, and the fourth will be identical to the second. But if a narrow categorization arises, then this will not be possible. Upon seeing the stacked triangles, one's thoughts may drift to the pyramids, or to high school geometry. The key is not only to abstract out what is essential about these objects, which is not difficult for anyone in possession of frontal lobes and an eighth-grade education, but not to perseverate on the natural associations one has with the inputs. The abstractions one forms must be context dependent, not association dependent. When this does happen, one will be able to produce new and useful information about one's environment that may also be generalizable to new situations.[4]

In summary, then, the two requirements of our thinking cap are as follows: a) first increase plasticity, up to a point, to facilitate learning and adaptation, and b) decrease coherence between brain regions, up to point, to facilitate flexibility of thought. While we have little idea of how to achieve either of these at this point via neuroengineering, neither is out of the range of speculation. We do know, for one, that plasticity and learning are facilitated by alertness. It should not be difficult to achieve greater levels of alertness given control of the brain either via deep transcranial magnetic stimulation (TMS) or deep brain simulation (DBS).

This, of course, is what we already do with the most widely used brain modification technique in the West, the ingestion of caffeine. Caffeine has probably had more of a profound effect on the development of Western civilization than any other substance—the skyline of New York would probably look radically different had not the Turks left behind this precious bean after their aborted seize of Vienna in 1683, and the outline of the intellectual landscape would likewise be altered. But the more precise stimulation afforded by neuroengineering will probably allow us much greater control in the future. With caffeine, and all drugs for that matter, one is fighting against two powerful forces in the brain. The first is short-term homeostasis, which wants to bring the brain back to its natural state, and often will overshoot this baseline in the process. The second is long-term habituation. We all know someone who sips coffee throughout the day without it appearing to have a perceptible effect.

Both of these effects may be fully or partially averted with future neuroengineering. As it turns out, the key to doing so will be to figuring out the precise salutary effects of sleep. We do know that sleep is different than mere rest, as is obvious. Lying in bed all night without actually entering into unconsciousness is never as satisfying as a full night's sleep. But whether it is dreaming, memory consolidation, rehearsing past events or preparing for new ones, or some other aspect more related to the psychic burden of the conscious state that requires all of us to hibernate each night is unknown. If, however, this benefit could be replaced artificially, a longer period or more alert period of awakeness could be reproduced. One clue may come from studying those rare people who seem

[4] Generalization seems to be the opposite of what is being recommended here, in that it does reduce multiple inputs to the same state. To maximize information inferences must be made with respect to similarity to past occurrences, and therefore some collapsing of states, as in the left of Figure 9.1, is necessary. But nimbleness of thought requires: a) that this generalization be highly context-dependent, and b) in most cases of interest, easy categorization be abandoned in favor of more model-driven and other abstract modes of processing.

to operate at full steam with what for most of us would constitute an extended nap. More generally, the circuitry behind arousal (reviewed in Chapter 2 and in the context of anesthesia and consciousness in Chapter 4) is gradually becoming more understood as the years pass, and it does not appear that there is any fundamental obstacle *in principle* from both modeling the circuitry and manipulating it.

Achieving our second and more important goal, that of decoherence between brain regions so as to increase cognitive complexity, is likely to prove more difficult, because in this case we are fighting stronger forces. This aspect of neural dynamics is a complex function of connectivity, synaptic efficacy, arousal itself, and the input to the brain. It may also be the case that if the aforementioned hardening of the categories is too far along, there may be nothing we can do. However, let us assume that we have a relatively young, plastic mind at our disposal and that we are attempting to increase the complexity of thought. One scenario for doing so could be achieved with the kind of closed loop brain stimulation (CLBS) described in the previous chapter. The two stages of this process would correspond to first detecting that the brain is entering too easily into a state in which brain regions are in coherence. This is not difficult to determine and in fact can be achieved currently with EEG. The second state would entail the production of decoherence. This could be achieved by a suitable blend of noise and inhibition administered to the overly compliant region via TMS or DBS. None of this is trivial. For example, the decohered regions still need to be in communication with their processing counterparts, and the fact that relatively low phase delay correlates with intelligence indicates that it would be detrimental to upset these channels in any way. It also goes without saying that additional thousands of scientist hours (and by extension, hundreds of thousands of graduate student hours) would be required in testing such a device once built.

But we have another completely different option at our disposal, to augment intelligence with an external device. We have already considered this possibility in the previous chapter, and also discussed the difference between a $\mathbf{C}^+$ and $\mathbf{C}^-$ device. Recall that the former intertwines with consciousness and alters the thought process, and the latter (merely) routes requests from the brain to a machine via a thought-reading device and then feeds the result back through the senses. Let us here consider in a bit more detail how to construct the former, as the latter is no different in principle than a very fast Internet connection. The increased bandwidth between man and machine may very well produce novel results via synergy but it will not create a fundamentally new type of being.

Let us therefore examine the $\mathbf{C}^+$ case. Let us also assume that we have an artificially intelligent device that we wish to feed directly into the psyche. We should also note that this device need not be necessarily smarter than its host in all its capacities, it just needs to form inferences that will complement the intelligence of the host, that is, to either think better or quicker in at least some domains. Thus, we will not need to wait until full-fledged robot intelligence of the science fiction variety to make this a reality, and in at least some areas, artificially intelligent agents already outperform natural human ones.

The key is to merge these capabilities of the former with latter. One clue as to how to do this comes from Baars' influential (1988) workspace theory of consciousness. In Chapter 4, we were critical of this account because it was a variation on the standard computational line, and thus inherits the difficulties of this form of functionalism. But this does not mean workspace models are not expressing an important truth, and it also does not mean that it cannot be retooled to conform more closely to a causally based theory of consciousness, such as rF. Recall that in this model the contents of consciousness are identified with a central working memory store. Various brain modules examine the current constituents of this store and (unconsciously) compete with each other to influence the next set of contents. The winner or winners get to act on the contents, and the store is then adjusted to reflect this action of these modules. The winning contents are then broadcast to all of the other modules once again; that is, they enjoy, in Dennett's term, cerebral celebrity. Translating into the casual terms of theory rF, the central store is identified with the kernel, which is a more dynamic structure that arises over possibly the entire brain but includes only cell populations in mutual and reciprocal influence. Activation spreads out from the kernel to other modules, which then act on these contents, and then rebroadcast the result back to the kernel.

It is clear then what needs to be done. The external device must be fed into the kernel, and must compete on its own terms with every other module vying for the "attention" of this structure. This may seem like a daunting task. Apart from the sheer mechanical barriers to achieving this, the kernel by construction is not identified with any part of the brain. It is more like a floating gas cloud, in constant transition as the processing task varies. However, there are likely to be some regions that are invariably associated with this cloud. If we were forced to lay down a bet the left dorsolateral frontal cortex (DLPFC) would probably be one of them. As previously noted, this area is identified with high-level planning and other executive tasks, including working memory, and the left hemisphere (in right-handed people) most probably houses what is sometimes called the stream of consciousness, although it is probably better to refer to this as the stream of verbal or conceptual consciousness. Tapping into this linguistically dominant hemisphere could very well give us what we want: the ability to directly manipulate the contents of consciousness, and in a such a way as to augment the intellectual capacity of the bearer of that consciousness.

Summarizing the difference, then, between a $\mathbf{C}^{+}$ and $\mathbf{C}^{-}$ device, we can say that the former will act more quickly and more directly on the contents of consciousness. Its output will not be routed into the visual field, for example, as in the latter case, and will not need to undergo the relatively slow processing associated with this process, in which it will fight with every other sensory input to capture attention. This does not mean that the external device will not need to compete, however. Echoing a very old concept in psychology, the pandemonium model, the various modules are like little devils all trying to outshout each other. The one who makes the most noise gets heard. What makes this model work is that a devil only gets excited when it has done something truly beneficial to the current contents of consciousness. The same will need to apply to an external device competing on equal terms. The external device needs to have a sense of the worth of its conceptual products, in order to know when to get excited, and when to lay low. In this it is no different than internal modules—this sense need not be perfect, just not so bad that it dominates consciousness when it would be better sitting quietly in the background and waiting for input to which it can make a significant contribution.

In summary, we will eventually have at least two means, possibly complementary, of augmenting intelligence at our disposal. The first is to take the existing brain and make it better, a more plastic learner, to increase the complexity of its processing, and perhaps most easily, to make it more awake more of the time, without the side effects of current arousing methods. The second is to add external devices, either by detecting the needs of the user of the device, and then routing information back through the senses, but more interestingly, to interact with the generation of conscious content directly, in order to more fully blend the benefits of AI into naturally intelligent systems. We now turn to the problem of making the mind more creative as well as more intelligent.

THE INDUCTION OF CREATIVITY

Although there is considerable debate regarding the *degree* of difference between creativity and intelligence, there is general agreement that there are important differences (Runco, 2004). This much is common sense. Let us take two people of relatively high but equal IQ, say on the order of 120. With apologies to our protagonist, Lenny from Long Island, the first could be an accountant, a field in which creativity is not entirely absent, but is certainly not overflowing, and in fact an abundance of which may be positively detrimental. The second could be a poet or visual artist. In dress, in mannerism, in lifestyle and, most importantly for the current discussion, in cognitive style, the two may be as different as night and day. The implication for our current purposes is clear: the thinking cap, and the so-called eureka cap, designed to provide inspiration on demand, will most likely require two different modes of action.

Exactly *how* intelligence and creativity differ is a much more difficult issue. If intelligence is conceived in the broadest sense as the successive application of operators to a problem space, then the difference is merely one of degree. For then creative approaches would be merely the application of more complex operators, such as "relax constraint," or "re-represent the problem"; as we will see, both of these are strategies for creative problem solving. But this is close to a vacuous statement, as we can always claim that the solution to a problem, or more generally, the creation of a novel work, artistic or otherwise, always involves a series of steps. The real question revolves around the nature of the differences between these steps. As we will see by examination of some representative examples, it is probably more profitable to keep the two modes of thought separate, and our two corresponding thought-enhancing devices separate, also. In the following sections, we will first try to determine what is unique about the creative process, then look at the neurophysiological evidence surrounding this concept, and finally, propose a means of artificially enhancing creativity.

Anecdotal and psychological evidence

We began our examination with the old chestnut illustrated in Figure 9.2, which may or may not be the origin of the phrase "to think outside the box." The task is to cover all nine squares with only four continuous lines and without lifting the pencil off the paper; the right half of the diagram shows one possible solution. All solutions, including this one, share one trait, namely, that some lines must go beyond the implicit boundary formed by the dots. The problem statement makes no mention of this boundary; rather it is imposed on the problem by the solver. This gives us our first, and perhaps most obvious insight into the creative process, namely, that it often entails the relaxation of one or more constraints that are falsely applied to the problem or situation.

Our second problem is less well-known and more difficult: How many total matches are played in a single-elimination n-player tournament? For example, what is the total number of matches played in Wimbledon singles, where the draw is 128 players. The interesting aspect of this problem is that it can be solved

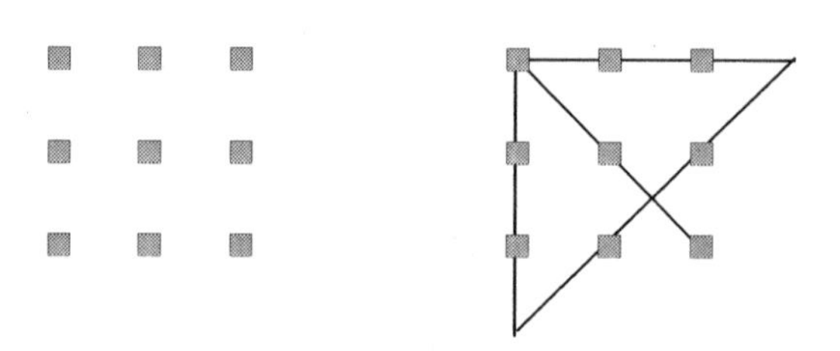

FIGURE 9.2 **A puzzle that requires the relaxation of an implicit constraint to solve.**

by brute force, either by summing 64 + 32 + 16 + 8 + 4 + 2 + 1 to obtain 127, or for the mathematically inclined, invoking the principle that

$$\sum_{i=0}^{n-1} 2^i = 2^n - 1$$

where n is the number of rounds in the tournament. However, there is an easier, insight-driven way of obtaining the same result. The total number of matches will be equivalent to the total number of losers (there is one loser for each match). All but the tournament winner loses once, and therefore the total number of matches is 128 – 1 = 127. This solution has the added bonus that it is immune to alterations in the nature of the tournament; it applies if there are byes, or if qualifying tournaments feed into some of the slots in the main draw. It also illustrates the principle that creative restructuring or re-representation of the problem will often turn a difficult problem into a simple one.

It is also the case that a similar restructuring will take a problem that is intractable or nearly impossible to solve by any means, including brute force calculation, and turn it into one that is soluble. As an example of this, consider the mutilated checkerboard problem in Figure 9.3, made famous because of its resistance to purely logic-driven solutions; apparently logic is not enough to reason well (sorry, Mr. Spock). The two corners are removed, and the task is to determine whether a set of dominos (a 2 × 1 object) can completely cover the checkerboard. An exhaustive search of all possible tilings would take an enormous amount of time; furthermore, we could make this search futile even for the fastest machines by making the board bigger. However, if we place alternating colors on the board, as in the right of the diagram, a little bit of insight quickly reveals the solution. Each domino must cover both a colored and clear square. But there are a total of 32 colored squares and only 30 clear ones; therefore there is no possible covering for the board.

While problems such as these are good examples of creativity in that they require insight to solve, or at least to solve quickly, they fare poorly in the laboratory.

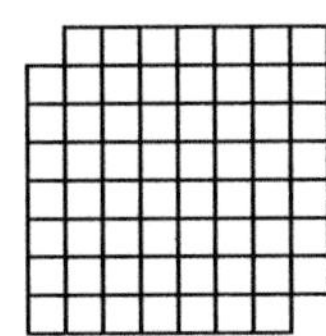

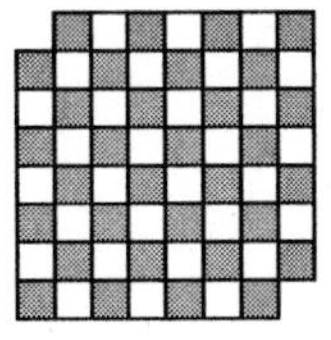

FIGURE 9.3 **In the original representation on the left, the mutilated checkerboard problem is nearly impossible to solve. Adding the alternating colors provides a valuable hint to the solution.**

It is difficult to generate large numbers of problems of this sort, and for all but the most creative thinkers a solution will not be forthcoming in a reasonable period of time. Psychologists have thus turned to a set of more manageable problems (well mindful that these simplified conundrums may leave out some of the more difficult elements of the creative process). We consider one such type of problem here, the Compound Remote Associate (CRA) problem (Bowden et al., 2005), and it will later serve as the basis for understanding how a eureka-producing device could work. A CRA problem consists of three words given to the subject, and the task is to find a fourth word that can be combined with all of the given words. For example, given French, car, and shoe, the answer would be horn. Problems in which the solution is both a prefix and suffix are somewhat more difficult (e.g., boot, summer, ground; answer: camp), but all problems are constructed so that a reasonable percentage of solutions (typically around 50%) are found within thirty seconds of contemplation. Furthermore, there are two paths to the solution, the first a more methodical exploration of the associates of one of the original words to see if it matches the other two, and the second, a more insight-driven approach in which the answer just pops into one's head, allowing an exploration of the neural correlates of the difference between these modes (see the following text).

It is in the latter mode that CRA problems, while not as complex as some other test problems, and certainly lacking in some components with respect to a full-fledged creative act, such as the generation of a scientific theory, highlight an important aspect of the creative process. This is the Chinese-finger-puzzleness of inspiration: the more one tries come up with a solution, the less likely that it will be forthcoming. An anecdotal corollary to this observation is that inspiration is fostered when one is in a dream or fugue-like state. There is the oft-told account of Kekule's realization of the structure of Benzene by dreaming of a snake swallowing its own tail (the Ouroboros of mythology), as well numerous claims by writers, scientists, poets, and others that it was only after they gave up trying to actively look for a solution to a difficult, nonlinear problem that it came to them. For example, this entire section in outline, and many of the individual sentences in detail (including this sentence itself!), came to the author in a period of no longer than thirty seconds while he was taking a shower.

The necessity of so-called divergent thinking, a semirelaxed state of mind in which many avenues are explored for creative thought is but one of a set of many closely linked ideas that course through the literature on creativity. A closely related concept is the idea of defocused attention (Martindale & Greenough, 1973), in which concentration is spread more thinly among competing conceptual nodes than in "normal" sequential thought processes. This state is thought to be facilitated by altered states that decrease frontal arousal (Martindale, 1977), such as the aforementioned dreaming and daydreaming.

We can also associate such states with their behavioral components. The creative mind will appear to others as one at play rather than at work, and the nimbleness of thought will often manifest itself in a child-like approach to the task at hand. Although we ourselves are in divergent mode here, simply trying to corral a number of unruly concepts into the conceptual stable for later treatment, it must be mentioned that the child-like adult at play is not at all like the child at play—children, especially young children, produce notoriously stereotypical creative works. It is sufficient to cite the refrigerators of grandparents throughout the world to justify this claim. Skipping ahead a bit into neurological territory, it appears that a well-developed prefrontal cortex, an area of the brain that reaches maturity last, in conjunction with an earlier flexibility of thought that is often lost with the innocence of childhood, *both* contribute to the creative process.

Our final strand of evidence comes from the much-noted connection between genius and madness. From van Gogh to John Nash, there is abundant anecdotal evidence the two are closely allied, and some have suggested that the latter is an unfortunate prerequisite for the former. More careful research has disputed this claim by pointing to numerous Nobel Prize winners and other clearly creative people in the arts and sciences as often being well-adjusted people with families, PTA meetings, and well-mowed lawns. This view would be consistent with the Maslovian claim (see Chapter 11) that the self-actualized person is the most productive and the most satisfied.

There is a compromise position between these extremes that was advocated by the late psychologist Hans Eysenck. Eysenck's three-factor model of personality comprises the three dimensions of neuroticism (N), extraversion (E), and psychoticism (P) Echoing Aristotle's claim that "no great genius has ever been without *some* madness" [author's emphasis], Eysenck (1995) proposal that "psychoticism in the absence of psychosis" may be critical to the generation of creativity[5]. As evidence, Eysenck cites the numerous studies linking P to creativity on standard tests of such, as well as more direct evidence showing that both male and female artists have higher P scores than nonartists. In a move that anticipates a key element of the neurally based theory of creativity discussed later, he also suggests that the reason P increases creativity is via its effect on associativity. The creative mind has a relatively flat associative gradient, possibly

[5] Eysenck has been deceased almost ten years but remains a controversial figure to this day because of his views on race and intelligence and also on subjects such as parapsychology. The devout student of these matters should not let this be a deterrent; the cited book is nothing short of masterful, and should be read by everyone who wishes to explore the psychological and neurophysiological roots of creative genius.

due to a reduction in lateral inhibition; that is, it is able to entertain many different concepts in unison relative to the more narrowly focused mind with low levels of P. However, when P is *too* high, one sees the over-inclusiveness of full-fledged psychotic thought, in which unrelated concepts are connected by a string of fanciful associations; this is clearly inimical to any sort of rational thought, creative or otherwise.

In summary, an initial examination of a few types of problems requiring creativity for their solutions and other trends in the literature on the subject allows one to come to the following tentative conclusions:

a) Creative problem solving often entails the relaxation of constraints, ignoring constraints that appear to apply but in reality do not, or restructuring the representation of the problem to make it tractably soluble.
b) Mood and attitude are crucial to creativity and insight. The creative mind is divergent, defocused, and will appear playful and nimble to the outside observer.
c) A moderate amount of the trait P, or psychoticism, is often found in those who are creative. This may be because lowered inhibition between ordinarily competing concepts or conceptual frameworks enables two or more of these to be entertained in unison.

It is easy to criticize any or all of these as being too vague to be useful, but it must be born in mind that creativity is a difficult beast to pin down, and in fact may be not a single type of thought but rather a constellation of similar modes of thinking all bearing a family resemblance to each other. However, if we are to construct a eureka-inducing device, or at least one that encourages so-called divergent thought, we cannot afford to wallow either in generalities or negativity about them—we must try to thrust to the center of this elusive phenomenon with whatever evidence is at hand. In order to complement the evidence generated so far, let us now examine two further items: first, the neurophysiological data on creativity, which, while not extensive, *is* instructive, and second, let us take a more in-depth theoretical approach to how neural processes could be structured so as to solve problems, such as CRA puzzles.

Neurophysiology and creativity

We begin our exploration of the neural basis of creativity by reconsidering the notion that that low cortical arousal is correlated with creativity (this is a relatively old hypothesis, see, e.g., Hull, 1943). This view is not distinct from the idea that the creative mind is one that throws a wider conceptual net and in fact goes hand in hand with it. Greater arousal will serve to make the currently dominant concept more dominant, via lateral inhibition to competing concepts; in contrast, lower arousal will lower activation of the dominant node, allowing greater activity among competing nodes.

Martindale (1977) has been the most forceful recent advocate of this position. In addition to providing empirical evidence consistent with the view that lowered cortical arousal increases associability, he has also attempted to resolve an apparent paradox. As it turns out, most creative people are introverts, which means that they have relatively high levels of internally generated tonic arousal levels—this is why they do not seek out stimulation from elsewhere. Thus, it appears that creativity is associated with high arousal. However, we must distinguish between two phases of the creative process, elaboration and inspiration. In the first, the problem is pored over intensively, without a solution forthcoming. In the second, the inspiration comes almost in a flash; this is the "Aha" moment. It is in the latter that both introverts and others demonstrate lowered cortical activity (Martindale and Hasenfus, 1978).

Another long-standing idea is that creativity is the province of the right hemisphere. Certainly, we have all been exposed to this idea in the popular literature, which encourages us to believe that the only thing that stands in the way of everyone being a creative genius is the big bad overly rational left brain. Despite the fact that this is a gross oversimplification—all thought, and especially complex generative thought, involves an interplay between the two hemispheres that is still not completely understood—there is some empirical support for this position. A number of results that measure hemispheric dominance show less left hemispheric dominance in creative people (Katz, 1994).

More direct results come from a fascinating recent study that purports to show that low-frequency (i.e., inhibitory) TMS applied to the left frontal-temporal region alters artistic production (Snyder, et al., 2003). Subjects drew pictures of dogs before and after stimulation; in many ways, the "TMS dogs" do seem to be more natural and more expressive. However, independent blinded judges did not find that the TMS-induced drawings were better artistically than the other drawings, although they did find them to be stylistically different. Unfortunately, given low subject numbers and the lack of true aesthetic progress under TMS limits the conclusions that one can draw from this study; more work along these lines is clearly needed. These researchers also looked at the ability to estimate the number of items on a computer screen in a brief glance (Snyder et al., 2006). It is also known that left anterior temporal lobe damage is implicated in this savant-like skill; hence they applied low-frequency rTMS to this region. They found an increased ability of normals to make this estimate (within 5 items) that was highly unlikely to be due to chance. While equally tantalizing to the drawing results, and better grounded statistically, it is unclear what these results say for creativity (or intelligence) for that matter. The authors speculate that the temporary lesion induced by TMS led to a better estimate by suppression of grouping mechanisms. However, such mechanisms are critical for both ordinary discursive and creative thought; in fact, the latter may be characterized as the result of novel but useful Gestalt-like processes.

Perhaps the best current evidence for distinct modes of processing between the two hemispheres, with the right better able to produce insight, comes from the Bowden et al. (2005) experiments with CRA problems. They tested the response time for the left and right hemispheres (by presenting words to the right and left visual fields, respectively) for solutions to unsolved CRA problems. Subjects responded faster when the target word was presented to their right hemisphere suggesting that the solution, although not at sufficient activation to produce a correct response, was more highly activated in this hemisphere. In other studies, this team also looked at EEG correlates of insight (Jung-Beeman et al., 2004). As previously mentioned, CRA problems can be solved with insight or by brute force. Those problems solved by the former and presumably more creative method, as indicated by subject self-reports, showed a characteristic increase in alpha activity at the parietal-occipital junction just prior to the solution, and an increase in right temporal gamma coincident with the solution. Again, the right hemisphere is implicated, although explaining the exact nature of these results is more difficult. The alpha burst reflects a decrease in visual activity, suggesting a turning down of the influences of sensory stimuli, and a turning inward to the problem at hand; the gamma burst is suggestive of a binding between the three words in the CRA problem and the desired solution.

Of special interest for the current purposes, the promotion of insight via mechanical means, is another study by the same authors of the neural activity prior to insightful solutions (Kounios et al., 2006). Before the problem was presented to the subject, both EEG and fMRI results suggest increased activity in anterior cingulate cortex (ACC) when insight was the mode of solution. The authors' interpretation of these results is that the ACC may be involved in suppressing irrelevant thoughts and thus allows the subject to approach the problem with a clean slate; this is consistent with the reduced arousal hypothesis discussed earlier.

Clearly, the current set of results, while intriguing, need to be augmented by further study, and in particular the role of the prefrontal cortex, the putative center of executive thought and therefore also for complex acts of creativity such as scientific hypothesis formation, musical composition, and the writing of poetry and prose, needs to be examined in more detail. Nevertheless, these studies provide important clues for our purposes, and it only remains to examine the dynamics of insight in more detail to bring us within sight of our goal.

Toward a eureka-inducing device

If we are to have any hope of constructing a eureka device we must look more closely at the fine-scale dynamics of inspiration. The CRA problems provide an excellent stepping-stone for this endeavor. Figure 9.4.A shows a simplified but representative network for the problem egg-step-speed (answer: goose). Each of the problem words is connected weakly to the solution, but more strongly to a primary associate (in the full network, there would be many more such associates). In addition, there is lateral inhibition between each of the problem words, and also between the associates (not shown). As discussed in Chapter 3, lateral inhibition enables the brain to concentrate on one concept at time in a sequential fashion; without it, pattern recognition and thought as whole would be a massed jumble of conflicting concepts. However, in this instance this inhibition prevents the solution. One or another of the provided words and its primary associate(s) will become dominant, and this will suppress the other concepts. The brain (or possibly the left brain alone) will go into a sequential mode, looking for associates of one or more of the clues at a time, and eventually coming to the answer. But this is a slow laborious process, not an insight-driven one.

To produce insight we may take a clue from the literature and lower tonic activation reaching this network (Figure 9.4.B). This seems reasonable; everything we know about inspiration implies that it flourishes in a defocused, low-arousal state such as daydreaming or the like. However, low arousal by itself will not produce a solution, because the dynamics of the network remain identical. One of the clue concepts and its associate will still dominate (step and child in the diagram), and the actual solution will still be below threshold. It seems that this does not help our situation at all.

We are at a crossroads, and it is now that we need to be heeding our own advice; it will take a bit of creativity to get past this dilemma. Let us assume that we have lowered arousal, *and* that we could somehow stimulate the solution alone (goose). This is seemingly absurd: the eureka machine does not know the solution to the problem, it is meant to help the mind solve it, and in any case we

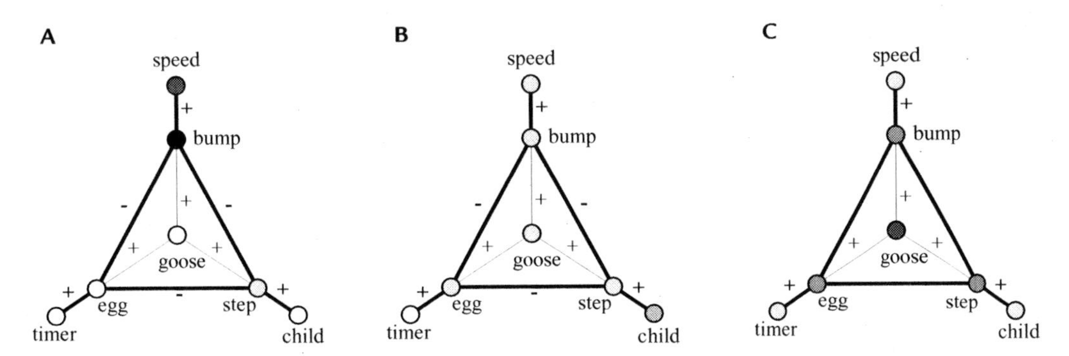

FIGURE 9.4 **A simplified network to solve a CRA problem. A: One possible state of the network after initial consideration. B: The network state if tonic activation is reduced. C: A possible state after reducing activation and adding noise. Eureka!**

have no idea of how to target a single concept in the brain. But let us not be so critical in this—the creative phase of our inquiry. If we *could* somehow activate the "goose" node, it would then trigger the three clues, and a stable resonance loop would rapidly form (we assume that the inhibitory connections are not so strong as to prevent this; otherwise the solution would never be found under any circumstances). In dynamical systems terms, we can say that the solution is a stable state of the system, albeit not the one the system wants to enter on its own; it needs a kick in that direction.

Of course, we can't activate the solution directly, but what this constraint-defying reasoning tells us is that if we could get the network into a state similar to the solution, it will do the rest of the work on its own.[6] To see how this might be possible, let us take a short detour into the field of stochastic resonance (SR). SR is the counterintuitive principle that random noise added to a weak signal may improve the signal-to-signal-to-noise ratio (Simonotto et al., 1997). Consider how this would work in the case of visual perception. Suppose the image is too faint for the eye to register, because the relevant features are below the detection threshold. Adding noise added to the image will bring these features above threshold, and if done judiciously, will not produce artifacts in the image.

How does SR apply to our situation? The natural counterpart to the perceptual case is to attempt to add a random signal to the low-arousal network (Figure 9.4.B) to encourage the solution to achieve threshold activation. This may not work, as the random noise may trigger one of the clue words and its associates as before. But in this case, we simply start over. Eventually, either the target word, and possibly some of the clue words, will be more activated than the rest, the excitatory subnetwork comprising the inner triangle in the figure will resonate, and the activation of the target will rise above the threshold for conscious detection. We also know that it should not take that long for this to happen because we are not trying to put the network artificially into a state it does not want to be in. We are simply nudging it one way or another until it gets there.

Thus, we have the starting ingredients of a recipe for inducing creative problem solving. Does this apply to our other situations? Let us look at the "out of the box" thinking needed to solve the dot problem posed in Figure 9.2 as a paradigmatic case. Strong activation of the dots will encourage the formation of a box and focus attention on the confines of that box in keeping with the Gestalt principle of closure. Weak activation will do so to a lesser extent, but will not necessarily encourage thinking outside these boundaries. Some noise, however,

---

[6] It would do so rapidly. This fast build-up of synchrony between ordinarily mutually exclusive concepts may be the reason for the flash of mini-euphoria that often accompanies insight (compare with the model of humor in the next chapter).

will trigger external elements, and this could lead to a solution. This is admittedly vague, but remember that for the current purposes we are attempting to encourage an insightful solution, *assuming that such a solution is already within the capabilities of the mind in question*. We are not trying, at least not yet, to make a problem-solving genius (but see the final section). If the capacity to imagine lines outside the box exists, a small amount of noise will encourage this, either by activating the appropriate perceptual cells directly, or by encouraging the abstract geometrical operations that allow this to happen.

In summary, our eureka device, then, needs to be constructed to do the following:

a) First, lower cortical activation, and then
b) Add a judicious amount of noise to encourage alternative solutions.

With respect to the first task, there are a number of means of achieving this goal, from biofeedback to inhibitory TMS to DBS applied to the appropriate cell populations. It may also be useful to experiment with different parameters applied to both the left and right hemispheres, in accord with the evidence presented previously suggestive of hemispheric differences. In either case, this is not an especially difficult effect to achieve, as evidenced by the fact that we already have a number of other techniques to encourage creativity from listening to music to a moderate amount of alcohol to simply popping into a hot bath. The key is not to sacrifice mental agility while defocusing attention; we do not want to create drowsiness, but rather a diffused intensity of thought.

With respect to our second requirement, the addition of noise, it is noteworthy that a number of researchers have postulated that ambient noise in the brain may be useful in both creative and ordinary problem solving (Ma et al., 2006). However, this preexisting noise may not be sufficient or be in the right location for our purposes. Again, we can conceive of a variety of means of introducing it. For example, the simplest thing we could try would be to present random images or sounds to the subject. The limitation with this method is that there will not necessarily be a direct correspondence in neural activity, especially in brain areas removed from primary sensory processing, to this input. To provide a more direct effect, we can again turn to TMS, but this time in excitatory fashion. A larger inhibitory signal may be used to "calm" a given region, and then an excitatory high-frequency signal of relatively low intensity could then be used to stimulate the region in question. Candidates for this region include the parietal-temporal junction, on the basis of the previous experimental results, but we should also not rule out the regions in the prefrontal cortex because of their central role in problem solving of all sorts.

Much experimental work needs to be done to see if this method is correct, and if it is, what parameters need to be used to optimize the effect. However, as always, we must return to faith in the regularity of the universe. Even such seemingly nebulous concepts as insight are governed by laws, and there is a reason why it happens to some people more than others, and why it happens more in certain situations for a given thinker than in others. There is a reason, for example, why insight comes in daydreams or the shower and not in concentrated attention or board meetings, and it is likely to be along the lines we described, or some suitable variant thereof. Thus, in the future we may not need to wait until our Muse deigns to favor us, and instead turn to technology to produce inspiration at will.

THE MANUFACTURED GENIUS?

No amount of insight will turn an undergraduate physics major into an Einstein, or a Salieri into a Mozart. To be a genius is to be more than creative; one needs the intellectual firepower to prepare the way for the eureka experience. These two aspects, high intelligence and high creativity, both in astoundingly short supply, make naturally occurring genius an exceedingly rare trait however loosely one defines it. We know, for example, of no cases of one genius begetting another genius; regression to the genetic mean prevents this from happening (Eysenck, 1995). Can we be so bold as to suggest that this apparent law of the psychological universe can be overturned with the appropriate use of technology?

It does not take a formidable intelligence to guess that we will answer this in the affirmative, although cautiously. These theoretical waters are uncharted, and we do not know nor can we predict what the application of a seemingly reasonable stimulation regime will be. It may be that a fundamental constancy law applies, such that increasing intelligence in one domain means that another domain must be borrowed from—this appears to be the case with so-called idiot savants. Still, with the addition of extra computing machinery to the brain we should, in principle, be able to achieve a superior form of sagacity, if not genius eventually, and the question remains how to do this.

One recurring theme in the literature on creativity is that novel accomplishments often proceed in two phases. Pasteur famously said that "chance favors the prepared mind," and it appears that an incubation period is often necessary before the Muse strikes. The first phase is characterized by furious rumination on the matter at hand. Avenues are explored, but they lead to dead ends, and other paths are pursued until an impasse is reached. Then, seemingly when no solution is possible, one sees an unusual cloud as one walks to give an unrelated seminar, or the wind blows a leaf in a funny way, and the answer flashes into one's mind, as if it was there all along. One curious aspect of this process is that one cannot simply stare at the sky or at the trees to begin with and hope to be inspired. Apparently, preparatory work is absolutely necessary, it is during this period that the problem takes form, representations are explored, and the obvious solutions are rejected. Without this phase, if one does become inspired, it will be by facile solutions that will turn out to be wrong after validation.

It is tempting to draw parallels between the two phases of genius, intellectual prowess and creative force, and it is also tempting to speculate that the same device may be used in different modes for both. The first phase is characterized by both psychic intensity and conceptual complexity. Recall that we concluded that a signifying aspect of the intelligent mind is the avoidance of easy attractors. Every situation for this mind is *sui generis*, and furthermore it is able to explore multiple consequences of this situation. In contrast, creativity is the product of the relaxed, defocused mind. In this stage, constraint-defying connections are made between seemingly incongruous items.

We now have a tentative design for a "genius cap," or a device that combines both aspects of the fertile mind. In the first, and longer lasting stage, it acts as a "decohencer," or device that leverages the full power of the multiple processing modules in the brain. This will eventually be augmented by artificially intelligent devices, wired directly into the consciousness of the user that will compete on equal terms with these native modules. After a suitable time, or perhaps when the device senses that the appropriate level of frustration has been reached, the device will throw a gentle wave of calm over the thinker's mind. This will allow greater synchrony between modules that are ordinarily in competition, and help to make connections between ideas that have heretofore been missed. Thus, the device will now act as a "coherencer." It may also turn out that a more rapid shuttling back and forth between these two modes may be useful in certain situations. In either case, this push-pull action acting together may achieve results not possible if either mode acts on its own.

Whether full-fledged Homo sapiens 2.0 will be the natural outcome of such a machine is difficult to say. However it is achieved, though, an increase in mental powers cannot come soon enough judging from the state of the world. It appears that evolution has provided us with just the right amount of intelligence to truly get us into trouble without the wisdom to be able to solve these difficulties. We can do better, and we will do better.

REFERENCES

Baars, B. (1988). *A Cognitive Theory of Consciousness*. Cambridge: Cambridge University Press.

Bowden, E., Jung-Beeman, M., Fleck, J., & Kounios, J. (2005). New approaches to demystifying insight. *Trends in Cognitive Sciences*, 9, 322–328.

Deary, I., & Caryl, P. (1997). Neuroscience and human intelligence differences. *Trends in Neurosciences, 20*, 365–371.

Eriksson, P. S., Perfilieva, E., Bjork-Eriksson, T., Alborn, A., Nordborg, C., A., P., et al. (1998). Neurogenesis in the adult human hippocampus. *Nature Medicine*, 1313–1317.

Eysenck, H. (1995). *Genius: The Natural History of Creativity.* Cambridge: Cambridge University Press.

Hull, C. (1943). *Principles of Behavior.* New York: Appleton-Century-Crofts.

Jung-Beeman, M., Bowden, E., Haberman, J., Frymiare, J., Arambel-Liu, S., Greenblatt, R., et al. (2004). Neural activity when people solve verbal problems with insight. *PLoS Biology*, 500–510.

Katz, A. Creativity of the cerebral hemispheres. In M. Runco, *Creativity Research Handbook.* Cresskill, N.J.: Hampton Press.

Kounios, J., Frymiare, J., Bowden, E., Fleck, J., Subramaniam, K., Parrish, T., et al. (2006). The prepared mind: Neural activity prior to problem presentation predicts subsequent solution by sudden insight. *Psychological Science*, 882–890.

Licklider, J. (1960). Man-computer symbiosis. *IRE Transactions on Human Factors in Electronics*, 4–11.

Ma, W., Beck, J., Lathan, P., & Pouget, A. (2006). Bayesian inference with probabilistic population codes. *Nature Neuroscience*, 1432–1438.

Martindale, C. (1977). Creativity, consciousness, and cortical arousal. *Journal of Altered States of Consciousness, 3*, 69–87.

Martindale, C., & Greenough, J. (1973). The differential effect of increased arousal on creative and intellectual performance. *Journal of Genetic Psychology, 123*, 329–335.

Martindale, C., & Hasenfus, N. (1978). EEG differences as a function of creativity, stage of the creative process, and effort to be original. *Biological Psychology, 6*, 157–167.

Neubauer, A. C., Grabner, R., Freudenthaler, H., Beckmann, J., & Guthke, J. (2004). Intelligence and individual differences in becoming neurally efficient. *Acta Psychologica, 116*, 55–74.

Reeves, L., & Schmorrow, D. (2007). Augmented Cognition Foundations and Future Directions - Enabling "Anyone, Anytime, Anywhere Applications. In C. Stepanidis, *Universal Access in HCI, Part 1* (pp. 262–272). Berlin: Springer-Verlag.

Reichle, E., Carpenter, P., & Just, M. (2000). The neural bases of strategy and skill in sentence-picture verification. *Cognitive Psychology, 40*, 261–295.

Runco, M. (2004). Creativity. *Annual Review of Psychology*, 657–687.

Russell, S., & Norvig, P. (2003). *Artificial Intelligence: A Modern Approach (2nd ed.).* Upper Saddle River, NJ: Prentice Hall.

Simonotto, E., Riani, M., Seife, C., Roberts, M., Twitty, J., & Moss, F. (1997). Visual perception of stochastic resonance. *Physical Review Letters, 78,* 1186–1189.

Snyder, A. W., Mulcahy, E., Taylor, J., Mitchell, D. J., Sachdev, P., and Gandevia, S. (2003). Savant-like skills exposed in normal people by supressing the left fronto-temporal lobe. *Journal of Integrative Neuroscience*, 149–158.

Snyder, A., Bahramali, H., Hawker, T., & Mitchell, D. (2006). Savant-like numerosity skills revealed in normal people by magnetic pulses. *Perception,* 837–845.

Thatcher, R., North, D., and Biver, C. (2005). EEG and intelligence: Relations betwen EEG coherence, EEG phase delay and power. *Clinical Neurophysiology*, *116*, 2129–2141.

Wallhovd, K., Fjell, A., Reinvang, I., Lundervold, A., Fischl, B., Salat, D., et al (2004). Cortical volume and speed-of-processing are complementary in prediction of performance intelligence. *Neuropsychlogia*, *43*, 704–713.

Chapter 10 HAPPY ON DEMAND

Although the pursuit of happiness is enshrined in the Declaration of Independence, in practice it takes a near perfect storm of events, internal and external, to make us happy for any length of time. Compare the plight of man with his faithful companion, the dog. To make Fido happy, just a few things are absolutely necessary. He must have a full stomach, he must have companionship in the form of fellow members of his species or human beings, he must have a park to roam in and do "his duty," and that's about it. In contrast, the purely physical needs alone of human beings are vastly greater. Our temperature regulation compared to most species is poor, and we need a variety of artificial coats to compensate for the lack of a natural one. Not only do we need three meals a day, we need meals that vary within the day and from day to day. Except in the worst weather, Fido could probably make do with a makeshift shelter and a soft pillow to lay his head on. We need not only houses but beds, pillows, high-thread-count sheets, televisions, iPods, computers, and cell phones, not to mention special rooms with special apparatuses to do our duties.

The sheer depth of our needs introduces a secondary problem, the need for one additional but hard to procure item, money, the lack of which will invariably lead to unhappiness (although the presence of oodles of cash rarely guarantees happiness and may actually prevent it; see the following discussion on the determinants of happiness). To make money we must work, work leads to reduced leisure and increased stress, and these, in many cases, all but prohibit happiness. To make matters worse, ones psychic bank account must be equal or greater than one's actual bank account. Work, for example, must jibe with our life goals; it must be intrinsically rewarding in addition to any income that it generates. In addition, while at work we must be comfortable and respected. That is, during the considerable period of time that we spend earning the money to pay for the things that may or may not lead to happiness, we must have also have both components of our needs met, physical and psychic. But the latter class of needs is usually not met through work alone. As the song says, you're nobody if nobody loves you. Current divorce rates and less formal observations, such as the behavior of couples in restaurants, suggests that love, at least in its most fulfilling incarnation, is both infrequent and fleeting.

And in the rare instance where it all comes together, the whole house of emotional cards can come tumbling down in an instant by something as trivial in the grand scheme of things as a persistent itch, a callous snub by a colleague, or the stub of a toe. The truth is that evolution cares not a whit about our happiness, except to the extent that: a) it uses ephemeral pleasures, such as orgasms, as motivators, and b) a terribly unhappy person will be poorly motivated and therefore less likely to survive. The brain is optimized not for happiness, but for survival, survival entails vigilance, and vigilance is at the opposite end of the emotional spectrum from contentment. Buss (2000), for example, discusses at length the importance of jealousy as an adaptive mechanism. Those without a strong proprietary feeling toward their mates were likely to lose them, and therefore less likely to pass on their genes. We are the end result of this evolutionary process, and almost all of us are subject to this motivating emotion that combines anger and possessiveness resulting in a distinctly unpleasant emotional cocktail.

Evolution also works its perverse ways by means of emotional homeostasis. Any upsurge in positive affect is soon drawn downward; likewise, transient disappointments fade by the kind action of forgetfulness. But evolution is most punishing toward any state that begins to hint at ecstatic fulfillment. This may not be obvious on the face of it, but this principle is in fact deeply encoded in human mythology, from Icarus's fall from grace, to modern slasher movies where the first victims are those having playful sex. Evolution simply cannot allow ecstasy of any sort to persist because it is at direct odds with adaptation. Extreme forms of content imply both a looking inward and a lack of striving, and this is not the kind of strategy that keeps one alive on the plains of Africa, nor in the contemporary boardroom.

Still, as the following discussion of happiness research indicates, once basic needs are met, a slim majority of people is not too terribly displeased with their lot. And somehow, a lucky few manage not merely to avoid evolution's strict calculus of emotions, but to revel in life and its vagaries. Just how they do so will be of great interest to the goals of this chapter, which is to show that technology will, in the not too distant future, not only enable more reliable and consistent moments of pleasure, but more importantly, lead to a consistent and fulfilling long-term contentment not only for the few but the many. As with many of the topics in this work, neural technology will enable an end around the normal dichotomies of quotidian existence. Happiness is not and should not be the opposite of adaptation; to the extent that it is so, it is only so for brains in their native state. How to carry them beyond these limitations awaits the survey of first the neural determinants of pleasure, and then the constituent factors of happiness.

PLEASURE AND THE BRAIN

Ideally, this section should be titled happiness and the brain, but so little is known about the neurophysiological determinants of short-term, let alone long-term determinants of this mental state that here we will concentrate on happiness's weaker and more ephemeral affective cousin, pleasure.

Self-stimulation and reward

In 1954, Olds and Milner made an astounding discovery, and one that has implications to this day for the affective structure of the brain, as well as for the neuroengineering of pleasure and happiness. They set out to confirm earlier work in which electrodes implanted into the midline of the brain of the rat produced aversive behavior; that is, the rat did not like being stimulated in this way. However, the techniques of the day were not as precise as they are currently, and many electrodes missed their mark. With one rat in particular, it was found that not only was the invasive stimulation not aversive, but that animal seemed to positively like being stimulated in this manner. After sacrificing the animal, it was found that the electrode had landed in the structure known as the nucleus accumbens, one of the important structures in the so-called reward circuit of the brain, as described further in the following text. Olds and Milner went on to confirm this result in many other rats, and to verify that this artificial stimulus generally will outweigh the reward value of any naturally occurring stimulus. In particular, whenever the animal was given a choice between the stimulation and food, it would invariably choose the former, regardless of its hunger. Most strikingly, if allowed to deliver the stimulation itself, by pressing a lever to deliver the electrical current to its brain, it would do so indefinitely until it fell into an exhausted sleep, ignoring all other forms of stimulation along the way. Once it awoke, it would go immediately for the bar, and continue where it had left off.

The rat brain is strikingly similar to the human brain in many ways, and this fact, along with the ease of care, combine to make it the animal of choice for neural experimentation. Nevertheless, there is a serious question as to whether such a result, intimately tied into the nature of pleasure and consciousness, would generalize to the more complex human psyche. In a series of bold, and many would say reckless experiments, the Tulane psychologist Robert Heath set out to examine this beginning in 1950. In a move that would not come close to meeting the stringent Internal Review Board criteria of today, Heath gathered a number of patients from mental wards in New Orleans and implanted electrodes in a number of brain areas implicated in reward, most notably the septum and ventral tegmental area (VTA) in an attempt to cure their illnesses.

In at least some of these cases, it appears that a similar effect was found as that of Olds and Milner on rats. Heath filmed many of his subjects, preserving many of his experiments for posterity. In one case a woman with blank affect is lying in her hospital gown on a cot in a soundproof room. We hear the words "sixty pulses" coming from the technician, and suddenly a smile appears on her face. Heath asks here, "Why are you smiling?"

"I don't know ... Are you doing something to me. I don't usually sit around and laugh at nothing. I must be laughing at something."

Then we hear the technician say, "One hundred forty," increasing the stimulation.

The patient giggles and proclaims, "What in the hell are you doing? You must be hitting some goody place."

Heath later claimed that his treatment had cured her. Heath also experimented with the possibility of self-stimulation. He gave a gay man, patient B-19, access to a button to deliver the electric pulse to his reward circuitry. According to Heath (1963), "During these sessions, B-19 stimulated himself to the point that he was experiencing an almost overwhelming euphoria and elation, and had to be disconnected, despite his vigorous protests."[1] Apparently, B-19 was experiencing something very similar to Olds and Milner's rats, although Heath claimed that humans are not nearly as compulsive as their animal counterparts. This is consistent with the more complex mediation of reward in humans by neocortical control. It is also consistent with heroin addiction, in which there is a marked shift from external to internal reward, but usually only a partial shunning of both gustatory and sexual pleasures.

We can place both the Olds/Milner and Heath results in perspective by examining what is now known as the reward circuitry of the brain as illustrated in Figure 10.1. The medial forebrain bundle (MFB) is a collection of fibers connecting to various midbrain, limbic and cortical structures. The two primary structures in this pathway are the VTA and the nucleus accumbens, with the septum and the prefrontal cortex also playing significant roles. The amygdala is also included in this circuitry, but plays a more prominent role in the antagonistic punishment system (see Chapter 2). Stimulation of the nucleus accumbens either directly

[1] In a test that manages to combine almost everything one should not do when working with human subjects into a single experiment, Heath also attempted to "cure" B-19 of his homosexuality by stimulating his reward center while he was engaged in sexual activity with a prostitute hired for this purpose. The idea was that the patient would then associate heterosexual activity with sexual reward. The experiment enabled B-19 to successfully achieve orgasm in this manner, but had no lasting effect on his sexual preference.

with invasive electrodes or by sensory stimuli results in a pleasurable feeling. The VTA is believed to be responsible for the processing of reward signals generated by the cortex, and communicates under normal (i.e., naturally rewarding) circumstances with the nucleus accumbens via dopaminergic fibers (i.e., those using the neurotransmitter dopamine). By stimulating these structures directly, we are in effect fooling the brain into thinking that it has performed some action deserving of reward.

This is also the mechanism of action of a third source of pleasurable reward, that of drugs. Chemical stimulation is a near-universal trait of mankind, and it is incumbent upon any neural model to account for not only the positive affective component of drug taking, but also the addictive property of many such agents. The latter will be discussed later in the context of avoiding homeostasis; here we simply note that most common psychoactive drugs have a direct or indirect effect on this system. For example, both cannabis and cocaine act directly on

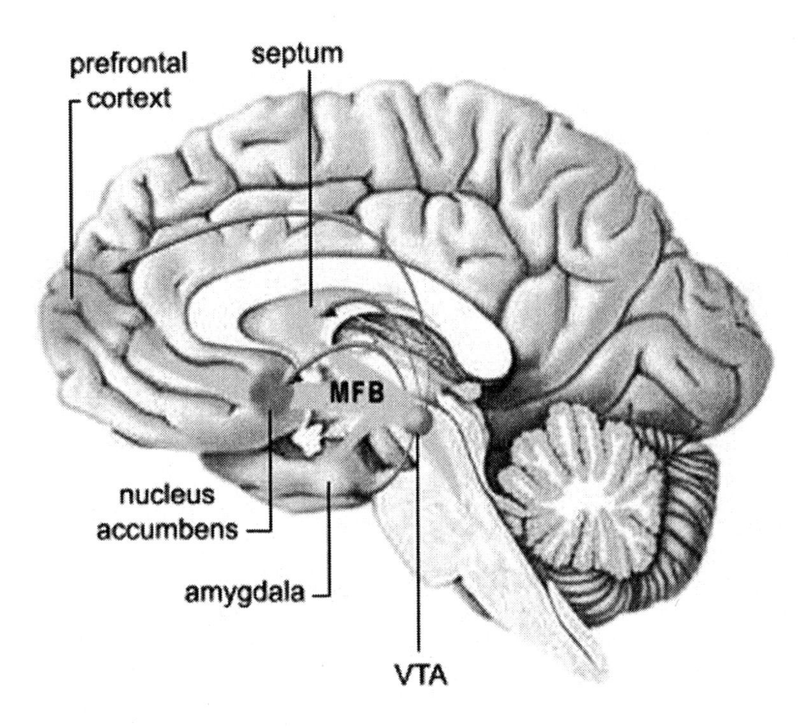

FIGURE 10.1 **The brain's reward circuitry. The medial forebrain bundle connects a number of structures, most notably the VTA and the nucleus accumbens. Reward is associated with activation of the nucleus accumbens either directly or indirectly.**

SOURCE: **Brain top to bottom.**

Bruno Dubuc. Used with permission. http://thebrain.mcgill.ca/

the nucleus accumbens. Opiates such heroin and morphine have a slightly more complex course of action. These act to inhibit the release of GABA, an inhibitory neurotransmitter acting in this structure. Endogenous opiates, such as the endorphin's associated with "runner's high," have a similar course of action, and have a similar etiology with respect to addiction. Other drugs, such as nicotine and alcohol, have multiple influences on the reward system, but all serve to increase the dopaminergic response.

Recently, Berridge (2003) has cast doubt on the generality of this model. He first claims that we cannot infer pleasure from desire. The fact that the rat continues to press the bar until exhaustion, and the fact that Heath's self-stimulating subjects were clearly being reinforced for their actions, does not necessarily mean that they liked this stimulation, in the purely phenomenological sense. For example, Berridge claimed that a careful analysis of the responses of Heath's subjects show that they were in a state of desire but were not necessarily achieving fulfillment of that desire. In conjunction with this claim, he notes that although rats in which the dopaminergic system was destroyed show no inclination to ingest sweet liquids, something they are ordinarily keen to do, if force fed sweets they manifest all the behavioral signs of pleasure, as judged by their facial expressions and other indicators. In other words, the reward circuitry may subserve motivation, but not necessarily pleasure.

Berridge's work points to a need to better understand the opioid system as well as the dopaminergic system in reward and pleasure. The former may very well have affective properties apart from its indirect effect on the latter, as in the case of morphine and the endogenous opiates. However, for current purposes, that of producing a device that increases pleasure and/or happiness, the current body of evidence and models thereof are sufficient to describe in broad outline how to achieve this. We know, for example, by direct experience that drugs such as alcohol and the opiates cause pleasure, not merely desire, although they often do this also. Is it such a stretch to conclude that the same states can be achieved by some means with nonchemical stimulation, invasive or otherwise? Furthermore, the data examined in the next two sections, on meditation and orgasm, will eventually allow us to make much more specific claims about what a pleasure or happiness machine would look like.

Imaging studies of meditation

There exists a minor cottage industry in hooking up holy men to machines to see what is going on in their heads. For example, Cahn and Polich's (2006) survey of the literature reveal no less than 65 EEG studies of meditators, with the majority taking place in the past 10 years. To a large extent, any such studies are going to be a highly indirect way of getting at the meditative experience, given: a) the inability of current imaging technology to reveal brain activity in sufficient spatiotemporal resolution, b) the imperfectly understood connection between neural activity and phenomenology, and c) the paradoxically low status of peak and mystical states in the Western intellectual framework, and the consequent poor understanding of these states. Furthermore, it is not at all clear that all meditative states are one and the same thing. For example, there is a fundamental distinction between traditions that encourage mindfulness, or the dispassionate observation of any thoughts that may arise without attending to any particular thought, and the concentrative traditions, such as yoga and transcendental meditation, that focus on the breath and a mantra respectively.

Nevertheless, there are a number of suggestive results that are summarized here.

i) EEG frequency and power

The single most consistent result in the literature is that meditation increases power in the alpha (8–12) Hz band (Cahn & Polich, 2006). It may be recalled from Chapter 6 that alpha is associated at the neural level with an increase in synchronization of firing between neurons and at the psychological level with an awake but relaxed state. The increase in alpha power appears to be stronger during meditation than in normal relaxation; however, the degree to which this is true may depend on the type of meditation. An advanced Kundalini meditator, for example showed a fourfold increase in alpha (Arambula et al., 2001). In addition to increasing power, meditation may also decrease the frequency of alpha. Transcendental meditation (TM) practitioners showed a 1 Hz lowering of alpha and there was a similar .8 Hz difference between novice and advanced Sahaja yoga meditators in the alpha band (Aftanas & Golocheikine, 2001).

These researcheres have also suggested that meditation increases theta activity (4–8 Hz). Theta is ordinarily interpreted as representing drowsiness, although it may reflect a different underlying mechanism in meditation given that the goal of this technique is often described as "relaxed alertness." In addition, there have been suggestions that meditative techniques increase gamma activity (30–80 Hz) (Cahn & Polich, 2006), indicative of long-range coherence over neural populations, and consistent with the coherence results discussed next. The difficulty of determining a single electroencephalogram (EEG) invariant of the meditative state is highlighted by one intriguing study showing that the center of gravity of gamma waves differed, *within a single meditator* (a Buddhist Lama) depending on what technique he was using (Lehmann, et al., 2001).

ii) EEG coherence
EEG coherence is a measure of communication between brain regions. Alternatively, we can speculate that increased coherence is an indirect measure of the size of the conscious kernel (Chapter 4), the set of reciprocally interacting neural populations. Coherence increases for meditation have been found as a function of both state and trait; i.e., meditators have found to have greater values of this measure than nonmeditators, and during meditation itself, coherence tends to increase (Aftanas & Golocheikine, 2001). The primary increases have been in the theta and alpha ranges, and can occur either within a hemisphere or across hemispheres. Fine-grained phase synchrony has also been studied. Herbert et al. (2005) found that TM practitioners showed greater phase synchrony between anterior and posterior brain regions during meditation relative to the rest condition and also relative to controls.

iii) Laterality
Despite some initial enthusiasm revolving around the notion that meditation may lead to greater right vs. left hemispheric activity, a number of recent studies have found little evidence for this effect (Cahn & Polich, 2006). If the phenomenological effects of meditation are primarily affective rather than cognitive, this is consistent with Murphy et al.'s (2003) meta-analysis of neuroimaging studies indicating that there is little evidence to suggest that the right hemisphere is dominant in either the generation or perception of emotion. However, more research is required before a definitive negative conclusion can be reached.

iv) Event-related potentials (ERPs)
Motivated by the idea that meditation reduces the effective background noise in their brains, a number of research teams have investigated the EEG response to sensory stimuli for these subjects. The quiescent mind should, in principle, be more responsive to external stimuli, as the time to overcome competing internal stimuli should be shorter. ERP studies offer partial support for this idea. Concentrative techniques produced shorter latencies and decreased amplitude for several stimulus modalities (Cahn & Polich 2006), and mindfulness techniques often produce lowered habituation (i.e., less attenuation of response over time). In a typical such study (Cranson et al., 1990), nonmeditators, novice meditators, and long-term meditators were studied for their passive response to an auditory tone with variable interstimulus intervals. The advanced meditators produced the shortest latency for the P300 signal, the novice meditators the next shortest, and the nonmeditators showed the longest latency ERP.

v) PET and Functional Magnetic Resonance Imaging (fMRI) studies
There is a growing body of evidence suggesting that meditation induces greater activity in the dorso-lateral prefrontal cortex (DLPFC) as well as in the anterior cingulated cortex (ACC) (Cahn & Polich, 2006). Consistent with this finding, activity in the ACC may also be implicated in feelings of love (Bartels & Zeki, 2000). The DLPFC is implicated in working memory, abstract thinking, and attentional control. As Dietrich (2003) notes, to some extent the finding of increased activity in this area contradicts the increase in frontal alpha with meditation, which is an indicator of reduced activity. However, it may well be that these two findings taken together are reflective of the unique nature of the meditative state. In ordinary consciousness, arousal and synchrony are at odds, with the former tending to reduce the latter. The goal of mediation, as previously stated, is to produce a paradoxical state of "relaxed alertness," and this may result in both increased activity and increased synchrony. We will return to this idea later.

Orgasm and euphoria

Orgasm is both the most powerful and the most beguiling of mental states. We might expect a reward for doing our genetic duty, but what a reward it is! There is little or no evidence, for example, that nonprimates experience anything more than the normal pleasure associated with satisfying a drive during coitus, such as drinking when thirsty. In humans, the reward value of orgasm surpasses that of almost any other experience, perhaps because our mental complexity means that we might desert this crucially adaptive activity if it did not overwhelm. Thus, one of the first things one might expect that neuroscience would investigate with imaging techniques would be the human climax. While we cannot rule out the possibility that an intrepid couple snuck into the lab late one night to "try out" the machine, a number of factors have curtailed the full-scale investigation of this phenomenon, including bias in funding toward pathology as opposed to normal states (especially pleasurable ones), natural reluctance to work in an area that is still taboo despite numerous attempts at societal liberation, and finally, and to a lesser extent, technical problems such as motion artifact associated with the climactic state.

The rapid propagation of techniques, such as fMRI, and related techniques combined with scientific curiosity has enable these factors to finally be eclipsed in recent years, and the few preliminary studies that have been run have generated useful results. A 2003 PET study (Holstege et al.) found that male ejaculation was correlated with increased activation in the VTA (cf. with Heath's self-stimulation results), as well as a number of other midbrain areas including the zona inserta, intralaminar thalamic nuclei, and most notably the cerebellum. Neocortical activity increases were found throughout the frontal, parietal, temporal, and occipital lobes, but exclusively on the right side. However, using a new and more accurate analysis technique that was better able to remove artifacts (Georgiadis, et al., 2007), the authors concluded that their earlier results were largely incorrect. Instead, the improved analysis revealed that the main neural correlates of ejaculation were cerebellar increases as before, but neocortical decreases in activation throughout, but especially in the prefrontal cortices.

Two research groups have studied the female orgasm, one in women with serious spinal cord injuries, and one in healthy subjects. In the former, fMRI revealed a number of areas of increased activity in the hypothalamus, the amygdala, the anterior cingulate, as well as the cerebellum and the frontal and parietal neocortical areas (Komisaruk, et al.). While remarkable in the sense that the subjects were able to experience orgasm at all (the authors speculate that the source of the sensation to the brain are the Vagus nerves, which bypass the injured spinal cord, as stated in Chapter 7), these results are of limited generality because of the small number of subjects used and because of the nature of the injury. A better study for current purposes is that of Georgiadis et al. (2006). As well as studying many more subjects, they provided two control conditions in addition to rest: clitoral stimulation without orgasm, and imitation of orgasm in the absence of sexual stimulation. The former can help distinguish between mere arousal and the release associated with orgasm, and the latter helps to eliminate purely motion-related neural changes. The main findings with respect to these control conditions are as follows: orgasm relative to imitation showed decreased temporal activation, and orgasm relative to stimulation showed marked decreased in prefrontal cortical activation. Taken together, these results suggest that temporal lobe deactivation is a correlate of sexual arousal, but that the main correlate of orgasm itself is prefrontal deactivation (we will return to this significant finding later; for now, we note that this was also found in the male ejaculation results). Furthermore, the cerebellar increases in activation were also obtained in imitation, suggesting that these are artifacts of motor control, rather than true concomitants of either arousal or climax.

Apart from sexual climax, the most common non-drug-induced euphoric experience is that provided by music. Blood and Zatorre (2001) had musicians choose passages that had previously given them "chills" and studied brain activity while they listened to their chosen music with PET. Among other things, they found increased activity in the insula, the cerebellum, and the right orbitofrontal cortex, and decreases in the amygdala and widespread cortical regions, including, most notably, the prefrontal cortex, in music that produced chills as opposed to music that did not.

The author, along with colleagues Allon Guez and Youngmoo Kim, decided in 2006 to investigate one aspect of these results in more detail. We used functional Near Infrared Spectroscopy (fNIR), to study the continuous neocortical response to music over time (recall from Chapter 6 that fNIR can only measure activity on the surface of the brain). In our first pilot study we simply exposed subjects to a single piece of music, "Won't get fooled again" by The Who, which contains a well-known climactic moment toward the end of the song in which an organ solo is followed by drums and Roger Daltry screaming "Yeaahhh!". We expected that if we were to find anything of significance, it would only be revealed after painstaking statistical analysis, perhaps by looking not only at the oxygenation signals directly but also how well they are synchronized between the measured brain areas. We were thus roundly disappointed to see the astonishingly strong and obvious results in Figure 10.2 on our very first test run. The graph shows inverted prefrontal oxygenation during the course of the song; the climax is marked by a rapid decrease of blood oxygenation in this region, almost as if the blood is being sucked out and diverted to other areas.

In order to confirm that this was not a fluke, and also to ensure that these results were not merely a confound of loudness, we studied four other songs in further detail. For each, we had subjects continuously rate how much they were enjoying the songs while we also recorded frontal and other cortical oxygenation levels. Once again we found moderate to strong correlations with subjects' ratings and decreases in neocortical activity, with the best results shown in the

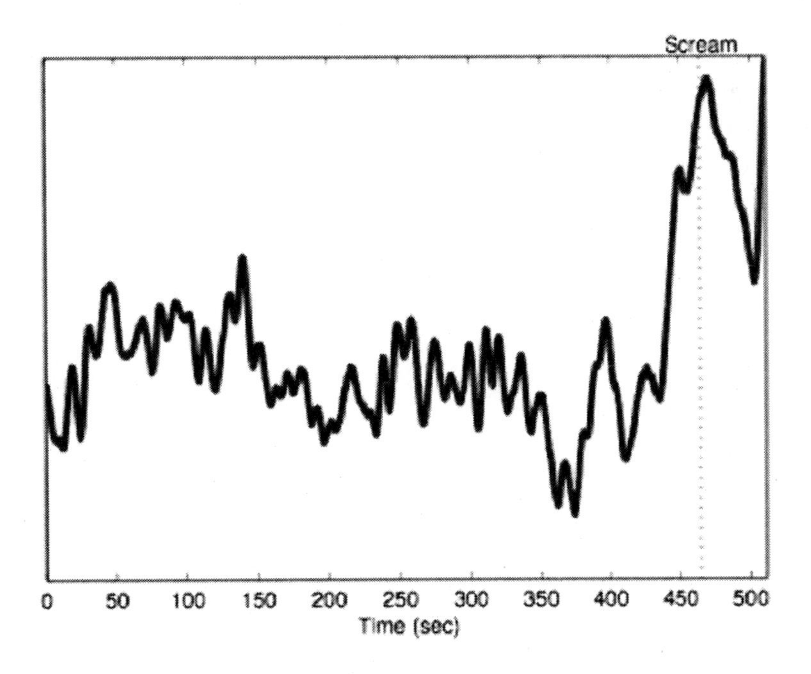

FIGURE 10.2 **Mean frontal oxygenation over eight subjects, inverted, recorded while listening to the song "Won't get fooled again."**

prefrontal areas. Figure 10.3 shows mean inverted oxygenation vs. the mean values of these ratings for the four songs in question. With the exception of "The Star Spangled Banner," there is a good correlation between these values. Note in particular that "Nessun Dorma" contains two climactic moments, one near the middle of the song, and one toward the end. This is mirrored in both the fNIR data and subjective ratings, and both curves are in almost lockstep with each other. It is rare to find such clean results with neuroimaging data, but before commenting further on this, let us turn to contemporary thinking on the nature and determinants of happiness.

THEORIES OF HAPPINESS

Comprehensive theories of happiness are almost as elusive as happiness itself; it is difficult enough to describe what happiness is let alone produce an account of

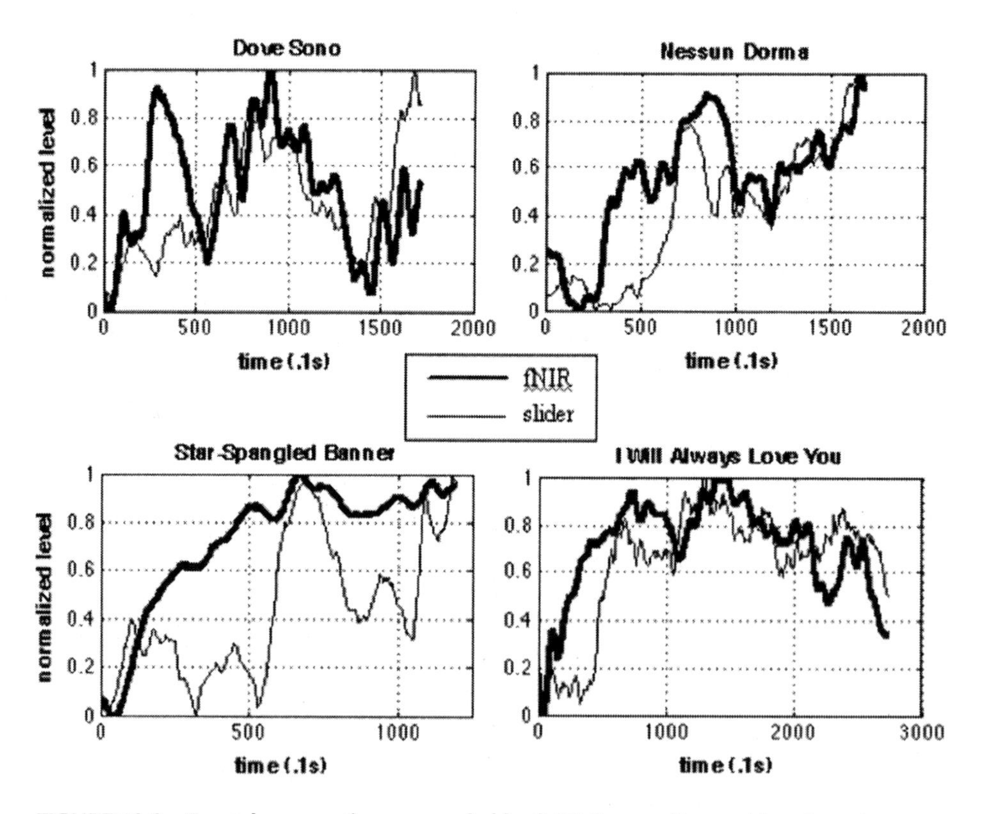

FIGURE 10.3 **Frontal oxygenation as revealed by fNIR (inverted) vs. subjects' continuous rating of song goodness for four songs.**

its causal determinants. This section will review some attempts to provide this phenomenon with an explanation; the final section of this chapter will synthesize many of these ideas and theories into a framework suitable for neuroengineering.

Oddly enough, one of the most commonly held theories of happiness, setpoint theory, turns out to be almost trivially wrong. According to this account, one's personality is more or less completely determined by one's genetic constitution and upbringing, and one's personality provides a fixed level of happiness that is largely independent of circumstance. This does not mean that one will not be affected by life events, but rather there will be a relatively swift return to a fixed level of happiness. In this view, if one is a jolly fellow, one will return to being jolly after being rejected by the girl next door, if one is neurotic and unhappy, one will be unhappy and neurotic after snagging the girl of one's dreams.

This theory is belied by both the experimental evidence and by common sense. While some adaptation to circumstance undoubtedly does occur—one will be more ecstatic immediately after winning a tournament than a week hence, for example—it is by no means complete, and the cumulative effect of events can turn one sour or joyful. Easterlin (2003) provides a comprehensive review of the flaws of setpoint theory; here we will summarize his findings. First, there is a clear correlation between health and happiness. If setpoint theory were true, one would eventually adapt to any disease, no matter how serious. Next, there is small but significant correlation between being married and being happy. Marriage has its travails, as all who have taken the plunge know, but it is generally preferable to facing life alone. Finally, a minimum level of wealth is required for both material comfort and contentment, and while an eventual downward adjustment in mental state to a monetary windfall undoubtedly occurs, an equally large upward adjustment to a fall in income is less likely.

Another equally flawed theory is based on the notion that we base our happiness by comparing what we have with others in our peer group. Again, there is undoubtedly a grain of truth in this notion. It is more than a little discouraging to see our high school foe pull up to the reunion in a late-model open-top Bentley. Conversely, Schadenfreude, the delight in others' misfortunes can, in the right circumstances, give off a satisfying if short-lived buzz of pleasure, as much as we may wish to deny that this is the case. However, this theory, in its pure form, is also at odds with the evidence. A rising standard of living in a given country is correlated with increased happiness, at least up to a critical point. If comparison theory were correct, then there should be no change in mean mental outlook, because on average a given member of this society would remain in the same place relative to his or her peers (Veenhoven, 1995). In fact, happiness is correlated with quality of life in differing countries, where this variable is measured by mean income as well as a number of other factors, such as political freedom.

This latter fact also presents difficulties for the so-called folklore theory, which presumes that national character largely determines individual happiness.

TABLE 10.1. Mean and standard deviation in life-satisfaction for ten countries in the early 1990s.

Nation	Mean	Standard deviation
Brazil	7.37	2.41
Britain	7.49	1.94
Canada	7.89	1.74
France	6.78	1.98
India	6.70	2.28
Japan	6.53	1.75
Nigeria	6.59	2.62
Russia	5.38	2.40
Sweden	7.97	1.74
USA	7.73	1.83

In fact, there is a substantial body of evidence that mean happiness does vary across nations. Table 10.1 shows the mean and standard deviations for life-satisfaction, a measure strongly correlated with happiness, across ten nations in the early 1990s (Veenhoven & Ouweneel, 1990). The problem with this sort of evidence is that to a large (but not complete) extent, it is not cultural outlook that determines mood, but quality-of-life factors within that nation. For example, there is something to be said for the notion that Russian culture breeds moodiness, but their low score in this survey is as likely to be as much the result of the turmoil that befell them during the fall of communism than any intrinsic factor.

Finally, we turn to an earlier and more comprehensive attempt to understand this often baffling phenomenon. Among traditional psychologists who have treated the fundamental problem of human happiness, as opposed to those such as Freud and Jung whose main concern was psychopathology and its treatment, one voice rings louder and more clearly than others. This is that of Abraham Maslow, one of the founders of transpersonal psychology, and the proponent of a theory that still has resonance to this day. In the 1950s Maslow proposed a so-called "hierarchy of needs," in which each level of the hierarchy must be satisfied before one can move to the next.

At the bottom of this pyramid are the four basic deficiency needs, or "D-needs," which must be met to stave off either misery or unhappiness. These include the purely physiological needs such as food, water, sex, and sleep. As Maslow stressed, these themselves are arranged in a hierarchy; if one is hungry enough, food is dominant over sexual needs in motivation, and both given short shrift in one has difficulty breathing. Next in the hierarchy is the need for safety, including the security of one's body and possessions. Maslow stressed the deep need for children to feel not only safe and protected, but to do so through structure in their environment. Once one feels safe and has one's immediate physiological needs met, one turns to interpersonal needs, such as love and friendship. It may not be readily apparent in our fractured society, but man is a deeply social organism—the worst punishment that can be inflicted on a prisoner is not the withholding of material goods, but the removal of human interaction via solitary confinement. The fourth level of the pyramid consists of esteem and related matters. Maslow distinguishes between two sublayers at this level, a lower one involving the adulation and respect of others, and a higher and more important one based on confidence built on genuine achievement.

At the top of the hierarchy is the need for self-actualization. In Maslow's words, "What a man can be, he must be." Self-actualization is a being need, or "B-need," and consists in coming to know one's true self, or the purpose for which one was put on earth, and then acting in accord with this realization. Maslow made a careful study of the lives of Einstein, Eleanor Roosevelt, and others who he considered self-actualized, and noted the prevalence in these people of freshness in perception, honesty, and candor in thought and deed, spontaneity in action, and a fundamental acceptance of the self and its limited place in the universe, among other things. Maslow's observations also caused him to refine his theory to include aesthetic and cognitive needs in addition to connative, or motivational ones in the set of B-needs. Finally, pace Hitchens (2007), Dennett (2007), and a recent spate of authors who in differing ways make the claim that the intellectually advanced individual is one free of religious sentiment, Maslow observed that highly successful, actualized beings were invariably deeply spiritual ones, with frequent peak or ecstatic experiences, and a transcendent or non-self-centered world view. In later life, Maslow flirted with some of the outgrowths of sixties counterculture, such as EST, that also were attempting to craft a new and more direct spirituality, but to his credit, found that these groups often devolved into as rigid an ideology as the straight-laced counterparts from which they sprung (Hoffman, 1988).

Maslow's schema is more than interesting, and indeed we will return to it when formulating a more general conception of happiness later. However, to the extent that true happiness is only available at the top level of the hierarchy, this conception passes the baton from one difficult-to-confine concept, happiness, to another concept that is equally problematic, self-actualization. How, for example, does one know what one's calling is? To a large extent, we infer this from what makes us happy, but then this leads to a circularity: we tend to be happy doing what makes us happy.

Given this limitation, and the limitations of the earlier theories, it is perhaps safest at this point to simply conclude that happiness is multidetermined. It is influenced by both national and individual character, and by an assortment of circumstances, both absolute and relative. Some sort of function, as yet unknown, and not necessarily of a simple character, combines to produce a level of satisfaction for each person in a given society. The next section will examine the most studied of such factors with a view toward forming a fuller understanding of happiness.

DETERMINANTS OF HAPPINESS

The past twenty-five years has seen a burgeoning of interest in the purely empirical study of happiness by two interacting groups of researchers. Economists are increasingly realizing that money and wealth are inadequate as measures of quality of life (QOL). To take a trivial example, if a man works eighty hours a week and earns twice as much as if he worked forty hours, he will not increase his QOL by a factor of two, and he may even decrease it. The same applies to a society as a whole; if we are truly interested in QOL, not only must we consider the quantity of goods produced by a society, but also the kind of goods, and the work required to make them. In general, if QOL is the main desideratum of societal engineering, it makes sense to speak not only of a per capita income, as a means of measuring productivity, but also a mean contentment index, gauging how well a society more directly meets the needs of its constituent members. The latter is much more difficult to measure than the former, and involves a number of extraeconomic factors, such as those discussed in this section, but if psychic wealth is ultimately what matters, there is little choice in the matter.

Among the factors influencing QOL, and by extension happiness, are the following. This should not be taken as a comprehensive list, but merely as those variables that have been studied the most, and/or variables that are known to have a substantial influence on happiness.

i) Income and wealth
There is an old Yiddish saying that "The heaviest burden is an empty pocket"[2]; the scientific results bear this out. Within poorer countries, income is a strong determinant of happiness, and within the developed nations, it is a weak or nonexistent factor. For example, in still-developing regions, material affluence can account for up to ½ of the variance in self-rated contentment scores (Veenhoven & Ouweneel, 1990). The simplest explanation for this data, and one that is also consistent with the data within individual countries, is that although money cannot buy

[2] *Der shversteh ol iz a laidikey keheneh.*

happiness, the *lack of* money invariably purchases misery. In poorer countries, the large number of people significantly below the poverty line have very low life satisfaction; in wealthier ones, happiness is spread more or less equally among the population, even when wealth is not distributed in this manner.

ii) Marriage
Contrary to popular conception, and the frequent laments of those dwelling in this noble institution, marriage does markedly contribute to individual happiness (Easterlin, 2003), as previously mentioned. Also, somewhat counterintuitively, the increase in contentment due to marriage is approximately four times that of cohabitation. The marriage effect appears to be a genuine one and is not simply the result of happier people being more likely to get married; this is supported by the fact that as people marry off in their twenties, the single population left behind does not show a decrease in happiness, as it would if marriage cleaved the population into two groups, one of happy and one of sad people. Rather it is the newly married that appear to get happier, and stay that way as long as the marriage lasts.

iii) Health
Health is another important factor with a significant contribution to happiness (Easterlin, 2003). Contrary to the predictions of setpoint theory, there is less than full adaptation to serious illness and disability, although in this instance, adaptation cannot be completely discounted. For example, there is a general decline of health with age, but not a general decline in happiness.

iv) Societal/political factors
Freedom at the societal level and at the individual level moderately to strongly correlates with happiness (Veenhoven & Ouweneel, 1990). The single most significant freedom turns out to be freedom at work; those with repetitive jobs or in low-ranking positions show greater stress and less contentment, as common sense would dictate. Physical safety and security is another important factor. Societies with a high murder rate or accident rate are significantly less happy.

v) Religiosity
Belief in God, pure and simple, appears to be a moderate factor in determining happiness. It is also roughly twice as strong a factor as identification with a particular organized religion (Veenhoven & Ouweneel, 1990). More research is needed to confirm the Maslovian hypothesis

stating that faith is a necessary concomitant of the fully actualized being, as opposed to being merely an elaborate way of convincing oneself that one is part of something larger that will persist after one's death.

vi) Meaning
The good life has often been equated with one that has a purpose that goes beyond one's immediate ego boundaries. Empirical research confirms that this is the case; those members of the population with a unifying purpose are significantly more content that those without (Easterlin, 2003). These results suggest, paradoxically but plausibly, that movement from one pleasurable experience to another can end up being distinctly unpleasant, and may even lead to depression. The circularity problem discussed earlier in the context of the Maslovian hierarchy still applies, in the sense that one often infers purpose from that which makes one happy, but it is still valuable to have empirical confirmation of the idea that pure hedonism does not suffice.

vii) Innate factors
Perhaps happiness is not so much a function of circumstance as of innate personality. There is strong evidence to suggest that this is the case. For example, half of the variance in subjective well being in separated twins is due to genetic factors (Lykken & Tellegen, 1996). One's mood, of course, varies during the course of the day and from day to day, but the mean mood for a given person is relatively stable over time (Diener, 2000). It also appears that temperament emerges early and is relatively constant throughout life (Goldsmith et al., 1987), as informal observation also suggests.

This final factor suggests that outlook has as much to do with happiness as does circumstance. The importance of this fact to the following discussion is as follows. Outlook, whether determined by innate factors or more local influences, is nothing more than a brain state. This in turn implies that if one could, artificially or otherwise, encourage the right sort of perception and thought, then one would naturally increase the possibility for happiness. One will always, of course, be battling to some degree with extraneural factors—it makes more sense, for example, to provide a starving man with food than to place him in a device that will allow him to ignore his hunger pangs. Likewise, dating will be a better long-term option for the unmarried than neural manipulation of the coupling drive. But, as with wealth, once a given threshold has been passed, it is one's attitude to the world that will determine one's level of happiness. Achieving the proper attitude, with or without technological help, is the subject of the next section.

CLEAR PERCEPTION, CLEAR ACTION: AN INTEGRATED VIEW OF HAPPINESS AND PLEASURE

In order to motivate the theory of happiness to be proposed, let us return to where we began, with Fido bounding through the park on his way to snatch a Frisbee. This feat, from a purely engineering standpoint, is quite remarkable, and involves an intense set of calculations that is, to some extent, still beyond current robotics (AIBO, Sony's robot dog, for example, cannot leave the ground let alone catch flying objects). But all of these calculations are entirely opaque to Fido qua doggy sensation experiencer; if he is thinking about anything, he is thinking about the wind on his coat, the play of the sun on the field, and the smells of the grass as he makes his run. More significant is what Fido is not thinking. He is not wondering whether it is appropriate for him be playing Frisbee in the afternoon when he should be gathering food, he is not worried that he may ruin the Frisbee by biting it too hard, and he is most definitely not thinking about how the stock market will react to the latest housing slump. In short, Fido is the breathing, living embodiment of the folk principle "Don't worry, be happy."

In order to develop this idea in neurological terms, and ultimately to show that it is consistent with the known results on pleasurable, meditative, or happy state, assuming that the broad features of these can be relegated to a single account, we must first disabuse ourselves of a dangerous although natural conception—that there will necessarily be a structural invariant of positive affective states (or any other kind of mental state). What this means is that we should not expect to find that a particular brain module is always active under these conditions or even that more general measures of physiology, such as the intensity of an EEG, within a given band must invariably hold. Why is this the case?

Recall from Chapter 4 that we concluded that consciousness supervenes on whatever the brain is doing at any one moment—anything else requires that the putative function that transforms brain states into mental states to not only have a memory, but be able to search through this memory to find the right items relevant to current computation. This would produce a theory of mind that while not completely out of the realm of possibility, would be far too complex to be understood, and unlike any other fundamental theory of the universe previously discovered. Given, then, that the underlying structure of the universe is simple, elegant, and comprehensible, we must conclude that a given mental state, pleasure, say, is a function of the instantaneous causal network as in the brain. Or as previously argued, the network formed by the conscious kernel, the set reciprocally interacting with neurons. But this network can be "plucked" from one of many structures or sets of interacting structures. In other words, it is not physiology that counts, but how that physiology implements a given set of causal interactions, and any particular set of network dynamics may be instantiated in any number of brain structures.

This is not to say that certain pleasurable states will not have stereotypical functional sequences. For example, reward may very well be characterized by a regular interaction between the nucleus accumbens and the frontal cortex, which in turn then affects other systems in more or less predictable ways. But other states, those arising from meditation, or those from aesthetic contemplation need not have a similar etiology let alone similar physiological characteristics to be either scientifically explicable or technologically manipulable, as long as the kernel maintains an invariance across these varied neural event sequences.

How, then, can we derive the dynamics of a network that represents the unquiet mind, or conversely, what are the characteristics of the carefree brain? The first and most obvious thing that we can say is that an unquiet mind is one that will either perseverate upon a given thought, or will cycle through a series of related thoughts. A mind in this state will be unable to give attention to items outside its source of worry, and will also not pay as much attention to purely sensory stimuli. This is not to say that these stimuli are not consciously perceived at all; simply that the phenomenal intensity of these sensations will be reduced. The extreme case of this qualitative state is mental anguish and physical pain, and it is noteworthy that these are both highly unpleasant and characterized, in the extreme, by the inability to entertain thoughts other than the pain itself.

The opposite holds for the carefree state. Here, the mind flows freely from one thought to the next, with little friction in this movement. Moreover, stimuli that are able to overcome inhibition in the brain, and enable synchrony between neural populations, are those that we tend to find enjoyable. This idea was first proposed by Colin Martindale (1984), the leading advocate for a scientific study of aesthetics for the past 30 years, and for many years the editor of the primary journal in this field.

As an example, the left-hand side of Figure 10.4 shows a perfect cadence, the most powerful chord sequence in the harmonic repertoire, and the way a typical classical piece concludes. The right-hand side of the figure shows a model of the processing of this sequence. Thicker lines mean stronger connections; hence, the tonic chord in C major is formed by a strong connection from the note "c," and weaker connections from the "e" and "g" notes, and likewise with the dominant chord (the strongest note for this chord is the "g"). Units in the input layer are also connected most strongly to their respective counterparts in the note layer, and more weakly to adjacent notes, reflecting a degree of perceptual overlap in processing of notes. The net effect of such a model (Katz, 2004) is to maintain input to the dominant V chord, after it is no longer being played and while the tonic I chord sounds; this effect is augmented by the proximity of the notes in the treble staff, as well as resonance between the chord and note layers. Other chord transitions do not produce the same degree of synchrony, and cadences that do not follow the principle of voice leading (that the notes between successive chords

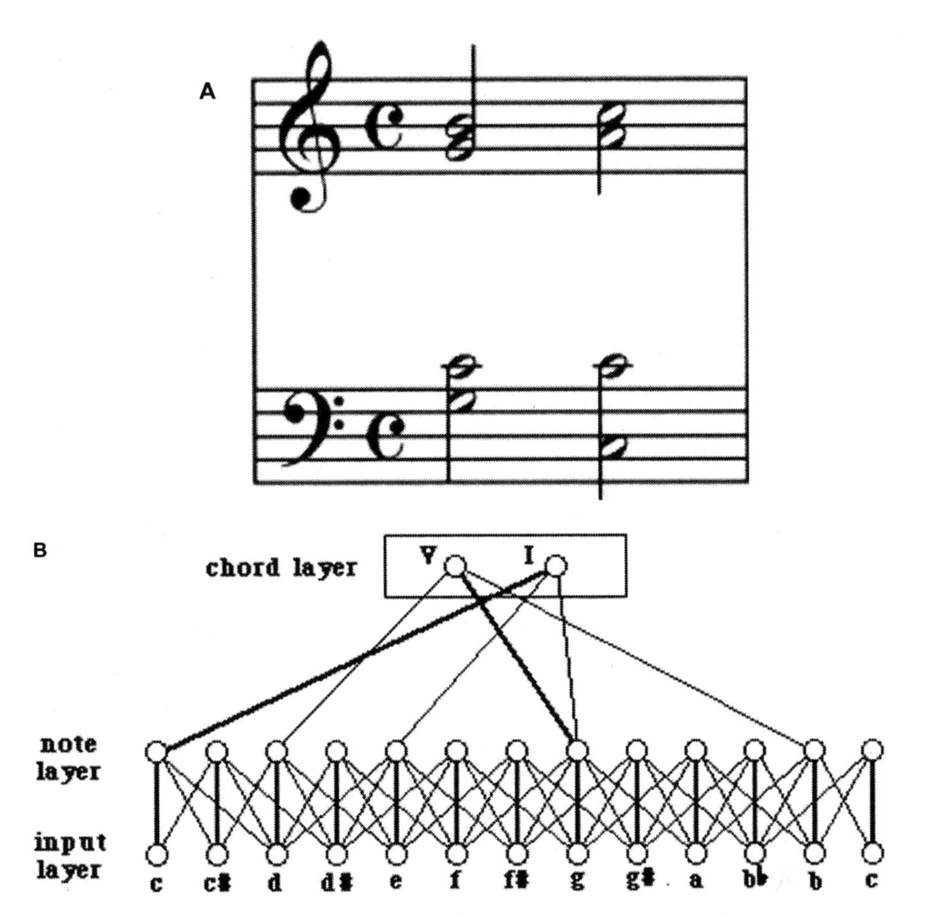

FIGURE 10.4 **A: A perfect cadence, or transition from the dominant to the tonic chord. The model in B: explains the power of this transition by noting the overlap between the structures, allowing the two chord units (V and I) to be synchronously active.**

should travel a minimal distance) as in this example also show a reduced degree of synchrony.

This unification effect is even more pronounced in the case of humor, which is characterized by a sudden sharp burst of pleasure. What can account for this? In accord with traditional models, it has been suggested (Katz, 1993) that humor works by the unification of opposing tendencies. However, these models do not explain a strong and sudden hedonic response to humor. Consider the following joke which will have resonance for those who interested in the kinds of issues that were raised in chapter 4 with respect to consciousness as well as those who have gotten their feet wet in academic waters: Two behaviorists meet at a psychology conference. They get drunk at a conference party and end up in bed. After making love, the man turns to the woman and asks, "That was great for you, how was it for me?" This of course is a reversal of the normal question in which the man inquires how it was for his partner. It is humorous not only because of the incongruity of this reversal, but because also because of its plausibility. Behaviorists deny the existence of inner states or qualia, or at the very least deny their importance, and must judge solely on actions. Hence, he must ask his partner to tell him what he would have felt if he felt anything by observing his behavior; he has no private access to these states (it hardly needs adding that analysis of this sort destroys any humor, but this is a price that humor theorists are used to paying in the service of science).

Humor, to be successful then, must contain two elements. An incongruity or reversal of expectation that must also be true to life or otherwise contain sufficient verisimilitude to be thought so upon reflection. Figure 10.5 shows how these two elements combine to produce a unity between two opposing concepts that would ordinarily not be able to be entertained in unison. The normal question appears on the left and ends with the first person pronoun; the identical question is on the right with the second person pronoun. The context, making love, triggers the expectation that the question on the left will be asked. Instead, we get incongruous one on the right, but this is supported by the additional fact that it is behaviorists who are tangling. In the model, this implies that both questions receive support, the first from expectations, and the second by inference. In short, two mutually inhibitory concepts receive sufficient support to briefly

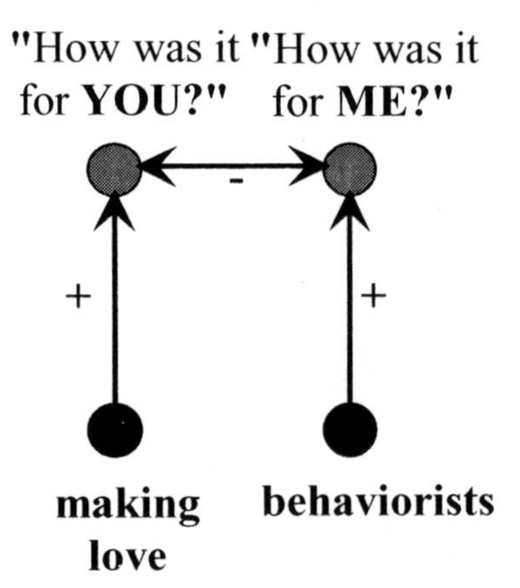

FIGURE 10.5 **The behaviorist joke rendered into a neural network. The joke works because the two ordinarily competing interpretations at the top are united by the expectation (making love) and the context (behaviorists making love).**

fuse them together, and this is hypothesized to be behind the sudden and large hedonic response to humor.[3]

Summarizing these models of music, humor, as well as earlier intuitive considerations, we can state that the state of displeasure is characterized by a rigid fixation on a single thought or closed sequence of thoughts; the state of pleasure is characterized by the free flow of thoughts, with frequent partial or full overlap between them. Together, these ideas motivate the driving principle of the rest of this chapter:

The Kernel Pleasure Principle (KPP)

> *The valence of phenomenal experience at any particular moment is proportional to the size of the concious kernel.*

There are two important aspects of this formula. First, as with all phenomenal experience, it claims that how good you are feeling is a function of what your brain is doing right now. Second, it claims that to the extent that your brain is able to unify its various sensations, whether they be externally or internally generated, and bring them into the conscious kernel, you will feel good, and to the extent that the kernel is narrowly focused, you will experience displeasure. In practice, these conditions imply the necessity of relatively large levels of synchrony between the neural populations processing a stimulus or set of stimuli in order to feel pleasure. In order for them to meet the kernel requirement of the KPP, they must be casually interacting and actively firing; in order to meet the temporal requirement, they must be firing in unison. These conditions are just the conditions for synchronicity.

Before showing how the KPP is broadly consistent with the results on pleasure and happiness previously discussed, let us examine one possible problem. This is that it appears, superficially at least, to suggest that intellectual activity is inherently painful, in that such activity will concentrate on a narrow set of thoughts of ideas. But we know that this is not the case, at least for some people some of the time. Game-playing must be intrinsically pleasurable, otherwise it would not be so popular, puzzles like Sudoku have become ubiquitous, and more abstract pursuits, such as scientific discovery, have at least some intensely pleasurable moments—consider the Aha! experience in the previous chapter.

[3] There are often other tendentious factors, such as aggression or sexual elements, which may serve to augment this boost in synchrony by generating a higher level of arousal. For example, the behaviorist joke is obviously sexual in nature, but it is also subtly aggressive, as it puts down those who believe qualia are outside the realm of scientific discourse.

The reason that these are not contradictions is that the having a eureka moment, and to a lesser extent ordinary problem solving, typically entails the psychological unity of heretofore disparate elements, as we have seen (think of the CRA problems). For example, in the case of scientific discovery, a number of unrelated pieces of data are shown to have an underlying unity, either by the supposition of a new law, or by bringing them into accord with existing theory. In the case of Sudoku, the finished puzzle is perceived as a completed work, with all the cells fitting together by virtue of the constraints on the placement of numbers. In both cases, there is the additional *extrinsic* reward value associated with completing a task; that is, a hedonic boost over and above that granted by the intrinsic properties associated with the processing of the given situation.[4]

We now turn to the consistency of the KPP with the data presented earlier in this chapter.

a) Ecstasy and euphoria
If forced to summarize the current results on these states, we come to an immediate conclusion: extreme states of pleasure are characterized by a reduction in neocortical activity and especially frontal activity. It is not an accident, then, that the French have dubbed orgasm "le petit mort"; it is as if higher-order functionality shuts down for a brief time. These results again seems to contradict the KPP until one realizes that cognition, except in certain unusual moments (insight, inspiration, and other moments of creativity) entails a narrow focus on a few well-defined thoughts. Categorical thought, by its nature, is reductionistic in that it takes a complex set of sensory stimuli and converts them into a small number of symbols. Hence, the activation of categorical thought implies a reduction in the overall quantity of thought; conversely, sufficiently strong pleasure will drive conceptual activity downward.

b) Meditation
The principle of unfocused attention or relaxed concentration can be thought of as the KPP translated into cognitive terms. As the data suggests, in the meditative state there is an increase of DLPFC activity, but unlike in focused cognition, also an increase in alpha and/or theta

[4] A similar distinction has traditionally been drawn between aesthetic and other more direct pleasures. The study of the former confines itself to understanding the direct properties of stimuli, such as beauty and elegance, independent of the associations these objects have. For example, if you are handed a check for a million dollars, you will not think that the check itself is beautiful, but you will be pleased with your new situation.

activity, indicating a slower more synchronous brain rhythm. Whether we should designate this semiparadoxical state of affairs as pleasurable is probably a moot point; it is clear that meditation is both soothing and stress relieving, and in the accomplished practitioner, may also be a window into the ineffable.

c) Determinants of Happiness
It is perhaps not an accident that the literature on the causes of happiness is relatively weak with respect to specific prescriptions; all studied determinants are, at best, facilitating rather than causal. In fact, if the KPP or some other extended formula for happiness is to be found in the brain, we should expect that external factors will only weakly correlate with happiness, as they only weakly determine internal neural dynamics. However, we can state that there do appear to be some *necessary* conditions for happiness, namely a life relatively free from financial and health concerns, a stable relationship, and a cheerful but not entirely anodyne outlook. In other words, at a minimum, one must be free from worry. But worry, whether justified or not, is the most narrow kind of thought that one can have, and one that if sufficiently strong will drive away other aspects of cognition and reduce perceptual capacity. An unpaid mortgage bill, for example, is like a thorn in one's side throughout the day, and will even penetrate one's dreams. Thus, the literature on the determinants of happiness is at least consistent with the KPP, in that it suggests that a minimum condition for long-term happiness is the relative absence of factors that serve to narrow cognition and partially shut down perception.

d) Maslow and self-actualization
Self-actualization is nothing more than living one's life fully, without unnecessary distractions, and unnecessary thoughts. As such, this process may be viewed as the KPP applied not to individual stimuli, but to one's entire environment. The unquiet mind acts a baffle chamber, where thoughts bound about, generating considerable mental heat, but ultimately of little consequence. In the self-actualized mind, one that knows what it wants and how to get there, the mind processes its environment with a minimum of friction, and its products reflect this directed energy. Furthermore, the clarity of such a mind allows them to pass through relatively unfettered, allowing all of the elements full play in the field of consciousness.

A useful, mathematically inspired analogy in this regard is that of a linear operator, such as a matrix, acting on a vector. For each such matrix, there is a characteristic vector known as the eigenvector, such

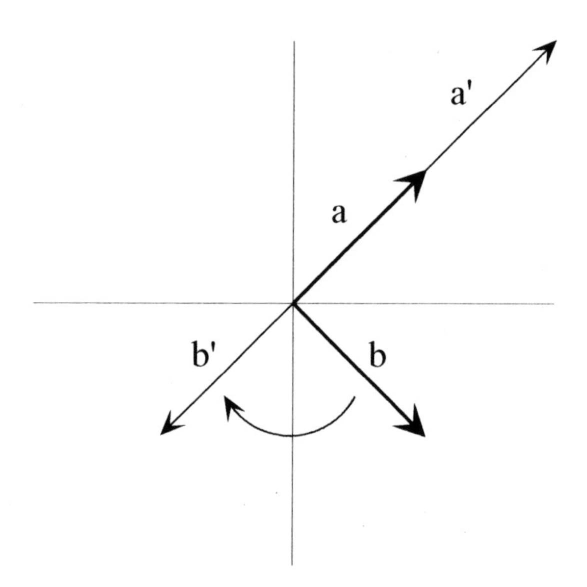

FIGURE 10.6 **The product of the matrix $\begin{bmatrix} 1/2 & 3/2 \\ 3/2 & 1/2 \end{bmatrix}$ and its eigenvector produces a new elongated vector** a' pointing **in the same direction. In contrast, when this matrix operates on the vector** b, **orthogonal to** a, **it produces a new vector b', which is rotated with respect to** b **but the same length as the original vector.**

that the matrix acting on this vector serves to both elongate it and preserve the direction of the vector; for all other vectors, at least some of the "energy" of the matrix goes into rotation rather than lengthening (see Figure 10.6). In this analogy, the mind is the matrix that operates on the vectors in one's environment. Becoming self-actualized means finding ones "eigencalling," that is, the relevant elements in one's environment that one can act upon with maximum effect. Until this happens, one is "spinning one's wheels," expending energy without generating anything of consequence.

Accompanying self-actualization will be clear perception and clear action; hence, the title of this section. When one is aligned with one's intrinsic nature, there will be a lessening in friction between perception and action. The reductionistic tendencies of the categorical mind, which serve to produce narrow thoughts, will instead be largely overcome, and abstract thinking, to the extent it *will* take place, will be a service of demonstrating a larger unity between events. In short, the actualized mind will see what there is to be seen, hear what there is to be heard, and when beauty does arise in its environs, whether it be music, song, art, or nature, it will be perceived fully and without unnecessary intellectual adornment. Just how technology can aid in achieving this state of mind is the subject of the next section.

HAPPINESS MACHINES: ORBS AND ORGASMATRONS

We begin by making a fundamental distinction, one presciently, if facetiously, raised over thirty years ago in Woody Allen's science-fiction movie "Sleeper." You may recall that Allen's character, Miles Monroe, a health food store owner from 1973, is cryogenically frozen and revived 200 years later by a rebel group. Because of his lack of identity in this future state, the rebels have chosen him to "assassinate" the nose of the dictator (the powers that be want to clone a new dictator from this sole remaining part). What concerns us here, however, is not the ridiculous plot, but the two devices in the house of Luna Schlosser (Diane Keaton), where Miles has (poorly) disguised himself as a robot to escape detection. The first device is an orgasmatron, which by an unknown technology produces orgasms when one enters a closet-like space—Miles discovers this by accident when he tries to hide there. The second device, the orb, has a more subtle effect, more akin to ordinary pleasure, but it also has its charms. When passed the orb, Miles end up "bogarting" the device, and the others have to pry it from his hands.[5]

We begin by concentrating on overpowering effects such as that delivered by the orgasmatron, and make the following prediction:

The extended bliss principle (EBP)

> *It will be possible in the not too distant future to place anyone, anywhere and at any time, and for as long as they wish to remain there, into a deliriously blissful state.*

The justification for the EBP is:

a) We can already do this for a short time albeit imperfectly with chemical agents,
b) With deep brain stimulation (DBS) we will be able to produce a stronger and more consistent effect, and moreover,

[5] Pop quiz: Using the prior theory of humor, discuss why this might be funny. Answer: Robots don't feel, hence the incongruity. However, this robot is really Mile Monroe, a needy nebbish from the past; hence the plausibility.

c) In conjunction with other technologies it will be possible to overcome the primary difficulty of chemical-induced highs, habituation and the accompanying cycle of addiction.

Finally, and despite b, and especially c, it will be argued that most people will not prefer to live in a permanent state of bliss, and will instead choose a modified orb-like device, described in more detail in the following.

The justification of a and b is straightforward, and has already been hinted at in the discussion of the brain's reward circuitry. We know that opiates and other substances, such as the noncoincidentally named ecstasy, have powerful emotional effects. Alcohol is often used in scientific experiments to produce euphoria, and of course, its use is not unknown outside the laboratory setting. Alcohol is a good example, however, of the limitations of conventional agents. The euphoric effect is fleeting, it is accompanied by a number of other less-desirable effects, both physical and mental, and if taken in sufficient quantities over a long period of time it will be harmful to the body.

It is natural, then, to ask whether there will be technological alternatives, such as DBS, to these crude agents, which act on a number of brain structures besides the intended target. The answer is trivially yes. It may very well be that Heath's subjects were not experiencing true pleasure but mere desire, although this is, to some extent, belied by their self-reports (perhaps the most interesting response came from one self-stimulator who said that it made him feel like "he had an upcoming date on Saturday night").

But given that we know that pharmacological agents can achieve these effects, there is only one reason that would stop direct stimulation from working, and that is if the effect was too global to be produced by a small number of electrodes. This is unlikely. While the KPP does suggest that pleasure is a "mass action" effect, it is also true that it can be initiated by a comparatively small number of neural events.

Consider the nature and diversity of the reward process. Virtually any stimulus can have a rewarding value. It can be money (pieces of paper), a grade (a mere mark on a piece of paper), praise (vibrations in the air), or an email from a love one (pixels on the screen). If each of these, and the many more types of reward, had the ability to directly produce a mass action in the brain, then the brain's circuitry would be incalculably complex. Instead, what needs to happen and likely does happen is that a stimulus, via conditioning, is associated with the triggering of a small set of reward structures. It is these, in turn, that initiate a global response, and if the KPP is correct, they do so by suppressing thoughts that inhibit the ability of neural populations to enter into global synchrony.

In summary, to create pleasure one merely needs to stimulate the central reward mechanism, rather than simulate a particular reward, or simulate the global effect of the reward system. Through DBS it will eventually be possible to do this in a much more precise fashion than is currently possible through chemical mediation. There is one remaining problem, however. This is that the brain, however stimulated, cannot sustain the reward state for any length of time, and in fact is explicitly designed via homeostasis to return to its default and presumably optimally adaptive state in short order.

To some extent, it is intellectually gratifying to know that we cannot be indefinitely and intensely gratified. Our current psychological universe is a balanced one, where there is a down for every up, an ebb and flow of emotions as with everything else. However, as with many of our tried and true notions, future neurotechnology will force us to rethink the very foundations of our conceptual outlook.

There are two forms of homeostasis that need to be overcome if pure and lasting bliss is to be achieved. In the short-term, the brain will produce less of the corresponding endogenous chemical to compensate for the artificially increased levels. For example, exogenous opiates, such as heroin and morphine, will cause the brain to produce less corresponding endorphins. This limits the effect of these substances, as well as resulting in a corresponding depression in affect when the effect of the exogenous substances wears off. In a second and more insidious long-term effect, the number of dopaminergic receptors in the brain's reward circuitry is reduced to compensate for the artificially increased levels (Volkow, et al., 2004). It is this process that is likely responsible for addiction; not only will it take more of the drug to achieve the same effect, when the user is not taking his or her drug of choice the tonic levels of dopamine in the reward system will be insufficient to stimulate the nucleus accumbers (NA). Thus, the addict will eventually need to use more just to maintain normal affective levels.

Can DBS or some other suitable technique overcome this homeostasis? Probably, if these are the only limiting factors. DBS, unlike drugs, does not work by modifying neurotransmitter levels; it works by stimulating the relevant structures themselves. For example, supposing that the NA is as hypothesized, the brain's reward center, stimulation of the cells in this structure directly bypasses the normal dendritic route; thus, dopamine levels, associated antagonistic agents, receptor count, or sensitivity within these dendrites should not matter. Whether other homeostatic mechanisms will arise is difficult to say at this point. If a mass action theory of pleasure is correct, then it is not the reward center per se that generates this emotion, but its action throughout the cortex. It may very well be that these other structures habituate to the increase in stimulation, therefore limiting the long-term prospects for bliss. However, there are at least three ways around this. First, it may be possible via chemical or other means to simply shut down habituating effects. Second, DBS and magnetic stimulation is not limited to the reward circuitry; it may be possible to stimulate multiple areas at once, and bypass homeostatic mechanisms in these places in the same way that DBS does so locally. Finally, the partial replacement of parts of the brain by artificial devices or the full virtualization of the brain, as discussed in the next chapter, will bypass entirely ordinary neural mechanisms. In these cases, once a state is realized, it will be able to be held in place indefinitely, if this is so desired.

Thus, extended bliss will be possible for any one and any time, as the EBP states, if not in the short-term, certainly within the next half century. This single fact, along with the following claims made regarding happiness induction, are perhaps the most significant of the all the claims made about the future of neurotechnology in this book. In many ways, all other drives are subsidiary to satisfaction, and if this can be achieved artificially, and without negative side-effects, then there can be no doubt that many will bypass the normal circuitous route to reward and choose to drink directly at the source, as it were.

Nevertheless, it is not likely that the vast majority of people will want to be in a continuous state of rapture. We may very well ask why we are not all hanging out in opium dens, or are all not heroin addicts. Contrary to the popular conception regarding the strength of heroin, a 1976 study found that only approximately 10% of casual users become addicts (Hunt & Chambers, 1976), a 1994 study places this somewhat higher at 23% (Warner et al., 1995). A comment by a working horticulturist, a recreational user in his forties is singularly instructive in this regard: "It was too nice. You're sort of not awake and you're not asleep, and you feel sort of like a baby in the cradle, with no worries, just floating in a comfortable cocoon. *That's an interesting place to be if you don't have anything else to do.*" (Sullum, 2003). The relevant passages are italicized. It is not that bliss is undesirable; it is that it is so desirable that it quells all other desires, and this paradoxically contains its downfall. We are not merely neo-behaviorist systems acting in a world that is a kind of giant, open-ended Skinner box, attempting to get the most reward for the least effort, but also personal identities, seeking to strive to grow, and to understand.

Future neurotechnology will not fundamentally change this state of affairs. While it is true that DBS or other similar means may overcome the considerable physical and mental problems associated with drug use, they do not alter the fact that the drive toward bliss does not overshadow all other drives, and that most will be not want to dwell in this most luxurious of psychic prisons indefinitely. What is more probable is that artificial stimulation will be used in precisely the way that Freud stated that psychoanalysis could not be used, that is, not so much to convert misery to ordinary unhappiness, but to take the latter and transform it into full-fledged happiness.

English, and probably most languages, are sorely lacking in the midrange affective names for this target state, but the terms exuberance and passion go some way toward capturing the nature of the intended emotion. One crucial difference between these terms and terms such as bliss, rapture, and ecstasy is that they entail an outward-looking component, whereas in the former the eyes are either literally or figuratively rolled upward, with attention facing inward rather toward the world. In contrast, in the "orb-induced" state, the fortunate recipient is still very much part of the world, social and otherwise. What changes is both their attitude and perception of this world. Similar to a caffeine high, but without the extraneous nervous energy, the person in this idealized state goes forth to greet each incoming stimulus, fully processing it and at the same time not becoming dominated by it. It is as if they are Andre Agassi let loose on the world at large, striking each ball on the rise, and with the keen vision of an eagle (Barbra Streisand once said of Agassi, "He plays like a Zen master out there").

In short, the trick is to produce, on a reliable basis across the populace, and on a consistent basis within a given individual, what Maslow would call a self-actualized being. Although the intensity of the desired effect is considerably less than that of euphoria, it is far more difficult to produce, because it is a function not merely of background affective level, but the way in which the brain interacts with its environment. Mildly stimulating the reward system is unlikely to have the desired result, because this will simply produce a kind of constant but reduced euphoria; this appears to be what Woody Allen had in mind with the orb. What is needed is a modified orb, specifically a device that takes into account both the context and the dynamics of processing. It must allow concentrative activity to occur as necessary, to permit problem solving and other frontal-related activities such as planning, and perhaps even a modicum of worrying. At the same time, it must restrict such activity to only that which is necessary, because the clumping of activity will reduce the size of the kernel, and in the long-term this will produce unhappiness.

As with the bliss device, we can divide technologies to achieve this into three camps: a) those that are created for brains as they currently exist, that is, as unadulterated wetware, b) those that are designed to work with augmented brains, and c) those that are designed to work with virtualized minds. As for the first case, we can envision a feedback system that would work as follows. Suppose the device noticed that for a given stimulus either an unusual amount of time was spent in a worrying state, characterized by amygdala or other activity in the aversion system plus frontal activity, or more directly, by noting that a particular brain state was being returned to again and again. In either case, the trick is the proper inhibition of activation clump that itself is inhibitory, either via magnetic means, such as TMS, or by DBS. Getting this to work just right will be no easy task, but should not be impossible. It amounts to producing a kind of meditation turned outward, and we already have examples of this, from Agassi's backhand to moments of flow in our own lives. If by no other means, it will be possible to produce these sorts of states by reverse engineering; that is, by studying those who have a gift for entering these attentive modes, and transferring the result to others that are less fortunate.

Thus, it ought to be possible to build a "worry zapper" that works in the unaltered brain. However, the cleanest form of control will occur when the brain is either augmented by external devices, or the mind is entirely virtualized. In these cases, and especially the latter, the course of thought will be able to be directed as surely as we can direct water through a pipe at the present. Like many ideas in neurotechnology, to some this may be a scary thought, but it must be recalled that the idea is not Big Electrode" (like Big Oil) will be sending signals to your brain to revere the Supreme Leader, or even to produce a specific thought sequence, but rather to keep the channels of thought free from congestion. When contemplating the beauty of a rose, the appropriately modulated mind will be doing that and nothing else, when working one's full attention and energy will be devoted to the task at hand, and as needed, the device will allow for both narrow forms of cognition and broad associative ones, of the sort discussed previously in the context of insight.

The goal will not be so much to eliminate all suffering, but to eliminate unneeded suffering. An analogy with physical pain is useful in this context. Pain serves an extremely important function, and those born without this sense, a rare but well-documented condition, are in constant danger of injuring themselves. But do we really need recurrent migraines, or the constant pain of arthritis, for example, to function better in the world? Likewise, the removal of unnecessary worry will not only make us feel better, it will make us more productive and more fully alive. Accordingly, the aims of the modified orb will be to guide the beneficiary away from the opposing poles of despair and euphoria, and toward the Maslovian actualized state.

Can it really be so easy to relegate human misery to a mere historical phenomenon? After all, a device protruding from the forehead cannot find a man a wife (and will probably repel a number of otherwise suitable candidates). Furthermore, one might wish to argue that depression and lesser forms of discontent are diseases of the soul, existential conditions that mere technology cannot address. Finally, there is the nagging worry that suffering itself is salutary, and that to take this away is to lessen the human experience.

To all of these concerns there is one simple response, namely, that one should not view neurotechnology as a conventional technology in any sense of the term. The neural dynamics of the current, evolution-designed brain generates its own set of checks and balances, such that the augmentation of one quantity leads to a diminishment of another equally desirable quantity. Thinking about how thought can be changed itself requires a new way of thinking, and one that shows that this is not inevitably true. The man in search of a suitable wife needs to first sort out who he is and what he wants, and only after he does this, either through conventional means, or via an external device, can he love fully. Some degree of mood swings may be desirable, but long-term depression is not, as anyone who has been afflicted will tell you. If anything, the ability to change, explore, and grow, necessary concomitants of truly relieving oneself of the tendency to fall into despair, requires an almost experimental, risk-averse attitude, something that is sorely lacking in the depressed state. Finally, suffering itself has traditionally been a necessary component of the human condition, but there is no law of the universe dictating that this must always be the case. It is no more inevitable than malaria before quinine, sweltering in the summer heat before air conditioning, or dealing with insurance agents before the Internet. For those of us raised in the preneurotechnology era, no doubt a life largely free of suffering will seem undeserved and therefore generate a great deal of guilt, but fortunately one day this guilt will also be zapped away, as surely as we will be able to eliminate unnecessary angst, ennui, and despair.

REFERENCES

Aftanas, L., & Golocheikine, S. (2001). Impact of regular mediation practice on EEG activity at rest and during evoked negative emotions. *International Journal of Neuroscience, 115*, 893–909.

Arambula, P., Peper, E., Kawakami, M., & Gibney, K. (2001). The Physiological Correlates of Kundalini Yoga Meditation: A Study of a Yoga Master. *Applied Psychophysiology and Biofeedback*, 147–153.

Bartels, B., & Zeki, S. (2000). The neural basis of romantic love. *Neuroreport, 11*, 3829–3934.

Berridge, K. (2003). Pleasures of the brain. *Brain and Cognition, 52*, 106–128.

Blood, A., & Zatorre, R. (2001). Intensely pleasurable responses to music correlate with activity in brain regions implicated in reward and emotion. *Proceedings of the National Academy of Sciences, USA, 98*, 11818–11823.

Buss, D. (2000). The Dangerous Passion: Why Jealousy is as Necessary as Love and Sex. New York: The Free Press.

Cahn, B., & Polich, J. (2006). Meditation states and traits: EEG, ERP, and neuroimaging Studies. *Psychological Bulletin*, 180–211.

Cranson, R., Goddard, P., & Orme-Johnson, D. (1990). P300 under conditions of temporal uncertainty and filter attenuation: Reduced latency in long-term practitioners of TMcy in long-term. *Psychophysiology, 27*.

Dennett, D. (2007). Breaking the Spell: Religion as a Natural Phenomenon. New York: Penguin.

Diener, E. (2000). Subjective well-being: The science of happiness and a proposal for a national index. *American Psychologist, 55*, 34–43.

Dietrich, A. (2003). Functional neuroanatomy of altered states of consciousness: The transient hypofrontality hypothesis. *Consciounsess and Cognition, 12*, 231–256.

Easterlin, R. (2003). Explaining happiness. *Proceedings of the National Academy of Sciences, USA, 100*, 11176–11183.

Georgiadis, J., Kortedkaas, R., Kuipers, R., Nieuwenburg, A., Pruim, J., Simone Reinders, A., et al. (2006). Regional cerebral blood flow changes associated with clitorally induced orgasm in healthy women. *European Journal of Neuroscience, 24*, 3305–3316.

Goldsmith, H., Buss, A., Plomin, R., Rothbart, M., Thomas, A., Chess, S., et al. (1987). Roundtable: What Is temperament? Four approaches. *Child Development, 58*, 505–529.

Heath, R. (1963). Electrical self-stimulation of the brain in man. *American Journal of Psychiatry, 120*, 571–570.

Herbert, R., Lehmann, D., Tan, G., Travis, F., & Arenander, A. (2005). Enhanced EEG alpha time-domain phase synchrony during Transcendental Meditation: Implications for cortical integration theory. *Signal processing, 85*, 2213–2232.

Hitchens, C. (2007). God is not great: How religion poisons everything. New York: Twelve Books.

Hoffman, E. (1988). The Right to be Human: A Biography of Abraham Maslow. New York: Tarcher.

Holstege, G., Georgiadis, J., Paans, A. J., Meiners, L., van der Graaf, F., & Simone Reinders, A. (2003). Brain activation during human male ejaculation. *Journal of Neuroscience, 23*, 9185–9193.

Hunt, L., & Chambers, C. (1976). The Heroin Epidemics: A Study of Heroin Use in the United States, 1965–75. New York: John Wiley and Sons.

Katz, B. (2004). A measure of musical preference. *Journal of Consciousness Studies*, 28–57.

Katz, B. (1993). A neural resolutiona of the incongruity-resolutiona and incongruity theories of humor. *Connection Science, 5*, 59–75.

Komisaruk, B., Whipple, B., Crawford, A., Grimes, S., Liu, W., Kalnin, A., et al. Brain activation during vaginocervical self-stimulation and orgasm in women with complete spinal cord injury: fMRI evidence of mediation by the Vagus nerves. *Brain Research, 1024*, 77–88.

Lehmann, D., Faber, P., Achermann, P., Jeanmonod, D., Gianotti, L., & D., P. (2001). Brain sources of EEG gamma frequency during volitionally meditation-induced, altered states of consciousness, and experience of the self. *Psychiatry Research; Neuroimaging, 108*, 111–121.

Lykken, D., & Tellegen, A. (1996). Happiness is a stochastic phenomenon. *Pscyhological Science, 7*, 186–189.

Martindale, C. (1984). The Pleasures of thought: A theory of cognitive heondics. *The Journal of Mind and Behavior, 5*, 49–80.

Murphy, F., I., N.-S., & A. D., L. (2003). Functional neuroanatomy of emotions: A meta-analysis. *Cognitive, Affective, and Behavioral Neuroscience, 3*, 207–233.

Olds, J., & Milner, P. (1954). Positive reinforcement produced by electrical stimulation of septal area and other regions of rat brain. *Journal of Comparative and Physiological Psychology, 47*, 419–427.

Sullum, J. (2003, June). H: the surprising truth about heroin and addiction. *Reason*.

Veenhoven, R., & Ehrhardt, J. (1995). The cross-national pattern of happiness: Test of predictions implied in three theories of happiness. *Social Indicators Research, 34*, 33–68.

Veenhoven, R., & Ouweneel, P. (1990). Cross-National Differences in Happiness: Cultural Bias or Societal Quality? In B. N., & P. Drenth, *Contemporary Issues in Cross-Cultural Psychology* (pp. 168–184).

Volkow, N., Fowler, J., Wang, G., & Swanson, J. (2004). Dopamine in drug abuse and addiction: results from imaging studies and treatment implications. *Molecular Psychiatry, 9*, 557–569.

Chapter 11

Uploading the Soul: Prospects and Mechanisms

The history of the attempt at self-preservation beyond bodily death reaches back at least as far as Benjamin Franklin's somewhat whimsical desire to be "immersed with a few friends in a cask of Madeira" upon his demise, possibly to be reanimated by once again coming in contact with lightning. Informed scientific inquiry into this possibility, however, awaited the technological advances of the second half of the twentieth century. An updated version of Franklin's goal, cryonic freezing of either the head or the whole body was initiated in the 1970s by the Alcor Life Extension Foundation and the Cryonics Institute. The idea is to preserve the brain in sufficient detail until the time that a future technology could bring it back to life. A more general movement, transhumanism, concerned with life-extension by any means and other technological enhancements to the native human condition, began about the same time (see Bostrom, 2005, for an excellent history of transhumanist thought). As characterized by FM-2030 (aka. F.M. Esfandiary[1]), a transhuman is a person in transition to the postevolutionary era, a period in which we, or our cyborg descendants, are in control of the development of the species (FM-2030, 1989). This brings up an extremely important and recurring theme in writings of transhumanist thinkers, that of the singularity (Vinge, 1993). A singularity is a radical discontinuity, such that the dynamics postsingularity are impossible to predict from those before this event. In this case, the discontinuity is between evolutionary development,

[1] Transhumanists tend to take on new and often imaginative names as they become involved in the movement to reflect the leaving behind of their (limiting) original human condition.

in which we have little or no control over the development of the species[2], and a later period in which we are in control of both our individual and collective destinies via technological means. We have already mentioned one of the reasons for believing that this will be a rapid shift rather than gradual transition is that one of the changes we will be effecting is to increase intelligence either by the enhancement of our native wetware, augmentation of such with hardware, or both. These new intelligent agents in turn will be better able to design means of enhancing intelligence, they will build even sharper aids to reasoning and thought, and so forth, leading to a positive feedback loop. Thus, we should see a transition from a very slow linear increase in intelligence due to ordinary evolutionary development to a rapid exponential increase in intelligence once these enhancements begin. The other primary reason for thinking posthumans will be unlike their organic ancestors, and the primary focus of this chapter, is the possibility of virtualization and the accompanying cessation of involuntary death. It is almost impossible to predict what life would be like once mortality loses its sting, and once we are no longer bound up with an organic substrate (although we will take a stab at addressing this in the next chapter).

Though the transhumanist community consists of a number of different and sometimes conflicting voices, almost all are unified in their ardent desire to have man move beyond his current limitations, a generally optimistic view that although this presents some dangers, they are minimal with respect to the potential gains, and a belief that when the changes do start they will happen suddenly and with a ferocity that will surprise even the most forward-thinking researchers. Naturally, the emphasis varies among thinkers on different aspects of this singularity. Kurzweil, for example, concentrates on his (2005) and subsequent analyses on the technology for creating Artificial Intelligences (AIs) of greater-than-human capacity (i.e., those that would pass the Turing test). Kurzweil calculates that this will occur on or near 2045, and depending on the turn of events, may be caused by runaway development on the part of machines acting independent of their human "masters," or as is more likely, on the part of man-machine hybrids or cyborgs. The extropy institute (http://www.extropy.org) was founded in the early 1990s with the goal of promoting the transhumanist cause; extropy is the approximate opposite of entropy and refers to the desire to fight what has heretofore been the inevitable decay with time of both the body and mind. The extropy institute closed in 2006 but continues as an online

---

[2] Somewhat surprisingly, this control is not nil; as we breed dogs for certain features, we breed ourselves by our preferences. A particularly interesting example is found in Turkish culture; there the trait of being slightly cross-eyed is considered desirable in a woman. This makes the prevalence of this trait higher (in both men and women) among the Turks, because a woman with this trait is more likely to have children and pass on this gene.

concern. A more recent and somewhat more academically minded organization, the World Transhumanist Association (http://www.transhumanism.org), founded by the Oxford philosopher Nick Bostrom and the utilitarian David Pearce, has attempted to bring renewed rigor to the field without sacrificing the original spirit of the movement.

While we are heavily indebted to these pioneers, the emphasis of this chapter will diverge somewhat from much of the previous work in this field. In particular, we will stress the primacy of a *theory* of personhood in determining the best technological means of an indefinite extension of life. In Chapter 3, it was stressed that knowing why is immeasurably more powerful than knowing that. This is true in this context, and then some. Simply put, without a proper theoretical underpinning, we will have no confidence in the results of any attempt at self-preservation in all but the extreme case of an atom-by-atom copy of the original.

Before justifying this statement, let us first entertain this very notion, which has been seriously suggested in the literature (Merkle, 1994) as a means of copying identities. To say that this idea is *prima facie* preposterous is probably overstating the case; it may be that some future engineering marvels that include nanotechnology and possibly other means could create tiny factories that build faithful replicas of their atomic surrounding. We can say, though, that the current state of the art gives no indication as to when, if ever, this will be achieved. More importantly, the very reason that some are thinking along these lines is because of the lack of theoretical understanding of the nature of the self and the nature of consciousness. One example illustrating this fact is worth many paragraphs of theorizing, or metatheorizing to be more exact. Let us say that an advanced civilization comes across the artifact shown in Figure 11.1.A in an archeological dig. They turn the device on and to their delight they find that it smiles at them, they can type on it, they can draw on it, etc. (their computers have long since reached sentience and are very demanding, unlike this pleasant little machine). Suppose they want to make more of these objects. They could if they really wanted, apply their advanced nanotechnology to create an exact duplicate of the machine (shown in Figure 11.1.B). Aside from the expense, this is of course an absurd proposition, and an extreme last resort. What they would likely do instead is attempt to reverse engineer the device, and in order to do that properly they will have to resurrect the notion of a computation, assuming that has not been lost in the mists of time. Armed with this understanding, not only will they be able to create facsimiles of the Mac plus, they will also be able to design a host of other similar devices that realize algorithms regardless of their appearance, their size, the material in the casing, and a host of other irrelevancies.

Likewise, it is only armed with an understanding of personhood that we will be able to extract out what is essential about a given brain that allows it to act as the substrate for that person's identity. With this knowledge, a far less costly

A

B

FIGURE 11.1 **A: Artifact found in an archeological dig. B: Picture of an exact duplicate created by a three-dimensional copying machine, down to the Apple logo.**
SOURCE: **http://commons.wikimedia.org/wiki/Image:Macintosh_classic.jpg.**

and difficult procedure could, in principle, extract the essence of the self from the brain. In addition, this theoretical knowledge will give us some confidence that the procedure was a success. It is not difficult to see how much is riding on not just the technology of uploading but on the theories of personhood and consciousness. Suppose, for the sake of argument, that a bodily view of identity held. Then identity uploads are by their nature not possible; you are only you by virtue of your organic instantiation. Suppose instead that something like the theory proposed in Chapter 5 holds, in which identity is a function of first-person construction of such. Then not only can we make statements about what needs to be copied, we can retain some degree of confidence that our procedure was successful in the absence of an absolutely faithful reproduction. Indeed, later it will be argued that one can restrict one's attention to a relatively narrow subset of the constituents of the brain and still have a successful upload.

REVIEW AND CLARIFICATION OF THE NECESSARY CONDITIONS FOR IDENTITY UPLOADING

Table 11.1 illustrates the minimal conditions necessary for uploading to be successful. If the vessel to be uploaded to is not capable of sustaining or otherwise producing true phenomenal content, then the upload will not and could not work. It is pointless to be uploaded as a zombie, because there is nothing "that which it is like" to be a zombie, regardless of how closely the behavior of this entity matches that of the original person. This row is equivalent to dying on the operating table. In the next row, we assume that genuine qualia are produced. If however, the transfer is not identity preserving, then a whole new

TABLE 11.1. Copy results as a function of the sustainability of consciousness and identity preservation (**de** = death equivalent for the original person). Identity must be preserved in a system that is capable of true feeling in order for the upload to be successful.

	Identity not Preserved	**Identity Preserved**
Consciousness not sustainable	Zombie with new persona **(de)**	Zombie mimic[3] **(de)**
Sustains consciousness	New sentience created **(de)**	Success!

[3] Strictly speaking, this cell is not a possibility given the first-person theory of identity developed in Chapter 6. In this case, you would not feel a sense of continuity with your former self because you could not feel anything at all. However, it is included for completeness and may be useful for theories that assume the independence of consciousness and identity.

sentience is created *ex nihilo*. This is akin to producing a baby, except that it is likely that the new entity will be fully mature—it just won't be the same as the old person. Finally, it is only the lower right cell that has any chance for success. In this case, qualitative content is associated with the new entity, and in addition, this content is such that new identity feels like the old person.

Thus, both preservation of identity and preservation of conscious content are necessary. Whether these two conditions are jointly sufficient to ensure success is a more difficult question. We are in uncharted metaphysical waters and drawing any conclusion with certainty may be rash. However, we may make better progress by turning the question on its side and asking the following: "What more could one possibly want to be included in the transfer?" If consciousness is preserved, then postupload we have a new sentient being for all intents and purposes. Furthermore, if that sentience preserves identity, then that sentience will be the same as the preload self. Invoking the notion discussed at length in Chapter 5 that continuity of identity is continuity of the *feeling of identity*, the new self will not be able to distinguish between the preupload and postupload state. In other words, if it did not open its robotic eyes and look and see its new mechanical form, it would not know that the upload had taken place.

Greater clarity still may be found by imaging what the experience of uploading would be like from the first-person point of view. You walk into the lab, naturally apprehensive, but excited about the possibilities that will be made possible by a virtualized existence (see the end of this chapter for a fuller discussion). The technician informs you that the procedure will take six hours, during which you will be unconscious. He tells you to start counting backward from 100 by threes. 100, 97, 94... And then you "wake up." Let us assume for simplicity that your sensory access is similar to your biological body and that your motor control is also similar (you are not seeing in the infrared range and you are not able to control a building crane with your thoughts), to minimize any period of confusion. What are you thinking as you come to life? Among other things you remember walking into the lab, signing the necessary papers, and going under the anesthetic. You remember your apartment, your digi-pet, a miniature giraffe, and the fact that your hover car needs to be inspected. Most importantly, you remember your goals in life, one of which was to try out virtualization like many of your friends. This is no different then from waking up from any other operation, or for that matter, sleep. The only way it would be different is if there was some essential oil of the soul that must be transferred, also. But if this is rejected as a prescientific hypothesis, as it must, we are left with only one conclusion. Your first-person illusion of continuity is just as good as the real thing, simply because there is nothing more. You will take up in your virtual life where you left off in your organic one, at least initially pursuing the same dreams, having the same likes and dislikes, and loving those who you love with the same intensity as before.

It is one thing to specify, in the abstract, the conditions for the success of an identity upload and quite another to give an actual procedure for this process. In the coming section, we look at making what must be regarded as the supreme technological feat, bar none, a tangible reality.

UPLOADING PROCEDURES

In rough order of plausibility, this section presents three uploading methods consistent with the previously developed theoretical constraints. None are likely to happen tomorrow, but none, on the face of it are completely infeasible. The last method in particular, gradual offloading, could be with us within fifty years, and with almost complete certainty in the next one hundred years. Paraphrasing Ray Kurzweil, the singularity is nearer than we think. The rapidly advancing pace of technology coupled with the fact that the universe is governed by laws, even when it comes to seemingly nebulous conceptions such as consciousness and identity, dictate that it be so.

Method 1 Thought and behavior replication

The first method to be considered works as follows. A device is constructed that takes two sources of information, a set of inputs and a set of outputs. The inputs consist of the collection of sensory inputs and the thought or thoughts that the subject is having at any given time. The outputs are what the subject thinks subsequently in this context, and any accompanying behaviors. Thoughts could be obtained by a mind-reading device of the sort described in Chapter 8; sensory inputs could be detected using cameras and the like. The purpose of the device is to construct a machine that will produce the same thoughts and behavior in the same context and also meets the constraint that the system is capable of full awareness. That is, it does not merely create an algorithm to translate the inputs into the outputs, but recreates the appropriate causal structures for generating conscious content.[4]

In other words, the device is a kind of universal emulator that builds a new you from the patterns of reactions of the old you. At first glance, this seems to be a preposterous way of going about things. After all, is not the emulator creating a copy of you, rather than transferring you to the new device that it creates? If you think that this objection is valid, it is probably because you are implicitly falling back on the natural but mistaken view of the self as a simple,

[4] If strong AI is true, contrary to what has been argued in Chapter 5, then this last extra constraint is not necessary.

unconstructed entity, that is magically attached to the body in which it lives. Once the self is conceived of as a psychological object, the reproduction of that psychology implies of necessity that the self is also reconstructed, as long as the facilitating condition of conscious content generation is also in place.

Once again, the easiest way to see this is to put yourself in the place of the uploadee. Let us assume that the emulator has done a superb job of capturing your thought patterns (by this we do not mean that it need be perfect, as the continuity condition of identity permits a wide degree of latitude, in keeping with the fact that the self is in a constant state of flux under normal circumstances). Because you don't trust this method, you request that your body be placed in a state of suspended animation, so that you may return if necessary. The emulator works right up to the time you go into this state, capturing your last trepidations, so that these thoughts can be remembered later. After you go to sleep, we turn on the new device. By construction, the conscious content experienced in this new machine is the same (or nearly the same) as if you yourself, that is your bodily you, had just woke up. If these contents are the same, then any illusion that the new entity has of continuity with your presleep self, will be the same illusion that you would have had if you were in your actual body. Thus, a full emulation of you is just you, as hard as that is to accept.

The real problem with this methodology lays not so much in clearing the philosophical hurdles, but in surmounting the considerable technological ones. The central problem is the sheer complexity, both numerical and structural, of the replication problem. The variety of possible responses to a given context is extremely large, and everyone will generate a unique response. Think, for example, of your acquaintances. Is there anyone you know who would react in the identical way to a given situation as anyone else? Of course, there will be some similarities in some contexts, but in general, no two people will behave in even approximately the same way over a small assortment of situations. Furthermore, in order to model this complexity, it is unlikely that we could get by with the kinds of correlational models discussed in Chapter 6 in the context of Brain-Machine Interface (BMI) adaptation. It will not suffice to simply connect inputs up to outputs because thoughts, like language, have a grammar, as stressed in Chapter 3. Therefore, we need a model, such as a finite state machine, that captures this structure, but learning in such a model is inherently a much more difficult task than forming a set of linear or nonlinear classifications.

There is really only hope, and that is that the apparent complexity of thought and behavior is just that. We know that systems can often exhibit complex behavior without having complex parts; cellular automata are just one of many such examples. In this case, simple rules govern how a cell updates itself on the basis of the values of the surrounding cells. If these rules can be induced, then the behavior of the system as a whole can be reproduced. Likewise, it would be impossible to model the rancorous discussion between your Uncle Frank and your Uncle Phil at the thanksgiving table as a series of speech acts, but an avuncular replication machine may be able to reduce your relatives to a set of much simpler parts. The apparent complexity of their conversation would then be not a function of the complexities of their respective minds, but the fact that the inputs to these models are highly varied.

One commonly used technique to reduce a system with a large number of dimensions to a smaller number is principal component analysis (PCA). PCA works by forming new dimensions from the original data so as to maximize the variance in the data along these factors; the collected set of new dimensions will often be far fewer than the original set. Table 11.2 shows the results of informally classifying the opinions of those who expressed themselves at the Thanksgiving dinner on a variety of political topics. Six dimensions of data are present, but after PCA this can be reduced to a little more than two. The first principal component corresponds to a traditional right-left division, corresponding to differences on the war, on health care, and increasing taxes. Frank, Dad, and the single guy are on the right, Phil and Estelle on the left. The second principal component corresponds roughly to an authoritarian-libertarian dimension, corresponding to the differences on the remaining three issues. Frank and Estelle are on the authoritarian side, and Phil and Dad on the libertarian side. Taken together, these two dimensions account for almost all the variance in the data (they don't quite explain everything because the anomalous single guy holds an authoritarian position on marijuana but a libertarian position on gay marriage), and thus a considerable reduction in complexity has been effected.

Do we have any confidence that PCA or similar techniques can reduce the apparent complexity inherent in personality as a whole to something more manageable? The most prominent account of personality based on this methodology is the Five Factor Model (FFM) (McCrae & Costa, 1997). These factors are openness, or the willingness to experience new ideas and

TABLE 11.2. Data extracted from Thanksgiving banter to put through a PCA.

	War in Iraq	Government Provided Health Care	Gay Marriage	Increase Taxes	Marijuana Decriminalized	Wiretaps for Noncitizens
Uncle Frank	+	–	–	–	–	+
Uncle Phil	–	+	+	+	+	–
Aunt Estelle	–	+	–	+	–	+
Dad	+	–	+	–	+	–
Single guy from Dad's office	+	–	+	–	–	–

situations; conscientiousness, as revealed by the need for achievement and attention to detail; extraversion, or the degree to which the company of others is enjoyed; agreeableness, or the tendency to be friendly and compassionate to others; and neuroticism, or the degree to which one reacts emotionally to a given situation. The factors are revealed, for example, by doing a PCA of a subject's ratings of themselves or others; despite the wide variety of such ratings, after careful analysis these five basic traits usually show up.

The FFM has been criticized as offering an overly condensed view of the variety inherent in personality, possibly because the factors are usually derived from artificial questions rather than naturally occurring situations. Still, if we could really reduce personality to 15 factors, or even 500, we would have a working model that could emulate anyone in any situation. The problem in the current context is that these sorts of analyses only account for the broad nature of responses. To capture Uncle Frank we need to more than capture his opinions, and possibly the extraverted manner in which he offered them, but also at least some aspects of the fine-grained nature of his responses, including the way he waves his cigar dismissively as soon as his table-side opponents begin their discourse. If we are to build a Frank emulator that is detailed enough to convince itself that it is indeed Frank (this is the identity requirement), then it needs to get not only these behavioral details largely correct, but also the thoughts that Frank is having that accompany his behavior. It turns out, for example, that he never liked Uncle Phil's wife Estelle, and every time he sees her he is reminded of the time she wouldn't let him smoke in her house. Unless this sort of thing is captured, we will have built an opinion-bot, but not a person.

The prospects, then, for the construction of sufficiently detailed replication are not good. It is unlikely that we could build an emulation of a cockroach merely by watching its behavior let alone something as complex as a human being. It is instructive to examine why this is the case. Our behavior and thoughts are not the result of a few underlying variables, or a few thousand, but ultimately 10^{11} variables, corresponding to the firing states of our neurons. Making allowance for the fact that not all are completely independent still leaves millions of determinants of our thoughts and behaviors. It is not simply the complexity of the environment that is responsible for this variety of personality, it is the complexity of the generator, our brains. If you were to place Uncles Frank and Phil in a plain white room with only a single chair, both would be annoyed (Frank more so), but both would have a completely different stream of thoughts. Memories and routines encoded in the hippocampus and neocortex, muscle routines saved in the cerebellum, cognitive routines stored in the frontal areas, cognitive styles reflected in the workings of the frontal and limbic systems, and additional factors more difficult to define combine to create a unique, and in many ways irreducible, personality. In the next section, we explore means of capturing this level of detail by copying not behavior but process.

Method 2 Synaptic copying

Suppose that instead of emulating the brain we wish to copy it. We can further break this down into two distinct questions: i) what parts of the brain must be copied to preserve consciousness and identity, and ii) to what level of granularity must they be copied? Let us approach the latter, and in many ways more critical question first. It is here that the previously developed theories of consciousness and identity will be most useful. The theory of conscious, reduced Functionalism (rF) states that it is the casual properties of the brain that are responsible for the generation of conscious content (Chapter 4).[5] Therefore, to produce a conscious machine we must produce something with similar causal properties to the brain. Coupled with the Neuron doctrine (Chapter 2), the idea that it is the interaction between interacting neurons, *and nothing else*, that is responsible for both the behavior of the organism and the internal dynamics we reach the conclusion that we need to construct a neural network along the lines of the brain in order to achieve a conscious machine. We do not need to go to a lower level of granularity than this, nor do we need to consider glia and other cells that (Chalmers, 1997) play a supporting role but do not influence the brain's dynamics.

The details of ionic transport are also unimportant for the current purposes. Though they are responsible for the firing of a cell, and therefore causally influence the behavior of the cell and by extension cells that it sends afferents to, the firing behavior of the cell can be completely understood in terms of the synaptic dynamics plus any other modulatory influences that may be present. This is a subtle but important point, and can be brought home by comparison to this absurdity: to build a conscious machine, we need to model the brain down to the level of the elementary particles, that is, the leptons, quarks, and bosons. If this were the case, then we would be in worse shape than if the predictions of the most ardent champion of nanotechnology came true (we would need pico or possibly femtotechnology in its stead). But we don't need this level of detail, because the casual behavior of the brain can be understood at the neural level.[6]

---

[5] Philosophical aside: The theory makes a weaker claim than this, namely, that the preservation of causal structure will also preserve conscious content; something else could be responsible for the actual generation. However, if we assume that this something else does not hold a preference for wetware over hardware, and there is no reason to believe it would, then causality embedded in any machine will do the trick.

[6] Another aside, with possibly larger implications: If causality were really the bedrock of qualia, as we have been suggesting, would not the interactions between elementary particles also sustain some sort of mental life? Chalmers (1997) appears to suggest so, and others have postulated a universe full of conscious content, even where no brains exist. But this would produce different content than the brain as a whole, and because of the lack of coordinated dynamics, of a different kind, possibly merely fleeting colors, sounds, or smells, but with nothing to fuse these sensations into a full-fledged phenomenal field.

The conclusion then, is that a synaptic copy will suffice as the foundation for a conscious machine; this much has been suggested many times in both the literature on consciousness and in the literature on creating posthuman, virtualized beings. Theory rF, however, imposes two additional constraints on any such procedure. We cannot achieve the same results by emulating the computations carried out by the copied network. As you may recall, the claim of this theory is that it is not that the set of algorithms that the brain instantiates that leads to mentality, but the causal properties of the brain that are responsible. In other words, the mind is not the software of the brain; mind is a product of causal interaction, not informational interaction, just like any other law that governs the physical behavior of the universe. For example, it is not sufficient to run a simulation of the neural network on a sequential machine. If we were to do so, the casual structure would be preserved, but the time-locked causal properties will be lost. For example, this simulation would ignore temporal conjunction, such as the overlap of spiking patterns, which theory rF deems crucial to phenomenal binding.

What about identity? If we made a veridical copy of the causal structure, identity would come for free. One would feel like one's self because the same causal structure that created the illusion of continuity in the body would now be in place in the virtual environment. Interestingly, however, we might make do with much less than a fully faithful rendition of our earlier self. To see this, consider the nature of continuity. The single most important aspect in fostering this sense is the join between the immediate past and the present. What happens before this join is much less important. The analogy we originally used when discussing identity was that of a line. If the two ends of the line meet up, or come close to doing so, it is perceived as a single line in a wide variety of cases, unless there is a radical change in the direction of the line.[7]

This translates into the perception of continuous identity in an uploaded sentience as follows. It may be sufficient to copy: a) the most recent memories, say stretching back six months, b) a smattering of memories before this, c) the major likes and dislikes of the entity, d) the hopes, goals, and dreams of the entity, at least in broad outline. The latter requirement, in particular, may be visualized as the equivalent of the direction of the line of continuity. If you wake up in the virtualized state with a memory of recent events, and therefore a memory of your recent goals, but then find that you have a completely new set of aims, then there is a chance that the Gestalt of continuity will suffer. Conversely, if you hit the digital ground running, picking up where you left off previrtualization, then it will appear that you are the same entity, especially when you can

[7] Mathematically, we can state that the line must be both quasi-continuous, and the first-order differential must also be quasi-continuous at the join.

remember with ease "your" most recent past. The fact that you have lost most of 1983 and your recollections of Duran Duran, for example, is unlikely to destroy your sense of identity (nor will it likely harm your musical development in your new state).

The practical consequences of this conclusion depend on the exact synaptic copying mechanism, but in general we can say that it will be easier for a device to record recent events than to do a full-scale synaptic copy. You could be fitted, for example, with a device that records synaptic changes over a period of a few months and then recreates this on the digital end. Whether this works or not is open to empirical study, but it may very well turn out that what is needed to convince you that you are yourself is considerably less than one might think.

In addition to making do with less than a full synaptic copy of our memories and other aspects of identity, we might make do without entire brain modules. There is little evidence that the primary visual cortex stores significant memories. This does not imply that this region: a) does not vary from person to person, and b) that it is not responsible in part or in full for the generation of conscious content (for example, it was argued in Chapter 4 that the primary visual cortex is necessary for visual consciousness because it is the only area that holds visual information at a sufficiently high resolution).

The claim is that to the extent that this region does vary between people, these differences are well below the threshold for the construction of identity. For example, let us say that color contrast (see Chapter 2) varies by 10% from person A to person B because of the degree of lateral inhibition in area V4. You are not going to feel like a different person if your visual field is altered by this amount or more; it may even be the case that you do not notice the difference. Thus, it suffices that for this region we use a standard pre-constructed module, and wire them into the regions that are of importance. These would include, at a minimum, the frontal lobes, large sections of the limbic system, association areas (but not primary sensory areas), and as yet to be determined chunk of the temporal and parietal lobes.

We have thus reduced the complexity of the synaptic copying problem by perhaps an order of magnitude, but it still remains formidable. Synapses are both small and structural rather than dynamic features of the brain, and are thus difficult to discern with current and most likely future methodologies. We can divide attempts to do so into two camps: nondestructive, which leaves the brain intact, and destructive, which does not. The former may, in turn, be divided into two categories, noninvasive and invasive. The prospects of a noninvasive means of sensing activity at the syntactic level, let alone reading off synaptic efficacy, are not good. Current neuroimaging resolution is, at best, on the order of 1 mm. The size of a synapse is on the order of 1 μm, or three orders of magnitude less. If there were a version of Moore's law for neuroimaging, stating, for example, that spatial resolution halves every n years, we would be able to wait 10 such cycles or so to get down to this level; however, improvements, if anything, are incremental rather than exponential, and may hit fundamental stumbling blocks. It is unlikely, for example, that fMRI will approach anything like the level of the individual neuron, because of signal-to-noise ratio limitations, and in any case the Blood Oxygen Level Dependent (BOLD) response is being measured rather than synaptic efficacy per se.

We have little reason to believe that invasive but nondestructive methods will fare any better. Certainly it is infeasible to measure anything more than a very small fraction of the brain's activity with deep electrodes, so this option is out of the question. The best bet, although still a long shot, is Kurzweil's (2005) proposal to send nanobots through the brain to monitor synaptic activity, possibly traveling along the brains extensive vasculature. They would then need to regroup somewhere outside the body and report their results, including the locations that they monitored and the synaptic strengths at these locations. While not out of the realm of possibility, this method is highly speculative, and the time line for this development is impossible to predict.

By far the best prospects are presented by destructive analysis of brain tissue slices. In this case, the analysis would take place after "death," although this should not matter if identity and consciousness can really be reconstructed. Some confidence that this is indeed a possibility is provided by the pioneering work already underway on the Blue Brain project, a collaboration between IBM, which supplies the computing power, and Henry Markam's Brain and Mind Institute at the École Polytechique in Lausanne. The goal of this project is not merely to form a map of a neocortical column, a basic building block of the neocortex comprising approximately 10,000 neurons in a cylinder 2 mm high and with a radius of .5 mm, but also the precise synaptic connectivity pattern in this structure (Markram, 2006). Among the many techniques at their disposable are multineuron patch clamp recordings to obtain these patterns, and a dedicated supercomputer (the Blue Gene/L) to confirm that artificial network behavior matches actual brain behavior. The former works by attaching a very small pipette (on the order of 1 micron) to the pre-and postsynaptic cells to determine how these cells interact.

As enticing as this project sounds, it is still a far cry from what would be needed to reproduce enough brain "stuff" to capture identity. If we make the generous assumption that we need to capture only 10% of the brain in order to achieve this goal, this still leaves us with 10^{10} cells. Thus, the Blue Brain Project, which aims to encode the behavior of 10^4 cells, even if successful would fall short of what is needed by a factor of 10^6. By way of comparison, the only organisms for which we have been able to produce full neural networks have been on the order of 10^3 neurons. For example, the network for the roundworm *C. elegans*, consisting of 302 cells, has been completely mapped, although it took ten years of intensive research to produce this result.[8] Again, current technology falls short by a factor of about a million.

This is not to say that the method of synaptic copying is without merit, merely that it will probably take at least a scientific generation before a vertebrate is fully mapped, and another few generations before a mammal or primate is mapped. The best hope for the virtualization of the soul to occur in the next half-century will therefore probably be found in our next and final method.

Method 3 Gradual offloading

Our final method not only stands the greatest chance of success, it surprisingly will involve the least amount of effort on our part. This is gradual offloading, or the transferal of identity to neural prostheses that were originally designed for other purposes. For example, suppose we begin modestly and install a chip that augments your hippocampal and other memory systems. Then we add a mood regulator, that gives you the sense of "effortless effort," the feeling of being in the flow whether at work or at play more of the time. Then we add another little chip that clarifies your intentions and desires, making you hew closer to your Maslovian destiny. These turn out to work well for you, as well they should: they were designed to overcome the numerous limitations of your evolution-designed neural machinery.

Then you find yourself wanting more and more. If a sharper memory is a good thing, why not run the whole thing off-brain, with a Wiki-like knowledge store at your cognitive fingertips? Why, for that matter, rely on the primitive emotional regulation system built into the brain, with its relentless pursuit of affective homeostasis, in the best case, or in the worst case, unstable response as in bipolar affective disorder? At some point, so much will be going on outside of the native wetware that more of "you" will be encased in the prosthesis than in your brain. When this happens, it should be possible to jettison the jelly, so to speak, and continue on as a virtualized self.[9]

Two factors will serve to bring this possibility closer to reality than it would be otherwise. The first follows from the nature of identity. As in the other upload

[8] Interestingly, it has been found to be a so-called "small-world" network; that is, each cell is only a few degrees of separation from each other, in the same way that there are supposedly six degrees of separation between any two people when the link in the chain is acquaintanceship. Small-world networks exhibit many desirable properties, including enhanced signal-propagation speed, computational power, and synchronizability (Watts and Strogatz, 1998).

[9] As always, a necessary precondition is that the stuff external to the brain be capable of sustaining consciousness. It will be no good to you if you turn into a zombie emulation of your former self when your brain is turned off.

mechanisms, 100% preservation is not necessary. We require just enough of you on the digital side to make you believe that you are you. Second, once the digital enhancements become sufficiently entwined with your original brain, it will become a more or less trivial matter for them to observe the workings of your brain. This means that they will be able to construct a simulacrum on the hardware side in preparation for virtualization. This emulation could proceed along two lines: a) an imitation of your behavior and thoughts, along the lines proposed in the first uploading possibility discussed earlier, or b) a fine-scale copying of the synaptic and causal lines of force in your brain, along the lines proposed in the second uploading possibility discussed previously. Thus, the three uploading possibilities are not necessarily mutually exclusive, and may proceed in consort to achieve a successful upload.

However, the current methodology retains one huge advantage over the prior methods in their pure form. The gradual nature of the process, whereby the enhancements slowly become a second cognitive skin, allows one to flip the switch at any time, turning off the brain and feeling what it would be like to be in the decerebrated state. If the new state is not to one's liking, for example, either because identity is not sufficiently preserved, or because the one's consciousness is dimmed or nonexistent, or because the experience is simply uncanny, then one can wait until more or better enhancements are added to your brain. You will not be forced to leave your body behind until you are fully comfortable with the result of doing so. Surely, this is the best of all possible worlds. One can dip one's toes in the brave new world of vitualized existence, confident that one can return to the shore if the water is to cold, or dive completely under when the completely natural fear of bodily termination is overcome.

WHO GETS TO BE VIRTUALIZED?

When will virtualization occur? As argued at the start of this chapter, this is very difficult to predict, as it depends not only on sheer computing power, and technological advancement, but to larger extent on fundamental breakthroughs in our understanding of both the nature of consciousness and the nature of personal identity, of which the speculations in this book may or may not play a significant part. However, we can speculate along with Kurzweil and other transhumanists that there will be a singularity, a massive nonlinearity before which you perish, and after which you enter the realm of the gods, as it were.

We can further speculate that there is a fundamental relation between time, wealth, and the possibility of living forever. The nonengineer may be surprised to learn that cost is a fundamental variable in almost every engineering pursuit: if money was no object, we could, for example, build a lightweight yet very sturdy bridge of almost arbitrary span out of titanium, or we could keep adding

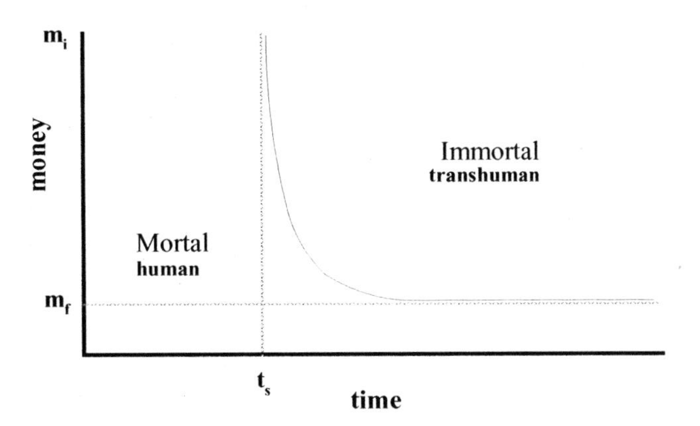

FIGURE 11.2 **One's chance for immortality will be dictated by both time and money. Before the singularity at t_s everyone will die; after, it will be a function of wealth.**

processors to a computer to achieve an arbitrary number of operations per second.[10] Uploading minds, while not your usual engineering fare, is no exception to this rule. The very first virtualization will be the result of a massive research effort, and may well cost into the billions. This, however, will come down considerably over time, most probably following something like the exponential decay shown in Figure 11.2. The singularity at time t_s indicates when the first upload will occur; no one who does not survive until this time will have any chance at immortality (barring metaphysical possibilities such as reincarnation, heavens and hells, and the like). After this point, it will depend on your ability to pay. The upload of the first soul will require a large initial investment m_i; after this, in keeping with most technological advances, the price will come rapidly down until it asymptotes at the final cost, m_f It is likely that m_f will either be within the means of most in the technologically developed world, or at the very least, most will be able to scrape together the cash to make this happen, given its supreme importance.[11]

In addition to the massive nonlinearity that occurs at time t_s in the graph in Figure 11.2, there are a number of other nonlinear factors implicit in the process of becoming an immortal; chief among these are those relating to medical advances of a more ordinary and corporeal sort. For example, let us say that the

[10] Not all operations are parallel, so this speedup proportional to the number of processors will not always be realized in every situation, but in many, such as image processing, it will be.

[11] 11 The mind boggles at the kind of financial deals one could arrange to "mortgage" one's soul. On the one hand, you might want to argue that you will have an infinite time to pay back the loan, so that you should unconditionally be given the opportunity to be virtualized. On the other hand, the lender will likely want to see some sort of collateral to protect his investment against the possibility of your deciding to spend the first few thousand years of your digital life immersed in a medieval fantasy game.

singularity happens in 2065, and in 2050 reliable artificial hearts with a lifetime of 20 years become widely available. If you suffer from heart disease and live to 2049, then you lose your chance to be immortal. However, if your heart does not give out until 2051, you will be able to have it replaced, which will then allow you to live to 2070, well past the singularity and into the transhuman stage (assuming sufficient wealth). Thus, we should probably replace Kurzweil's and Grossman's (2004) rallying cry, "Live long enough to live forever," with the more awkward but more accurate "Live long enough to take advantage of medical advances that keep your husk of a body alive just long enough to be finally free of it."

In summary, if you have already received a membership form for the AARP, then it is unlikely that you will reach immortality, at least via technological means. If you are between 25 and 55, then there is a chance, but you'd better make sure your investments on the stock market pan out. Younger than that, and it is anybody's guess, but virtualization may well be within your grasp, regardless of your means. If this all seems more than a bit capricious, there is not much else to say other than that is the way it is. If life is unfair then the laws governing eternal life are, in a strict mathematical sense, infinitely unfair. Either you get to zig and zag for four score or so years, and then become abruptly terminated, or you get to float onto a substrate that preserves your ego for a Methuselahn span or into eternity, if that is your desire. Just what it would be like in this disembodied state, or more properly, released from the birth-bodied state, is a topic that we will introduce now, and discuss in more detail in the next and final chapter.

LOOK, MA. NO BODY!

The prospect presented by extracorporeal existence is at once both frightening and exhilarating. If it feels strange when the dentist deadens one's mouth with novocain, how much stranger must it feel to have one's entire body deadened, to leave what Hans Moravec calls the jelly behind, and to float untethered in the digital realm. We know from sensory deprivation experiments what is likely to happen; very soon after being cut off from stimulation, the brain starts to produce vivid hallucinations akin to the dream state. The mind, it appears, craves stimulation, and when it is not present externally it will be generated internally. Of course, being digitally ensconced need not mean being cut off from the world. The default transfer would not be to a Dell laptop, but to a robot body with full sensory input—visual, auditory, and tactile, and possibly also taste and smell.

More to the point, in the same way that the computer created a revolution in data manipulation, virtualization opens up a host of alternative possibilities that will begin to be realizable with the addition of external chips to the brain, but will only reach their full culmination in this state. Some of our strengths as humans do stem from the semiarbitrary nature of the body that evolution has designed for us, but almost all of our limitations derive from this contingency. These limitations are twofold: first and most obviously purely physical constraints in that we can act only in ways that our bodies allow us to act, and second, and perhaps more essentially, mental, in that our brains are housed in a finite casing, and a rather small one at that. One can agree with corporo-centrists that the body is both a temple and a figure of grace, incredible in its complexity and artful and subtle in its movements, and at the same time hold that there remains an infinite collection of ways of being that we do not ordinarily begin to consider because of our physical parochialism. Here are a few such possibilities:

i) Polymorphically unbounded sexuality

At a transhumanist conference Natasha Vita More, president of the extropy institute, described a future containing, *inter alia*, negsexuals, technosexuals, postsexuals, VRsexuals and retrosexuals (Vita-More, 1997). Sexuality, for better or worse, will be one of the first areas that will be exploited with neurotechnology, and in fact may end up driving many of the innovations in the field. This trend can only be amplified once virtualization occurs. Celibacy is as unlikely in this state as in our current embodiment (of course, if one wishes to have sexuality extracted from one's digital identity, then this will be possible, but we suspect that most will leave it in place, at least initially). Sexual arousal is so powerful that we almost have to reach back to the years prior to puberty to convince ourselves of one important fact: the attraction to a confluence of body parts, gestures, and movements is a completely arbitrary affair, arranged by evolution for the propagation of the species. Virtualization will permit the complete disentanglement of biology and sensuality; everything and anything will be able to carry an erotic charge, by fiat, including all parts of the body, tables, chairs, tennis balls, houses, streams, and possibly vapors (although how this last one would work is difficult to say). Moreover, we will be able to take any form, including bizarre combinations of male and female, animal and man, machine and animal, that we will leave to the reader's imagination. Finally, and perhaps most importantly, as Vita More has argued, when making love it will be possible to feel not only one's own sensations, but also those of one's partner. The fundamental things still apply in this strange new world: it is the connection between consciousnesses that will end up counting as much if not more so than the devices that bring us together.

ii) Alternative incarnations

As already mentioned, being without a *fixed* body does not necessarily mean being disembodied; it simply means that one can be injected into alternative forms. These will likely include animals, real and imagined; robots, mechanical and bioinspired; historical figures, famous or otherwise; and a host of possibilities as yet unimagined. Whatever the form, the existence of proprioceptive and haptic feedback from the boundaries of that form will serve to define the physical identity of the sentience.

We can justify this by invoking an extension of the principle of transparency of action discussed in Chapter 6. That principle stated that whatever thoughts originally allow a BMI to drive an external device (such as a robot arm), after training one will simply have the intention to control the device in the desired way, and it will happen; there will be no need for cognitive intermediaries. This is what happens with our bodies as they are now; you do not think of moving muscles when lifting your arm, nor do you need to think of moving handlebars to turn a bike. Likewise, we can state that after sufficient time (which may not be very long at all), it will feel not as if you are receiving signals from your prehensile tail, say, but as if the tail is part of you and marks your boundary with the world. The same process of automatization that takes place with action, only in reverse, will eliminate the need to consciously process tactile signals, and allow any external parts or distortions of one's original form to be incorporated into one's body image.

This does not mean that it will not feel unusual to have a tail, or the neck of a giraffe, or other more extreme morphologies. Right now, we walk around feeling that we "live" inside our heads, but as philosophers have stressed at least since the time of Descartes, consciousness is not spatially extended nor physically located. The illusion comes about primarily because vision is our most prominent sense, and our eyes are located in front of our heads, and to a lesser extent because the facial nerves richly innervate the somatosensory cortex; roughly one third of this area is devoted to facial input (see the homunculus in Figure 2.11.B). Just what it would be like to lose this illusion, or to adopt an alternative one, is something that we will need to experience to specify in further detail.

iii) Emotional amplification and alteration

It is remarkable how much effort modern academic psychology has put into understanding cognition at the expense of emotion, although the efforts of scientists such as Damasio (1994) and LeDoux, (1998) and others are largely changing this. What is still not fully appreciated in some scientific circles, but has been known for centuries by dramatists (see Aristotle's theory of tragedy) is that the need to feel is fundamental, regardless of the valence or other quality of the emotion. It is as if we are not fully alive until we react, whether that reaction be positive or not.

If a play or movie produces tears on faces, it will also produce bums on seats.

Thus, we can postulate that once the phenomenal basis of emotion is understood, which will very likely precede virtualization, we will be able to turn up the pathetic, ironic, or horrific crank at will, possibly in conjunction with a dramatic spectacle. Movies (projected, naturally, directly into one's consciousness) will come not only with action and dialogue, but with an emotional annotation, detailing how the audience can be emotionally tweaked. Nothing like a shot of virtual adrenaline to heighten the suspense of a mystery or thriller. If this sounds crass, it need not be; the effect can be subtle, and may be all the more effective for that. In addition, a clearing of the emotional/cognitive landscape may be in order for the less dramatic arts, such as music. How Beethoven would sound when the background radiation of the mind is turned down, when microdisturbances of both physical and psychic origin do not interfere, is something only the gods have known until now; we may one day enjoy that privilege also.

iv) Lucid "waking"

The concept of lucid dreaming dates back at least as far as St. Augustine of Hippo in 415 AD. It was popularized, however, by the powerful but possibly apocryphal stories of the Yaqui shaman Don Juan, as recorded by the anthropologist Carlos Castaneda (*See The Art of Dreaming*, 1993). In these tales, Don Juan encourages his disciple to attempt to cultivate the same kind of awareness that he has in waking life in his dreams, and also to attempt to harness them for spiritual purposes. Regardless of the truth of these accounts, lucid dreaming has received considerable academic interest, and appears to be a genuine phenomenon, accessible to most if not all (la Berge, 1980). It is possible to become a fully conscious agent in your dreams, and not simply be a passive (Eliot, 1994) dream body subject to the phantasmagorical universe constructed by your unconscious fears and desires.

There are numerous problems standing in the way of creating an entire fantasy world in your sleep. First, lucid dreaming is difficult to induce, although some methods work better than others. For example, the wake-back-to-bed technique entails waking up five hours after you go to sleep, at presumably the best REM cycle of the night. As you fall asleep, you then try to remember to "wake up" within your dreams. Second, control in the best of lucid dreams is limited. You may be able to will yourself to fly, but the landscape that you find yourself hovering above is not generated by your now lucid self. To some extent this is a good thing, as it combines some special powers, with a degree of unpredictability. However, you cannot decide in advance which aspects that are under your volition, and which are to be independently generated. Finally, obtaining a high degree of awareness is often antithetical to lucidity; to the extent you enter into "normal" consciousness, the dream world becomes fuzzy and vague.

Needless to say, all of these problems are surmountable in a universe in which not only is the background wallpaper virtualized, but the self is also. The machine that is running everything is also running you; hence, it knows your thoughts and desires. Imagine something and it will be so: the genie has granted you an infinite number of wishes. Unlike ordinary reality, in which you navigate through a world largely not of your own making, with lucid waking you are both the creator and the responder. Of course, to make it interesting you may either limit your powers or encounter other beings with independent wills who have become similarly ensconced, as in *Second Life*. The key difference between this situation and earlier technologies, apart from the much finer resolution that the future will afford, is that your will and your thoughts are entangled in the virtual mesh. It is very possible that you will become so firmly embedded in the virtual universe that you forget the "real" world from which you sprung.

v) The key to all mythologies

The Reverend Edward Causabon, the husband of the heroine Dorothea Brooke in Eliot's *Middlemarch*, was not quite able to formulate his comprehensive system of human knowledge before he died. He was beset by a number of problems, including a limited set of resources, a fundamentally faulty approach, a beautiful and demanding wife, and most fatally, a subpar intellect, a fact that became painfully obvious to Dorothea only after she stated her vows. One tool, crucial for research that was missing in Victorian times was the Internet. The impact of this technology cannot be underestimated. Searching through musty library shelves for the appropriate book or journal is time-consuming and often futile. Academic life has been greatly invigorated by the online alternative, where one can not only download journal papers and other resources with a mouse click, one can easily acquire related research that was not known of before the search commenced. What we are seeing now is nothing short of an exponential explosion in the growth of knowledge, where ease of access accelerates ease of production, which in turn makes access easier, etc.

One thing is still missing though. We still have the same old brains in the same old casing. It is doubtful that Causabon would have come any closer to completing his "Key to All Mythologies" had Eliot included a fiber optic line running into his country pile. The Internet has allowed us to penetrate deeper without necessarily giving us the ability to go wider, to integrate and combine disparate knowledge sources to yield a novel synthesis. To do this, we need to start doing more of kinds of things that we mentioned in Chapter 9, including amplifying both intelligence and creativity.

As enticing as these possibilities are, they will remain awkward compared to what will be possible postvirtualization. To take a rough analogy, the difference is between enhancing a batter's arm with steroids and creating a robot player. The former will produce limited (though still controversial) gains, the latter will allow us eventually bat 1,000. The digitized entity will be able to be given not merely the raw knowledge of a field, but the expert's understanding of that field. This is hard to envision, because the expert's knowledge comes at the cost of many years of toil. But it is not the struggling per se that produces expertise, it is the end product of that struggle that contains the wisdom, and there is no reason why that product could not be captured, bottled, and injected into an identity, once that identity is completely under the aegis of a digital machine. As hard as this is to imagine, it is harder to accept. We are taught, rightly, to admire the hurdles as well as the achievement. Still, how much more fulfilled would the good Reverend's life had been had he been able to produce something on par with Joseph Campbell's work, and how much more admiring would Dorothea have been.

All of these possibilities and more are likely to be explored in what may be termed the digital adolescence of the soul, that time of exploration when the self allows itself to experiment and change. The difference between this period and normal adolescence is that this will be a chosen time, not one that is thrust upon one by hormones, and in addition it is likely to last much longer—possibly many thousands of years, depending on the sentience. The next and final chapter considers a set of potentially more radical changes that may take place as one approaches digital adulthood, including the possibility that the repellant force that drove mankind to virtualization in the first place, death and the termination of the self, paradoxically becomes the very state that is sought.

REFERENCES

Bostrom, N. (2005). A history of transhumanist thought. *Journal of Evolution and Technology, 14.*

Castaneda, C. (1993). *The Art of Dreaming.* New York: HarperCollins.

Chalmers, D. (1997). *The Conscious Mind: In Search of a Fundamental Theory.* Oxford: Oxford University Press.

Damasio, A. (1994). *Descartes' Error: Emotion, Reason, and the Human Brain.* Putnam.

Eliot, G. (1994). *Middlemarch.* New York: Penguin Poplular Classics.

Kurzweil, R. (2005). *The Singularity is Near.* New York: Viking.

Kurzweil, R., & Grossman, T. (2004). *Fantastic Voyage: Live Long Enough to Live Forever.* Emmaus, PA: Rodale Books.

la Berge, S. (1980). Lucid dreaming as a learnable skill: A case study. *Perceptual and Motor Skills, 51*, 1039–1042.

LeDoux, J. (1998). *The Emotional Brain: The Mysterious Underpinnings of Emotional Life.* New York: Touchstone.

Markram, H. (2006). The blue brain project. *Nature Neuroscience Review, 7*, 153–160.

McCrae, R. R., & Costa, P. T. (1997). Personality trait structure as a human universal. *American Psychologist, 52*, 509–516.

Merkle, R. (1994). The molecular repair of the brain. *Cryonics, 15*.

Vinge, V. (1993). The coming technological singularity: How to survive in the post-human era. *Vision-21: Interdisciplinary Science & Engineering in the Era of CyberSpace.*, NASA.

Vita-More, N. (1997). The Future of Gender and Sexuality. *EXTRO3.* San Jose.

Watts, D., & Strogatz, S. (1998). Collective dynamics of 'small-world' networks. *Nature, 4*, 409–410.

Chapter 12 The Mind, Unbound

The fundamental and recurring theme of this work is that neuroengineering will produce changes that are unlike any other technology, and ones that may very well be beyond the scope of our current neurally imposed abilities to imagine. This chapter is the culmimation of that claim: Here, we will examine not merely devices that enable us to think better and to feel more content but devices that change the very nature of our identity, possibly by simply throwing a switch, or clicking on an icon. It is impossible to predict precisely when the kinds of advances discussed here will occur, as extrapolation breaks down when looking so far ahead. But happen they will, and it can only to be to our advantage to examine them in full.

We will first treat the most obvious benefit of the transmission of identity from its current hidebound existence to a digital home, the shift from the extreme volatility of the former to the in principle eternal persistence of the latter. Freed from the second law of thermodynamics, that all (material) things decay, there is no reason why anyone could not live forever, or at least until the heat death of the universe. We will next consider the kinds of things that we are likely to do in this virtualized state. Apart from the obvious short-term gratifications available to a being in this form, there will be deeper longings that the unbound soul will seek to pursue. Chief among these will be the desire to experience, from the inside out, the minds and perspectives of other people and other cultures. This is considered in the section on psychic holidays. It will be argued, however, that this expansion of the mind's horizons will itself ultimately prove to be as hollow as the short-term hedonic gratifications offered by virtualization, and to a lesser extent by earlier more "conventional" neurotechnologies. Ultimately the soul will seek to be incorporated into something larger than itself; as will be discussed, if adopted, this will necessarily entail the loss of identity, and therefore will terminate the prior immortality of the self. But this will be a comparatively gentle landing into the afterlife, as these states of being will only be achievable by discarding the shell of the little birth self for something more spacious and accommodating to the mind in restless pursuit of truth. Among the possibilities only realizable by this process will be the fully luminous soul, one with no corridors or passages that slow the transit of thought and by such friction decrease the mind's full energy. This final possibility will close our discussion.

TEMPORARY IMMORTALITY

Nobody wants to die, except for those whose lives have become physically or mentally unbearable. What is less obvious is whether everyone would benefit from an extended lifespan. Regardless of one's beliefs about the afterlife, leaving this world is never easy, but at the very least the pain of departure is eased by knowing that one has achieved what one has set out to do.

Consider a typical life. An infant gradually grows from a bundle of primitive needs that must be met, to a being that starts to appreciate that there is a world around it. At this time the infant also begins to sense that there are other beings around it that provide a different but no less important form of nourishment, that of personal interaction. This includes, of course, the mother and father, but also friends, and the pleasures and travails of these relationships will be repeated many times through the in lifespan. The infant grows into a child and then moves into an extended educational phase, a phase that usually acts as a kind of holding tank until the child can be released into the wider and more confusing world.

The post-adolescent scenario has many variations, but typically involves finding a new psychic center after a considerable struggle (in many cases never fully resolved), and a form of tentative stasis ensues. Cupid's arrow strikes, some more deeply than others, and one learns that the unloved life is not worth living. One's own children may come along at this point, and through them one relives one's earlier development, a recapitulation as it were, with the added benefit of greater experience and greater detachment. One then achieves what one hopes to achieve or one does not; usually some median between these extremes is found. Marital discontent may turn into divorce or it may quicken into a deeper bond, less exciting than courtship but in many ways more fulfilling. Careers take their natural course and then wind down or sputter out. Minor ailments turn into operations, the mind and body slows, and before you know it, you and your spouse are queuing for the early bird special and maintaining muscle tone by lifing small weights in the community pool.

This is to no means either denigrate such a life nor deny its pleasures, both fleeting and of a deeper sort such as that found through personal sacrifice for the good of others. It is to suggest that in many cases, despite many tribulations, life has a kind of fullness to it. Nature, it would seem, has providentially provided just the right lifespan for those lucky enough to escape early disease or death by accident. It is almost as if it is saying, "What more can you possibly want: I have given you, in these threescore years and ten, loves and feuds, failures and triumphs, sunsets, moonlight, and picnics in the shade, climbs up mountains and trips down rivers, mojitos, burritos and taquitos, chickens molé and paprika, work, play, work that feels like play and play that resembles work, anguish, joy, and ecstasy, tens of thousands of sleeps fillled with both sweet dreams and nightmares, and now, in the end, it is time for a sleep that has neither."

Is it not the ultimate hubris to attempt to subvert this course of events? Well, not entirely. For one thing the typical life course is not the only life course. It is one thing to request more time to spend with one's grandchildren, and then their children, and so forth. This would certainly be nice, but whether substantial additional discovery or growth is possible in this context is doubtful. It is quite another to request more time to fulfill a scientific dream, an artistic vision, or to explore new dimensions in any type of creative endeavor. Architects, for example, often do not reach their prime until their sixties, leaving them a scant twenty years to express their vision if they are lucky. Einstein, it is said, was still working on a unified field theory on his death bed. Given his lack of productivity in his later years it is unlikely he would have benefitted from a reprieve from our common fate, but remember, we are now in the neuroengineering future, where in addition to keeping Einstein's mind alive, we can also rejuvenate it, and return him to the fertile intellectual years of his youth. One is also reminded of the itinerant Hungarian mathematician Paul Erdös, who like Einstein, died "in battle" while attending a conference in Warsaw. It has been said that Erdös imbibed generous amounts of amphetemines in order to accomplish as much as possible in his allotted time span. He died in 1996 at the age of 83 as the second most prolific mathematician in pages produced after Euler, and no one who knew him doubted that he would have continued publishing for many more years were he able.

Extraordinary people, then, almost always require more time to fulfill their vision. The discovery cycle is longer, at times by a significant amount, then the period of greatest productivity in a normal life, which usually does not exceed 35 years. In addition, why should we stop at a single cycle of discovery? Why should not our ubermensch pursue mathematics first, and then music, and let his "lifetime" of achievement in the former influence his compostions in the latter. To be sure, currently not everyone will have such a desire, but remember, the average Joe of the future, or the average Lenny, our accountant and protagonist from the first chapter has now been uploaded into a machine. The so-called ordinary man will, by virtue of neural technology, be transformed into an extraordinary one. What separates Einstein and Moscowitz is merely brain power, nothing more. After Moscowitz virtualizes, it will be a matter of a few tweaks here and there to give him the intellectual background, the intellectual power, and above all the intellectual curiosity of an Einstein if he so desires. He may not wish to devote his extended lifespan to physics, but one thing is almost certain: it is unlikely he will use his extra time to fill in tax returns for the next twelve centuries.

It may also be that we need not wait until full-scale virtualization is possible before this scenario unfolds. Augmented cognition holds the promise of increasing one's intellectual and creative capacity well before fine-grained manipulation can be carried out on the purely informational structure of the brain. In this case everyone, or at least everyone with access to this technology, will be extraordinary, and will require many lifetimes to fulfill their potential.

Just how many lifetimes will be necessary? The oxymoronic title of this section is not an accident; it is unlikely that given the chance, anyone will be want to live forever. Remember, eternity is a long time, and even Erdös may tire of his mathematical adventures after a few thousand lifetimes. Among the reasons a quasi-immortal being may choose to self-terminate are the following:

i) A temporary state of suicidal depression
This may seem unlikely, given that mood-control technology will go hand in hand with virtualization technology, and may well precede it. However, the intrepid mental explorer will likely visit many states, at least some of which, for want of a better phrase, are those that end up being complete downers. This may occur infrequently but it only need happen once in the course of forever for life to be over.

ii) Reaching the end of the line
A virtual Einstein, like an extended Erdös, may simply achieve all that he set out to do. Let us say that Einstein does discover a unified field theory, and it turns out to be everything he expects: elegant, comprehensive, and providing a direct window into the mind of the Creator. He could, of course, choose to tackle another problem for the next 10,000 years (consciousness, for example), or could simply say "enough is enough already, time for others to throw their weight into the mix." Our instincts tell us that were we in his shoes, we would take the former route, but we haven't been thinking for the past 800 years, and we haven't seen winter follow autumn and lead to spring and summer for that many years either.

iii) Absorption into a larger being
Neurotechnology opens up the possibility of a different kind of death, one which involves the termination of identity but not necessarily the termination of all of the information contained in that identity. For

reasons that will be discussed more fully in the section on soul amalgams, it is probable that what may be called super souls will emerge in the future, swallowing the contents of a given identity into its greater self and thereby vanquishing it. Recall from the chapter on personal identity that the notion of self is dependent on the mind's ability to create the illusion of continuity over time. There are no hard and fast rules about how much "mind stuff" has to persist for this illusion to hold, but if too little remains, then the construction of self cannot be accomplished. For example, personal identity could not possibly survive the absorption into a larger soul that itself is composed of a dozen other souls. The new emergent being may have some of the same memories and the same talents of its constituent minds, but it would not feel like any single one of them.

iv) A different view of death
Finally, death itself, and especially one's view of it, is not necessarily a constant. In our present embodied state, we have a strong evolutionary drive to survive at almost any cost. This drive may not be instilled in a virtual being, or it may be suspended on purpose. Moreover, as the soul matures over a period of centuries, it may come to view death in an entirely new way. From our limited perspective, death is the ultimate nonlinearity, a complete end to everything we know and to consciousness itself. This may not be the view of a more advanced being, or of the kind of soul described in the final section of this chapter.

Well before the souls' self-inflected termination, however, it is likely that it will decide to take full advantage of what virtualization can provide. A few of these possibilities are discussed in the following sections.

NEUROHOLIDAYS

As currently constituted, the mind encounters two primary impediments to its true freedom. These jointly interact to confine it to a relatively narrow and delimited subsection of psychic space, thereby shielding it from a much larger set of larger possibilities, both frightening and exhilarating. The first such barrier is imposed by society. Unless one is a hermit, one cannot help but be influenced by one's cultural milieu. It will creep into one's consciousness directly through the media, through personal interaction, it is embedded in the language that we speak, the way we hold our bodies, the foods that we eat, and the music and art that we produce. There is a cultural ego as well as an individual one, and if anything, the former is more ruthless at paring away possibilities beyond its collective comprehension.

In so doing, as it takes one cognitive step forward, it lurches backward in another complementary dimension. A gain in rationality invariably leads to a loss of simple and direct faith, the elevation of linguistic expression means a suppression of the intuitive arts, and above all, the celebration of individuality means the loss of both collective wisdom and collective sentiment (and vice versa, in all cases; we are not merely lamenting the ills of postenlightenment Western culture, but speaking of cultural trends in general). Whether there can be real cultural progress, i.e., progress that does not merely supercede an earlier gain but incorporates this prior wisdom into current thought, is an open question. Certainly, however, we can all admit that if and when this sort of progress does happen, it moves at a snail's pace, and tentatively and with many false steps: a Hegelian dialectician would not have kind words, for example, about the degeneration of a few noble aspirations in sixties counterculture into the leisure suit and disco music of the ensuing decade.

The other barrier to psychic freedom is the self, which by construction, dictates what can be understood and what is outside of its ken. One is bound by one's experience, which, at least in the preextended life prior to full-scale virtualization is usually extremely limited. Playing the piano, and playing it well, can occupy much of one's leisure during childhood and a smaller although significant portion of time into adulthood. There are no great pianists who are also chess players, there are few if any mathematicians with political skills, and the average person is lucky to be able to speak more than two or three languages well in a lifetime. If there is only one life to live, only one life's worth of experience can infuse into the ego. This is a trememdous limitation, although one which is not often contemplated, most likely because our expectations of what life is and should be are dictated by what is feasible and not on ultimate needs or desires.

Kahneman (2003) and others following in his intellectual footsteps have made much of the concept of bounded rationality. The structure of our brains (see Chapters 2 and 3) means that we are as much emotional and associational agents as reasoning ones, and it is very easy, even for the well-educated, to fall prey to elementary fallacies. But really this phenomenon is part of a wider set of limitations. It is not so much that the mind is bounded in its ability to think, it is bounded full stop. The day laborer in India cannot know what it is like to be an American scientist, and the American scientist cannot know what it it is like to hawk dumplings on the streets of Shanghai. And as we get older, the ruts of thought grow deeper, making it less likely that we can think any other way than the way we have been thinking all along.

It is possible to currently pursue, to a limited extent, the thought patterns of other minds, but only by indirect means. We will here briefly discuss one of the most significant such methods: artistic appreciation. Art has many functions, but among its most seductive is that it can act as a window into an alternative way of thinking. Figure 12.1 illustrates two *objects d'art* exemplifying this process.

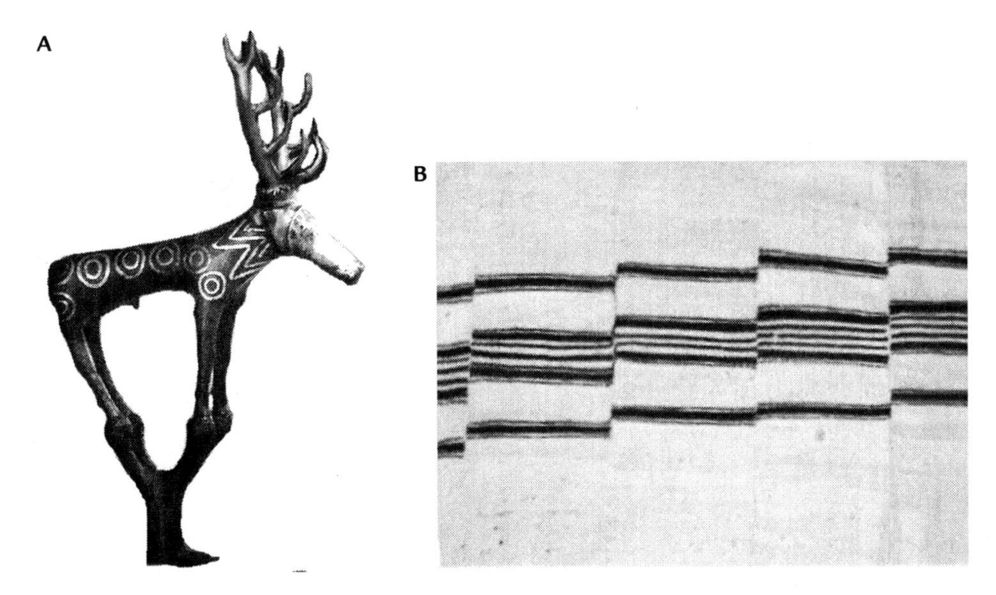

FIGURE 12.1 **A: Burial object, Alaca Höyük (photo, Georges Jansoone). B: Ghanese cloth, contemporary (courtesy of the collection of Samy Rabinovic).**
SOURCE: **A: http://nl.wikipedia.org/wiki/Museum_van_Anatolische_beschavingen. B: Samy Rabinovic.**

The left panel shows a silver stag found in the tombs of Alaca Höyük in central Anatolia, circa 2500–2300 BC. The dead of this proto-Hittite civilization were buried in a fetal position and with numerous artifacts including bulls, stags, jewelry, weapons, and abstract designs.

For these Hattian people, as with many premonotheistic cultures, such as American Indians, nature was a living, breathing thing; not so much an object to be bended to one's will, but an object to be revered and worshiped. This is clearly seen in the pictured silver figure, most obviously from the fact that it is situated on the pedestal, but also from the design itself. This is not so much a stag as the essence of a stag, rendered into pure form by the consummate skill of an artist whose vision rivals that of anything in the modern canon. The mind that has been cultured in the putatively superior monotheistic Weltanshaung has great difficulty in perceiving the vivacity of nonhuman forms, animate and inamate: its spiritual orientation, if it exists, is one that is directed to the source of all things and not the things themselves. Thus, the silver stag allows the modern mind a glimpse, however restricted, into this pagan mode of thought.

Likewise, the cotton textile fragment from modern Ghana shown in Figure 12.1.B provides an alternative path for the Western mind to follow, in this case literally so. Occidental thought patterns run in lines or curves, and even Picasso's most expressionistic work follows this rule. Here, however, we have a radical break with the very principle of continuity. The artist, for whatever reason, simply saw no need for the sets of indigo stripes to align, and the end result is a pleasing if somewhat cognitively jarring pattern.

These examples, however powerful, and however influential, cannot compare to breathing in the full effect of a culture dissonant to one's own. One may as well compare ordering sushi with living and working in Japan. Furthermore, even the latter only allows one to acquire the patina of Japaneseness if done after the age of twelve, that is, after one's personality is fixed in stone. For the occidental to become Japanese, to understand what it would be like to revere the animal kingdom, to be completely comfortable with noncontinuity, something along the lines of a psychic holiday is called for.

How can a future neurotechnology achieve this radical goal? Assuming that identity virtualization has already been achieved, a la the mechanisms presented in the previous chapter, we can conceive of two basic mechanisms:

i) Psychomorphing
One way of taking such a "trip" is to simply blend two virtualized minds together. By analogy with the graphical combining of two images, let us designate this as psychomorphing. Normally, morphing consists of two successive processes. First, the images are reshaped to achieve a form intermediate between the two. Then, the pixels in the images are altered to achieve a compromise color; if one was red and the other yellow, for example, a 50/50 morph would show orange. By direct analogy to this process, general attitudes and predispositions are the shape of the soul, memories its color content. The former are extremely important, because they might not exist at all, or exist to any appreciable extent in the mind to be morphed. You may never, for example, have composed a song or let alone played an instrument. If we were to morph you into a jazz composer, it would not be a matter of simply blending your musical styles, but more giving you first capacity to write music, and then to mix in your personality with that of the composer.

How this would work in practice is best illustrated by example. We will take the our neuro-hero from earlier chapters, Lenny Moscowitz, who has already been virtualized as a means of saving him from a certain bodily death, and attempt to morph him into his mental opposite, an ashik. An ashik is a wondering minstrel with mystical leanings and figures prominently in the Armenian, Azeri, Iranian, and Turkish literary/poetic traditions. The work ashik literally means lover (Turkish aşık, Arabic عاشق), and in this case the feelings are directed primarily to Allah, although romantic love may metaphorically stand in for this higher bond in poem and song. Lenny's primary connection to music was through the easy listening station he had on at work throughout the day;

he and Mrs. Moscowitz also made the trip to "the city" once or twice a year to catch the newest musical on Broadway before he became ill.

Mrs. Moscowitz was, of course, overjoyed just to have the old Lenny back, after years of illness and his final slide toward death in the past few months. The new virtualized Lenny is also happy just to be alive, but he has also been starting to take advantage of everything his new virtualized state has to offer. The doctors have installed all the standard acoutrements of the unbound mind, including a heads-up display in Lenny's upper right visual field that provides instant search and annotation on Lenny's thoughts if he should so desire, augmented cognitive and intuitive faculties, and a bright and purposeful disposition, in accord with the mechanisms outlined in Chapter 10. It is, in fact, the latter that is causing the problem, at least from his wife's viewpoint. Lenny is not so much seizing the day as seizing the universe. His cursory examination of all that is on offer has convinced him that his sheltered Long Island existence, pleasant enough before his illness, somehow fails to satisfy his new clear-eyed self.

In the past few days, he has seen banner ads in his internal display for something called a neuroholiday. Intrigued, he scans the list of possibilities and settles on an infusion of Ashik, noting both its reasonable cost (he is after all still an accountant, despite his cerebral tune-up) and the fact that he has always wanted to play an instrument; his decision is also influenced by the fact that he suddenly feels the need to experience something completely new. The holiday company performs a quick analysis of Lenny's constitution and comes up with following transformation:

$$\text{Lenny}_{[90\%]} \otimes \text{Ashik X}_{[10\%]} \Rightarrow \text{Ashik Lenny}$$

In this formula, the left-hand side represents an admixture of 9 parts Lenny to 1 part of an unnamed Ashik who has been compensated for the dissemination of his mind, and on the right-hand side we designate the resulting person by placing the honorific Ashik in front of the first name in the traditional style. One of the main considerations of the algorithm was how much "otherness" to add in. Too little and the customer will feel like he has not left his old personality, but too much and the resultant entity will no longer be themselves—this is virtual death.

In Lenny's case it decides that anything in excess of 10% would be taking too big of a risk. In particular, it extracts from Ashik X just the essential framework for Ashik-hood: the ability to play the saz, or long-necked lute with a somewhat metallic twang favored by these itinerant minstrels; a limited ability to improvise (always a weak point for the old Lenny), both melodically and lyrically; and most importantly, a a dash of Ashik X's motivation, which is to celebrate heaven and earth through song. These correspond to the stretching aspects of morphing, twisting, and bending Lenny into a new form, being careful not to introduce too much shear in the process. The blending aspect of morphing consists of adding long-term autobiographical memories. These include the basic canon of songs, and in addition the memory of sleeping one night in a cemetary, and dreaming of being called to the saz by a descending angel. Traditionally, Ashik's learn whether they have been chosen for their profession in this way, and Ashik X can recall this experience from his adolescence. It is extremely important to realize that this is now Lenny's memory also. He does not remember the dream as if it were a movie; he remembers having the dream himself, and this dream now colors how he thinks about himself and his relation to the world as a whole.

It is this essential radicalness of the change that contributes to his wife's consternation as she visits him in the hospital the next day. He is sitting in his bed, strumming the instrument which was delivered that morning, improvising on a standard tune about a bülbül (nightingale) singing for its mate. As she enters the room, he quickly alters the song to praise her beauty, something she has not heard since before they were married. She tells him that she is flattered, but in truth, she is more than a little disturbed by the praise. She thinks of running to get a nurse, but his strong robotic arm restrains her, and then a more conventional meeting of souls takes place. Afterward, the latent accountant in Lenny decides that the neuroholiday was indeed well worth the expense.

ii) Psychoflavoring

The great advantage of psychomorphing over other approaches is that, apart from the admittedly advanced technology of virtualization, it requires little additional theoretical support. To take on another character, one merely needs to partially incorporate into one's nature the personality of another virtualized being. However, the disadvantage of the prior approach is that the psychic traveller may in the process acquire a set of quirks and idiosyncracies of his blend partner. These may be more distracting than they are worth. Suppose, for example, that Ashik X's improvisatory ability is intimately tied up with the emotions he felt at the loss of a good friend. Grief may be beneficial to the soul in the long run, but this is unlikely the trip that Lenny thought he was purchasing.

The other alternative, then, is to add a distillation of the essence of a particular national character, as it where, to one's own. This essence would contain only those traits shared by all members of a given subset, and leave aside undesirable components. Thus, the new formula would look something like the following:

$$\text{Lenny}[90\%] \otimes (\text{Ashik essence})[10\%] \Rightarrow \text{Ashik Lenny}$$

How then to capture the essence of Ashik, or more generally, the essence of a given national character? The notion of a distinctive national style has long been an attractive one to those who take an analytic approach to culture (Lin Yutang's justly famous book *The Importance of Living* from 1937 is a prime example); it is only recently, however, that this matter has been investigated with scientific rigor. A number of different cultural variables have been investigated, but the most extensively studied are those of the NEO-PRI-R system, comprising extraversion (E), conscientiousness (C), openness (O), neuroticism (N), and agreeableness (A) (that is, the same variables we looked earlier at in conjunction with personality). These variables emerge as the key dimensions of culture as determined by Principal Component Analysis (PCA) (other variable systems, such as that of Hofstede & McRae (2004) are strongly correlated with one or a combination of these variables).

Summarizing a large body of data based on these and similar variables, the following results are evident (McCrae and Terracciano, 2005):

i) There are reliable and replicable differences in the characters of nations.
ii) These differences sometimes coincide with the stereotypes about the culture, as judged by those external to the given culture, but more often they do not.
iii) The magnitude of these differences are dwarfed by differences between people within a culture (i.e., national character, to the extent that it exists, is not nearly as significant as individual character).

Figure 12.2 shows the extraversion and neuroticism of 51 cultures on a two-dimensional plot; similar cultures, according to these measures, thus appear proximally, and distinct cultures are distant. Given these results, how seriously should we take this methodology? The plot provides ample evidence that these results are interesting, valid, and unfortunately for our purposes, largely meaningless. Evidence for the validity of the graph comes from the fact that cultures with similar origins tend to be grouped together. For example, there is a cluster on the center right of Anglo-Saxon cultures: the English, Australians, New Zealanders, Canadians, and Americans. That the results are meaningless for us, however, follows from the fact that they are context-independent, and thus only weakly describe a given national character.

To take just one example, Japanese appear to the left, that is, on the introverted side of the graph. This is what one would expect of supposedly insular island nation. But these results are belied by any of a number of examples. Walk into any yakitori joint, or McDonald's in Tokyo for that matter, and you are

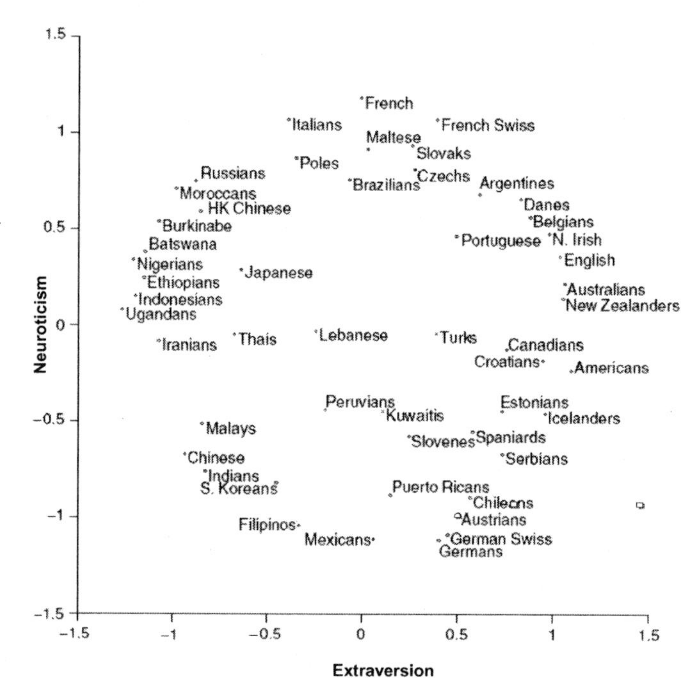

FIGURE 12.2 **Extraversion (horizontal axis) and neuroticism (vertical axis) for 51 cultures (adapted from (McCrae & Terracciano, 2005)).**
SOURCE: **McCrae, R.R., Terracciano, A., & 79 members of the Personality Profiles of Cultures Project. (2005). Personality profiles of cultures: Aggregate personality traits. Journal of Personality and Social Psychology, 89, 407–425.**
Used with permission of the American Psychological Association.

likely to find that Japanese waitstaff are in general much friendlier, and much more genuinely eager to please, than those in their Anglo-Saxon counterparts. Japanese introversion, to the extent that it is real, is a complex, and most importantly, a context-dependent effect, and one that is colored by every apect of life from work to family life to picking up a container of sushi at the train station.

To put this another way, you cannot make an Australian Japanese by placing them into a machine that makes them more introverted; what you will get is a quiet Australian. Nor can you make Lenny Japanese or an Ashik by tweaking a few parameters. It may, in principle, be possible to capture Japaneseness or Ashikness with a superior PCA based on a larger set of notions. But this is true only in a mathematical sense: to be an Ashik is to live a certain life, and through that life to love, transcendentally and otherwise; these are not easy elements to capture.

There is an alternative, though, that may allow us to capture the essential nature of character of a nation or subgroup without inheriting idiosyncratic characteristics. What if, instead of crossing Lenny with an ashik essence we morphed him with the cross of numerous ashiks? This is represented by the following formula.

$$\text{Lenny}_{[90\%]} \otimes (\text{Ashik}_1 \otimes \text{Ashik}_2 \otimes \ldots \text{Ashik}_n)_{[10\%]} \Rightarrow \text{Ashik Lenny}$$

In this case, in the process of forming the ashik blend, the characteristics of the individual ashiks will cancel out, and the residue will approximate ashik essence, in the same way that whiskey distillers blend out the quirks of each cask to achieve a uniform and smooth result. "Single cask" Ashik Lenny may have more character, but as in the case of whiskey, it will also be more variable, and contain extra notes that Lenny may not have bargained for. In small doses, it may even have the effect of creating an entirely new identity, by virtue of violating continuity of personality. This is something that the casual tourist would definitely try to avoid, although, as the next section details, may be the goal of the more intrepid neurotraveller.

SOUL AMALGAMS

One recurring motif througout this work has been that, contrary to intuition, there can be a free neurological lunch, or at least a heavily discounted one. At some point, it will be as easy to give oneself extra memory as it is to place a RAM chip inside one's PC. Intelligence will expandable, in the same way that today we can magnify physical strength, and even something as ill-understood as creative genius will be accessable to all. Finally, and perhaps most importantly, in the long-term, happiness will be a sustained and universal property of mankind. (Kurzweil, 2005)

At this juncture, however, we run into an important exception to this principle. As has been discussed in Chapter 5, personal identity is malleable, and in fact is subject to constant fluctuation, but it can only be stretched so far before the original identity is lost. This point, as was previously argued, will occur just when the new self no longer views itself as a continuation of the old self. The problem, then, is as follows. The clearest and shortest path for the self to expand its psychic horizons will be to be incorporated into a larger collection of selves that have already agluttinated for a similar purpose; let us call such an entity a soul amalgam. However, it is unlikely that identity will survive such an incorporation, especially if it is joining up with a large amalgam. Furthermore, once it allows itself to be swallowed, it will not determine the future of the amalgam, and it is likely that other souls will be ingested at a later time, reducing it to a small fraction of the overall whole. The situation is somewhat akin to a country stew, where the leftovers from the day before form the basis of the next day's ragout, and so on for the next day. Any given days' addition will eventually be reduced to mere homeopathic influence on the taste of the future stew.

Likewise, by choosing to join a larger community of souls you are also choosing a world in which the you that made the original choice no longer exists. Fundamentally, there is no difference between dying a traditional, physical death and losing one's identity in an identity dilution. In both cases, there is no continuation of a single being from the past to future. If this is so, then why will people (old-fashioned conjunctions of bodies and minds or virtualized selves) choose to have this done? The purpose of this section is to argue that though this leap into the unknown appears to be the height of irrationality, it will in fact prove to be the ultimate and most meaningful development in the long line of neural developments that began with cochlear implants and simple sensory modification, continued through to intellectual and emotional enhancement, and now ends with a procedure that can be construed either as suicide or rebirth depending on one's point of view.

First, it should be mentioned that we are constantly going through little deaths, not just of the "petite mort," or personal hedonic varitery, but also in our relations to others. Kurzweil (2005), for example, in his discussion of the same topic, mentions that romantic love is at least a partial ceding of the self to the needs of the other. If love does not entail self-sacrifice, it is not worthy of the name, and of course the placing of the needs of the loved above ones own are not limited to romantic love. The willingness to either delay or entirely give up personal ambition is an almost universal attribute in the set of complex emotional states that arises in association with love, or more properly *agape*, for one's children. There is also the more generalized form of giving toward others such as charity and the like; increasingly research is showing that such altruism is hardwired into the human condition, and is not simply the side-effect of an expectation to be compensated in return in similar currency (Moll et. al. 2006).

Still, it must be admitted that these sorts of examples fall more properly under the umbrella of self-sacrifice rather than total absorption of the self into a larger or greater being. Being in love may be wonderful, it may be bewildering, and it is almost always bittersweet, but it does not entail loss of memory or loss of other characteristics necessary to maintain personal identity; it is more of a broadening rather than an elimination of the soul. Likewise, there can be no doubt that being encased in a throng cheering for the home team entails a distinctly existential commitment to an external abstraction, distinct from and greater than the ego (although most sports fans would not put it as such), but does not require the abnegation of the self. To advance the notion that full-scale envelopment will be desired we need to turn to a stronger, more profound desire.

One natural place to start is with Freud's death instinct. In *Beyond the Pleasure Principle* (Freud, 1961), he describes the limitations on describing the human condition as one in which the organism seeks to maximize hedonic utility. Among the many classes of behaviors that he could not explain within his original framework were the existence of repetitive and compulsive behavior, masochism and sadism, and the recurrence of nightmares and other unpleasant dreams, as well as the more glaring examples of self-destructivenss inherent in the human conditions, such as war. More generally, Freud believed that there was a fundamental drive to return to a preexistent inanimate state, in which anxiety is suppressed by the complete cessation of thought, that is, the state of nonexistence.

Although Freud chided his colleagues for their refusal to accept what he thought was an uncomfortable but important truth, we wish to suggest that it was Freud's own pessimistic nature about the human condition[1] that prohibited him from seeing the stronger implication of his claim. To Freud, the death instinct was not something to embrace, but rather just one more unsavory fact in the litany of such truths regarding the human condition. Furthermore, it stands counter to the life instict, causing humans to create with the one hand, and to destroy with the other, a gloomy picture of mankind if there ever was one. But this is a mistake, a false dichotomy. In fact, the death instinct stands not in opposition to pleasure and the life-giving forces, but is part and parcel of them.

We can see by reexamination of the relevant current neurophysiological evidence, admittedly off-limits to Freud. Recall in the discussion of happiness in Chapter 10 that there were two main findings with respect to the positive affect and the brain: a) the worry-free brain is one in which neocortical activity in general, and especially frontal activity, is reduced relative to the baseline state, and b) the meditative brain is one in which there is a higher-than-average synchrony between regions that are active. Both sets of results are in need of more confirmation, but there is a growing body of evidence in their favor. Taken together they suggest there is a single, unified drive away from concentration on the particular and toward the universal.

Furthermore, depending on context and perspective, the cortex with reduced albeit synchronized activity may be construed either as a pleasure drive or drive toward annihilation. It is pleasurable, first and foremost because this is what the data tell us. We can also, however, think about this in more intuitive terms: the synchronized brain that is free from exective control is one with little or no presence of the little voice in the head worrying about where to park, when to pick up the dry cleaning, as well as deeper existential ponderings. However, it is also a death drive because it is a move toward reduced neocortical activity and thus to the death of identity.

[1]There can be no clearer evidence of this than Freud's previously discussed claim that the aim of psychoanalysis was to convert "hysterical misery to ordinary unhappiness."

We can also look at this in psychological rather than neural terms. Both large-scale synhronization and reduced frontal activity lead to a loss of distinction between perceptual and conceptual objects; pattern recognition depends on one set of neurons rather than another being active. Recall from prior discussion that the self is a construction, like all other concepts, albeit one with a special place and with greater generative force than probably any other concept in the brain. Thus, it will, like other concepts, first become fuzzy as one enters into the pleasurable state, and then completely disssolve into a sea of distributed activation as one enters euphoria.

To die a partial death, then, is not something we want, but something we seek anyway. If this view is correct, then the real opposition is not so much between the pleasure principle and the death instinct as between the survival instinct and both of these taken together. Survival requires both clear boundaries between concepts and especially between the self and other; pleasure delights in the breaking down of barriers, and the reduction of specific thoughts in favor of a more distributed buzz of activation.

What, then, does the union of the erotic with the thanatic have to do with a future technology that will allow the wholesale swallowing of a digitized soul by a larger composite virtualized being? The answer is surprisingly straightforward. At first the soul will revel in the possibilities created by a virtualized existence. The fruits of this techological tree will be rich and varied, and are beyond any analogy we could make as bound beings. The best we could do is to imagine our finest day, a day in when we were not coming down with a flu, where the boss complimented our work, when the children were well-behaved, when we came home and found an unexpected check in the mail, and at night found a truly good French movie on television—one where the romantic denoument is not so subtle as to be barely perceptable, but still unimistakingly Gallic in spirit. Then multiply the affective clarity of this day by a couple orders of magnitude, add in a virtual tete-a-tete with a reconstructed figure from history, throw in a kaleidoscopic light show that responds to one's immediate imaginings in the twilight of sleep, and you will have some idea of what life will eventually be like.

But after a number of days, months, and years of this psychic feast, the drive to move further will still be there, because this is part of the fundamental human condition (it may be part of the fundamental condition of any sentient being, although this is difficult to say). One will not want to experience more things, but to experience them differently, and to this will mean signing on the dotted line of the contract that takes one's self and injects it into the soul amalgam. This is akin to joining an exclusive club, with a few slight differences. One must do more than make a stiff deposit and a commitment to to eat so many meals a month on the premises; instead one must make the Mephistopelean deposit of one's soul, and promise to reside forever within the personality of one's fellow members.

This prospect would be fearful except for one key fact—what is being requested is an alignment with the natural inclinations of the self. The self, paradoxically, wants to transition to a state where it no longer is a distinct and unique entity. This is the meaning of the death/pleasure instinct, and highlights the contrast between it and the survival instinct. The fundamental instinct of the body, an evolutionary artifact, is to survive and regenerate; the mind, especially the fully self-aware mind of human beings, is at once more conflicted and hesitant, wanting to endure but also aware of the limitations of endurance for its own sake. Personal identity, in the current nonvirtualized world, already undergoes a never-ending series of revisions by choice or chance; virtualization will accelerate and magnify this process, and open up the possibility that identity can be sacrificed in order for greater growth to occur.

That there is no direct analogy to this process in our current state does not diminish the force of this claim. Even Freud, the cigar-smoking rationalist with both feet firmly planted on the ground, implicitly realized this in his direct alignment of the death principle with the notion of Nirvana of Eastern religions. Freud had little truck with faith and spirituality except as explanatory tools, but the sheer quantity of the evidence convinced him, especially in his later writings, that there is a powerful force that stands in contradiction to the conventional scientific view of the organism as something that wishes only to adapt and persist in its environmental niche. Our final section examines this in more detail.

THE FUNDAMENTAL LUMINOSITY OF MIND

In the same way that the union of colored lights combine to produce a single white beam, as numerous mulitple souls come together to form a psychic unity, the ensuing form will begin to achieve what Buddhists call the fundamental luminosity of mind. It may seem odd that technology could produce such an exalted state, but as this book has stressed, neural technologies are not ordinary technologies. Not only do they have the ability to alter the self, they have the ability to create new kinds of self, including, in principle, beings that prior to such advances visited our planet only once a millenium as prophets and seers. However, being a prophet or the like is, if such a concept is meaningful in any scientific sense whatsoever, a state of mind, and not a metaphysical designation as determined by a Seraphic board on high. By joining together a set of disparate souls, all contributing their colliding perspectives, and harnessing the intellectual power that will be possible with auxiliary computational devices to winnow from this collection impurities and irrationalities, prophets, seers, and Buddhas can be manufactured, just as surely as we make cars and houses and planes today.

Whether this is a good or bad thing is difficult to say, and perhaps we, as non luminous souls, or more aptly as souls that partake of luminosity only in the odd moment of grace, cannot be relied upon to make such a judgment. We can, however, get a glimpse of what such a being would be like by virtue of its compositional process. It would be, above all, a being without division, and one in which the ordinary dichotomies would be erased. It would not favor one food or music over another, it would be the true multicultural entity, and it would excel in all forms of endeavor, from fencing to mathematics to music composition. It would be self-contained, and not seek to look beyond itself either for intellectual or emotional satisfaction.

At its pinnacle, it would not seek at all, because it would know the outcome of seeking, and know that that object, whatever it was, was already firmly ensconsed within its being. There would be no subject and no object, because this distinction would be lost in the purely luminous state; all would be within, and nothing without. Again, we cannot and should not judge this state, but what we can say is that without subject and object there could be no movement and no striving. Without these there could be discovery, there could be no meeting of minds, and there could be no admiration or worship. Such a being, for example, would not even be aware of the numerous suboptimal souls gathered at its virtual feet, fawning over its perfection. It would also have no psychic trajectory, nothing to shoot for, and nothing to gain, because there would be no direction in which it could possibly improve.

Perhaps, though, in its inner core, among the many thoughts that is has amalgamated, would be the smallest realization that without differences, its perfection is incomplete and thereby marred. Perhaps it would therefore voluntarily choose to fracture into multiple souls in multiple bodies, in a kind of Big Bang of the mind, giving rise to among other things, Lenny Moscowitz, the accountant from Long Island, his family, his friends, and their families. We also cannot rule out the possibility that we ourselves are the product of an earlier such fracturing, and via the manipulation of the mind by physical devices, we are about to recapitulate the entire cycle once again.

REFERENCES

Freud, S. (1961). *Beyond the Pleasure Principle*. (J. Strachey, Ford.) New York: W W Norton.

Hofstede, G., and McRae, R. (2004). Personality and culture revisited: Linking traits and dimensions of culture. *Cross-Cultural Research, 38*, 52–88.

Kahneman, D. (2003). Maps of bounded rationality: psychology for behavioral economics. *The American Economic Review, 93* (5), 1449–1475.

Kurzweil, R. (2005). *The Singularity is Near*. New York: Viking.

Moll, J., Krueger, F., Zahn, R., Pardini, M., Oliveira-Souza, R., and Grafman, J. (2006). Human fronto-mesolimbic networks guide decisions about charitable donation. *Proceedings of the National Academy of Sciences, 103*, 15623–15628.

R.R., M., and Terracciano, A. (2005). Personality Profiles of Cultures: Aggregate Personality Traits. *Journal of Personality and Social Psychology, 89*, 407–425.

Yutang, L. (1937). *The Importance of Living*. New York: Reynal and Hitchcock Inc.

INDEX

I

J

K

L

M